Dietrich Schlichthärle

Digital Filters

Springer

Berlin
Heidelberg
New York
Barcelona
Hong Kong
London
Milan
Paris
Singapore
Tokyo

Dietrich Schlichthärle

Digital Filters

Basics and Design

With 257 Figures and 34 Tables

Springer

Dr.-Ing. Dietrich Schlichthärle
Bosch Telecom
D-60277 Frankfurt/Main

Library of Congress Cataloging-in-Publication Data

Schlichthärle, Dietrich
Digital filters: basics and design / Dietrich Schlichthärle.
ISBN 3-540-66841-1
1. Signal processing--Mathematics. 2. Digital filters (Mathematics)
I. Title
TK 5102.9. S367 2000
621.382'2--dc21

ISBN 3-540-66841-1 Springer-Verlag Berlin Heidelberg New York

Springer-Verlag is a company in the BertelsmannSpringer publishing group
© Springer-Verlag Berlin Heidelberg 2000
Printed in Germany

Typesetting: camera ready copy from author
Cover-Design: MEDIO Innovative Medien Service GmbH Berlin
Printed on acid-free paper   SPIN 10752879   62/3020   5 4 3 2 1 0

To my father

Preface

Today, there is an ever increasing number of applications that make use of digital filters or digital signal processing (DSP) algorithms in general. In electronic equipment of any kind, analog functions have been increasingly replaced by digital algorithms. This is true for instance for audio and video equipment, for communication, control and radar systems and in medical applications. Continuing improvement of CMOS technology has allowed higher complexity and increasing processing speed of dedicated integrated circuits and programmable signal processors. Higher clock speeds and parallel processing with the associated higher throughput opens the door for more and more real-time applications.

The advantages of digital signal processing are based on the fact that the performance of the applied algorithms is always predictable. There is no dependence on the tolerances of electrical components as in analog systems. This allows the implementation of interference-proof and long-term stable systems. The adjustment of components which is often necessary in analog circuits becomes obsolete. Also, extremely low frequencies can be processed without the problems which occur in analog systems.

In the field of consumer applications, digital signal processing has obtained wide application in the storage and reproduction of audio and video signals. In paticular, CD technology, which has entered most households today, has to be mentioned. Compared with the previous vinyl technology, the reproduction of audio signals may take place arbitrarily often without any loss of quality. Besides the real-time processing of signals, simulation techniques which allow the imitation of physical processes in arbitrary time scales on computers gain more and more importance.

It is the aim of this textbook to give insight into the characteristics and the design of digital filters. The starting point for the design of recursive filters is the theory of continuous-time systems (Chap. 1) and the design methods for analog filters as introduced in Chap. 2 which are based on Butterworth, Chebyshev and Bessel polynomials as well as on elliptic functions. After the introduction of the theory of discrete-time systems (Chap. 3) and the basic structures of digital filters in Chap. 5, we derive the rules which allow the calculation of the coefficients of recursive digital filters from those of continuous-time reference filters in Chap. 6.

An important part of this textbook is devoted to the design of nonrecursive filters which cannot be treated with the classical filter design methods. Approximation criteria such as the minimisation of the mean-square or maximum error and the maximally flat design are presented in Chap. 7. Of foremost significance is the Remez exchange algorithm which is widely used in commercial filter design software.

The connecting point between the world of the physical signals which are considered continuous-time and the world of number series and mathematical algorithms which are discrete-time by nature is the sampling theorem, which is treated in Chap. 4.

Chapter 8 is devoted to all those effects which are due to the finite register lengths in real software or hardware implementations. Both signal samples and filter coefficients can be represented with limited accuracy only. The consequences are deviations from the design target, additive noise and – under certain circumstances – the occurrence of instabilities.

The final chapter gives an introduction to terms like oversampling and noise shaping. These are techniques which are primarily used to reduce the complexity of analog components in analog-to-digital and digital-to-analog converters.

In the Appendix, the textbook is completed by a selection of filter design tables for Butterworth, Chebyshev, Cauer and Bessel filters. Furthermore, the interested reader will find several subroutines written in Turbo Pascal which can be used in their own filter design programs. These may be easily translated into other dialects or programming languages.

The study of this book requires the knowledge of the theory of continuous-time systems. In this context, Chap. 1 gives only a short overview of terms and relations which are used in later chapters. A basic knowledge of algebra, infinitesimal calculus and the mathematics of complex numbers is also needed.

Maintal, February 2000   Dietrich Schlichthärle

Contents

1 Continuous-Time Systems

The relationships between input and output signals of analog systems which are introduced in this chapter are treated using the theory of continuous-time systems. For a certain class of systems which have the properties of linearity and time-invariance (LTI systems), this theory provides relatively simple mathematical relationships. The present textbook is limited to the treatment of filters which possess these two properties.

Linearity means that two superimposed signals pass through a system without mutual interference. Suppose that $y_1(t)$ is the response of the system to $x_1(t)$ and $y_2(t)$ the response to $x_2(t)$. Applying a linear combination of both signals $x = a\,x_1\,(t) + b\,x_2(t)$ to the input results in $y = a\,y_1(t) + b\,y_2(t)$ at the output.

Time-invariance means that the characteristics of the system do not change with time. A given input signal results in the same output signal at any time. So the property of time-invariance implies that if $y(t)$ is the response to $x(t)$, then the response to $x(t-t_0)$ is simply $y(t-t_0)$.

If these conditions are fulfilled, the characteristics of the system can be fully described by its reaction to simple test signals such as pulses, steps and harmonic functions.

1.1 Relations in the Time Domain

One of these possible test signals is the so-called delta function $\delta(t)$. The delta function is the limit of an impulse occurring at $t = 0$ with area 1 and a temporal width which tends to zero. The filter reacts to this signal with the impulse response $h(t)$. As, in practice, we deal with causal systems in which the reaction follows the cause in time, $h(t)$ will disappear for negative t. If the impulse response of a system is known, it is possible to calculate the reaction $y(t)$ of the system to any arbitrary input signal $x(t)$.

$$y(t) = \int_{-\infty}^{+\infty} x(t-\tau)\,h(\tau)\,\mathrm{d}\tau = \int_{-\infty}^{+\infty} x(\tau)\,h(t-\tau)\,\mathrm{d}\tau = x(t) \,*\, h(t) \tag{1.1}$$

Relationship (1.1) is called the convolution integral: $y(t)$ is the convolution of $x(t)$ with $h(t)$. Both integrals are fully equivalent, as can be shown by a simple substitution of variables. The symbol for the convolution is an asterisk.

A further possible test signal is the step function $u(t)$. The reaction of the system to this signal is called the step response $a(t)$. The step function is defined as follows:

$$u(t) = \begin{cases} 0 \text{ for } t < 0 \\ 1/2 \text{ for } t = 0 \\ 1 \text{ for } t > 0 \end{cases}.$$

Also, $a(t)$ will disappear for negative t in causal systems.

$$a(t) = 0 \qquad \text{for } t < 0$$

If the step response of the system is known, the input and output signals of the system are related by the following equation:

$$y(t) = \int_{-\infty}^{+\infty} a(t-\tau)\frac{dx(\tau)}{d\tau}d\tau = -\int_{-\infty}^{+\infty} a(\tau)\frac{dx(t-\tau)}{d\tau}d\tau. \tag{1.2}$$

In this case, the output signal is the convolution of the step response of the system with the time derivative of the input signal.

Delta function and unit step are related as follows:

$$\delta(t) = \frac{du(t)}{dt} \qquad\qquad \text{and} \tag{1.3a}$$

$$u(t) = \int_{-\infty}^{t} \delta(\tau)d\tau \qquad\qquad \text{vice versa.} \tag{1.3b}$$

For LTI systems, similar relations exist between impulse and step response:

$$h(t) = \frac{da(t)}{dt} \qquad\qquad \text{and} \tag{1.4a}$$

$$a(t) = \int_{-\infty}^{t} h(\tau)d\tau \qquad\qquad \text{vice versa.} \tag{1.4b}$$

1.2 Relations in the Frequency Domain

In the frequency domain, the system behaviour can be analysed by applying an exponential function to the input of the form

$$x(t) = e^{pt}, \tag{1.5}$$

where p stands for the complex frequency. p consists of a real part σ, which causes an increasing or decreasing amplitude, depending on the sign of σ, and an imaginary part ω, which is the frequency of a harmonic oscillation.

$$p = \sigma + j\omega$$

The corresponding real oscillation can be derived by forming the real part of (1.5).

$$\operatorname{Re} x(t) = e^{\sigma t} \cos \omega t$$

Substitution of (1.5) as the input signal in the convolution integral (1.1) gives the reaction of the system to the complex exponential stimulus.

$$y(t) = \int_{-\infty}^{+\infty} e^{p(t-\tau)} h(\tau)\,d\tau$$

$$y(t) = e^{pt} \int_{-\infty}^{+\infty} h(\tau)e^{-p\tau}\,d\tau \tag{1.6}$$

The integral in (1.6) yields a complex-valued expression which is a function of the complex frequency p.

$$H(p) = \int_{-\infty}^{+\infty} h(t)e^{-pt}\,dt \tag{1.7}$$

Consequently, the output signal $y(t)$ is also an exponential function which we obtain by multiplying the input signal by the complex factor $H(p)$. Thus, input and output signals differ in amplitude and phase.

$$y(t) = e^{pt} H(p)$$

$$\operatorname{Re} y(t) = |H(p)|e^{\sigma t} \cos[\omega t + \arg H(p)] \tag{1.8}$$

The function $H(p)$ is called the system or transfer function as it characterises the system completely in the frequency domain.

1.2.1 The Laplace Transform

The integral relationship (1.7) is called the Laplace transform. Thus, the transfer function $H(p)$ is the Laplace transform of the impulse response $h(t)$ of the system. In practice, a special form of the Laplace transform is used which is only defined for the period $t \geq 0$, the so-called one-sided Laplace transform.

Table 1-1 Table of selected Laplace and Fourier transforms

Time domain	Laplace transform	Fourier transform
$\delta(t)$	1	1
$u(t)$	$\dfrac{1}{p}$	$\dfrac{1}{j\omega} + \pi\delta(\omega)$
$t\,u(t)$	$\dfrac{1}{p^2}$	$j\pi\dfrac{d\delta(\omega)}{d\omega} - \dfrac{1}{\omega^2}$
$e^{-\alpha t}\,u(t)$	$\dfrac{1}{p+\alpha}$	$\dfrac{1}{j\omega+\alpha}$
$(1-e^{-\alpha t})\,u(t)$	$\dfrac{\alpha}{p(p+\alpha)}$	$\pi\delta(\omega) + \dfrac{\alpha}{j\omega(j\omega+\alpha)}$
$e^{j\omega_0 t}\,u(t)$	$\dfrac{1}{p-j\omega_0}$	$\pi\delta(\omega-\omega_0) + \dfrac{1}{j\omega-\omega_0}$
$\cos\omega_0 t\,u(t)$	$\dfrac{p}{p^2+\omega_0^2}$	$\dfrac{\pi}{2}\left[\delta(\omega-\omega_0)+\delta(\omega+\omega_0)\right] + \dfrac{j\omega}{\omega_0^2-\omega^2}$
$\sin\omega_0 t\,u(t)$	$\dfrac{\omega_0}{p^2+\omega_0^2}$	$\dfrac{\pi}{2j}\left[\delta(\omega-\omega_0)-\delta(\omega+\omega_0)\right] + \dfrac{\omega_0}{\omega_0^2-\omega^2}$
$\cos\omega_0 t\,e^{-\alpha t}\,u(t)$	$\dfrac{p+\alpha}{(p+\alpha)^2+\omega_0^2}$	$\dfrac{\alpha+j\omega}{(\alpha+j\omega)^2+\omega_0^2}$
$\sin\omega_0 t\,e^{-\alpha t}\,u(t)$	$\dfrac{\omega_0}{(p+\alpha)^2+\omega_0^2}$	$\dfrac{\omega_0}{(\alpha+j\omega)^2+\omega_0^2}$

$$F(p) = \int_{0}^{+\infty} f(t)\,e^{-pt}\,dt \tag{1.9a}$$

If we apply this transform to the impulse response of a system, it does not impose any limitation as the impulse response disappears anyway for $t < 0$ as pointed out in Sect. 1.1. When calculating the integral (1.9a), the real part of $p\,(\sigma)$ is considered a parameter which can be freely chosen such that the integral converges. Table 1-1 shows some examples and Table 1-2 shows the most important rules of the Laplace transform. The inverse transform from the frequency into the time domain reads as follows:

$$f(t) = \frac{1}{2\pi j} \int_{\sigma-j\infty}^{\sigma+j\infty} F(p)\,e^{pt}\,dp. \tag{1.9b}$$

Equation (1.9b) is a line integral in the complex p-plane in which the integration path is chosen with an appropriate σ such that the integral converges.

Table 1-2 Rules of the Laplace and Fourier transforms

Time domain	Laplace transform	Fourier transform
$f(t-t_0)$	$F(p)\mathrm{e}^{-pt_0}$	$F(\mathrm{j}\omega)\mathrm{e}^{-\mathrm{j}\omega t_0}$
$f(at)$	$\dfrac{1}{\lvert a\rvert}F(p/a)$	$\dfrac{1}{\lvert a\rvert}F(\mathrm{j}\omega/a)$
$\dfrac{\mathrm{d}f(t)}{\mathrm{d}t}$	$pF(p)-f(0)$	$\mathrm{j}\omega\,F(\mathrm{j}\omega)$
$\displaystyle\int_0^t f(\tau)\,\mathrm{d}\tau$	$\dfrac{1}{p}F(p)$	————
$\displaystyle\int_{-\infty}^t f(\tau)\,\mathrm{d}\tau$	————	$\dfrac{F(\mathrm{j}\omega)}{\mathrm{j}\omega}+\pi\,F(0)\delta(\omega)$
$f(t)*g(t)$	$F(p)\times G(p)$	$F(\mathrm{j}\omega)\times G(\mathrm{j}\omega)$

In this context, the Laplace transform has an interesting property which is important for the characterisation of the behaviour of LTI systems in the frequency domain. The convolution of two functions $x_1(t) * x_2(t)$ in the time domain is equivalent to the multiplication of the Laplace transforms of both functions $X_1(p) \times X_2(p)$ in the frequency domain. For LTI systems, the output signal is the convolution of the input signal and the impulse response. In the frequency domain, the Laplace transform of the output signal is therefore simply the Laplace transform of the input signal multiplied by the system function $H(p)$.

$$Y(p) = H(p)\,X(p) \tag{1.10}$$

Hence (1.10) establishes a relation between input and output signal in the frequency domain and is therefore equivalent to (1.1), which relates input and output signal in the time domain.

1.2.2 The Fourier Transform

With $\sigma = 0$ or $p = \mathrm{j}\omega$ in (1.7), we obtain a special case of the Laplace transform.

$$F(\mathrm{j}\omega) = \int_{-\infty}^{+\infty} f(t)\mathrm{e}^{-\mathrm{j}\omega t}\,\mathrm{d}t \tag{1.11a}$$

This integral relationship is called the Fourier transform. Unlike the Laplace transform, (1.11a) does not contain a damping factor which enforces convergence. So the Fourier integral only converges if the signal is limited in time or decays fast enough. Otherwise, $F(\mathrm{j}\omega)$ will contain singularities which can be treated using the concept of distributions. These singularities are often called spectral lines, which are mathematically represented as delta functions. The concept of distributions removes the named limitations of the Fourier integral and extends its applicability to periodic functions with infinite energy. Table 1-1 shows some examples of Fourier transformed functions and allows a direct comparison with the corresponding Laplace transform. The simple transition between Laplace transform $F(p)$ and Fourier transform $F(j\omega)$ by interchanging p and $j\omega$ is only possible if the Fourier integral converges for all ω, which means that no spectral lines may exist in the frequency domain. The stable filters which are the subject of this textbook have transfer functions with this property, so that $H(j\omega)$ can be obtained from $H(p)$ by the simple substitution $p = j\omega$. The inverse Fourier transform is given by

$$f(t) = \frac{1}{2\pi} \int_{-\infty}^{+\infty} F(j\omega)e^{j\omega t} d\omega . \tag{1.11b}$$

Table 1-2 shows some further properties of the Fourier transform.

The Fourier transform $H(j\omega)$ of the impulse response $h(t)$ is also called the system or transfer function. Evaluation of (1.8) for $\sigma = 0$ or $p = j\omega$ shows that $H(j\omega)$ describes the behaviour of the system for the case of harmonic input signals in the steady state.

$$\text{Re } y(t) = |H(j\omega)|\cos(\omega t + \arg H(j\omega))$$

$H(j\omega)$ is therefore the frequency response of the system. The magnitude of $H(j\omega)$ is called the amplitude or magnitude response which is often written in the form

$$a(\omega) = -20\log|H(j\omega)| , \tag{1.12a}$$

where $a(\omega)$ is the logarithmic attenuation of the system given in decibels (dB). The negative argument of the frequency response is the phase response of the system.

$$b(\omega) = -\arg H(j\omega) = -\arctan\left(\frac{\text{Im } H(j\omega)}{\text{Re } H(j\omega)}\right) \tag{1.12b}$$

The phase response determines the phase shift between the harmonic oscillation at the input and at the output of the system with the frequency ω as parameter. The derivative of the phase response is the group delay τ_g of the system.

$$\tau_g(\omega) = \frac{db(\omega)}{d\omega} \tag{1.12c}$$

If preservation of the temporal shape of the signal is an issue, the group delay has to be flat in the passband of the filter; this is equivalent to a linear-phase behaviour. This guarantees that all parts of the spectrum experience the same delay when passing through the system. As an example, consider a simple first-order low-pass filter which is characterised by the following transfer function:

$$H(p) = \frac{a}{a+p} \qquad \text{with } a > 0 .$$

The substitution $p = j\omega$ yields the frequency response of the filter.

$$H(j\omega) = \frac{a}{a+j\omega} \qquad \text{with } a > 0$$

The magnitude of this expression is the amplitude response.

$$|H(j\omega)| = \frac{a}{|a+j\omega|} = \frac{a}{\sqrt{a^2 + \omega^2}} = \frac{1}{\sqrt{1 + \omega^2 / a^2}}$$

The phase response is the negative argument of the frequency response.

$$b(\omega) = -\arg H(j\omega) = \arg \frac{1}{H(j\omega)} = \arctan(\omega / a)$$

By differentiation of the phase response, we obtain the group delay of the filter.

$$\tau_g(\omega) = \frac{d\, b(\omega)}{d\omega} = \frac{1 / a}{1 + \omega^2 / a^2}$$

1.3 Transfer Functions

The transfer behaviour of networks consisting of lumped elements such as inductors, capacitors, resistors, transformers and amplifiers is described by linear differential equations with real coefficients. Equation (1.13) shows the general form of this differential equation with $x(t)$ representing the input signal and $y(t)$ the output signal of the system.

$$a_N \frac{dy^{(N)}}{dt^N} + a_{N-1} \frac{dy^{(N-1)}}{dt^{N-1}} + \ldots + a_1 \frac{dy}{dt} + a_0 y$$
$$= b_M \frac{dx^{(M)}}{dt^M} + b_{M-1} \frac{dx^{(M-1)}}{dt^{M-1}} + \ldots + b_1 \frac{dx}{dt} + b_0 x \tag{1.13}$$

The Laplace transform of (1.13) yields an algebraic equation which characterises the system completely in the frequency domain. The following relationship assumes that the system is at rest for $t = 0$. For an electrical network, this means that the stored energy in all capacitors and inductors is zero.

$$a_N p^N Y + a_{N-1} p^{N-1} Y + \ldots + a_1 p Y + a_0 Y$$
$$= b_M p^M X + b_{M-1} p^{M-1} X + \ldots + b_1 p X + b_0 X$$

Re-ordering of the above relation gives the transfer function of the network, which is the quotient of the Laplace transforms of the output and the input signal.

$$H(p) = \frac{Y(p)}{X(p)} = \frac{b_M p^M + b_{M-1} p^{M-1} + \ldots + b_1 p + b_0}{a_N p^N + a_{N-1} p^{N-1} + \ldots + a_1 p + a_0} \tag{1.14}$$

Equation (1.14) is a rational fractional function of the frequency variable p which has the following properties:
- The coefficients a_i and b_i are real.
- The degree of the denominator N which equals the order of the filter corresponds to the number of energy stores (capacitors and inductors) in the network.
- The degree of the numerator must be smaller or equal to the degree of the denominator. Otherwise the system would contain ideal differentiators which are not realisable.

Beside this rational fractional representation of the transfer function, which is closely related to the differential equation of the system, there are two further forms which are of practical importance: pole/zero representation and partial-fraction expansion.

If the numerator and denominator polynomials in (1.14) are decomposed into the respective zero terms, we obtain the pole/zero representation.

$$H(p) = \frac{Y(p)}{X(p)} = \frac{b_M}{a_N} \frac{(p - p_{01}) \cdots (p - p_{0(M-1)})(p - p_{0M})}{(p - p_{\infty 1}) \cdots (p - p_{\infty(N-1)})(p - p_{\infty N})} \tag{1.15}$$

The zeros of the numerator polynomial are the zeros of the transfer function, the zeros of the denominator polynomial are the poles. Poles and zeros have the following properties:
- Poles and zeros are either real or appear as conjugate-complex pairs. This property is forced by the real coefficients of the numerator and denominator polynomials.
- The real part of the poles has to be negative, which means that the poles of the system must lie in the left half of the p-plane. This condition guarantees stability of the system. It ensures that any finite input signal results in a finite output signal. When the input is set to zero, all internal signals and the output signal will converge to zero.

The pole/zero representation has special importance in the cases where given transfer functions are realised as a cascade of first- and second-order subsystems (Fig. 1-1a). First-order sections are used to realise real poles, second-order sections implement conjugate-complex pole pairs.

Assuming ideal noiseless amplifiers, the pole/zero pairing to create second-order filter blocks and the sequence of these subfilters can be chosen arbitrarily. As the operational amplifiers used in practice generate noise and have a finite dynamic range, the pole/zero assignment and the choice of sequence are degrees of freedom within which to optimise the noise behaviour of the overall system.

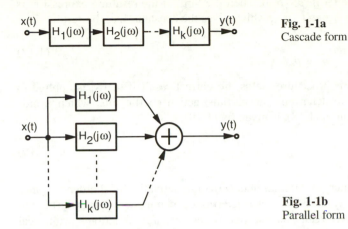

Fig. 1-1a
Cascade form

Fig. 1-1b
Parallel form

Example

The example shows the transition from a rational fractional transfer function to the pole/zero representation.

$$H(p) = \frac{p^3 + 9p^2 + 20p + 16}{p^3 + 5p^2 + 8p + 6} \tag{1.16a}$$

$$H(p) = \frac{(p + 6.18)(p + 1.41 + j0.78)(p + 1.41 + j0.78)}{(p + 3)(p + 1 + j)(p + 1 - j)} \tag{1.16b}$$

$$H(p) = \frac{(p + 6.18)(p^2 + 2.82p + 2.46)}{(p + 3)(p^2 + 2p + 2)} \tag{1.16c}$$

Equation (1.16b) shows the decomposition of the numerator and denominator polynomials into real and complex zeros. In (1.16c) the complex pole and zero pairs are combined to second-order terms with real coefficients. The given transfer function can therefore be realised by cascading a first-order and a second-order filter block.

Unlike the pole/zero representation of the transfer function, which consists of the product of first- and second-order terms, the partial-fraction expansion is composed of a sum of terms (1.17).

$$H(p) = \frac{A_1}{p - p_{\infty 1}} + \frac{A_2}{p - p_{\infty 2}} + \ldots + \frac{A_{N-1}}{p - p_{\infty(N-1)}} + \frac{A_N}{p - p_{\infty N}} \tag{1.17}$$

This representation is closely related to a possible realisation of the transfer function by a parallel combination of first- and second-order systems (Fig. 1-1b).

The coefficients A_i of the partial-fraction expansion can be easily derived on condition that the poles are known. First of all, the transfer function $H(p)$ is multiplied by the pole term $(p - p_{\infty i})$. Then the limit of the resulting expression is determined for p converging to $p_{\infty i}$, which gives the constant A_i as the result.

$$A_i = H(p)(p - p_{\infty i})\Big|_{p \to p_{\infty i}} \tag{1.18}$$

Understandably, this procedure yields the correct result if (1.18) is applied to (1.17). Another way to determine the partial-fraction coefficients, which avoids the forming of the limit in (1.18), is given by (1.19):

$$A_i = \frac{N(p_{\infty i})}{D'(p_{\infty i})}. \tag{1.19}$$

$N(p)$ is the numerator polynomial and $D'(p)$ the derivative of the denominator polynomial [9]. If the coefficients A_i are determined, conjugate-complex terms can be combined to give realisable second-order transfer functions with real coefficients. It must be noted here that the partial-fraction expansion requires the degree of the numerator to be less than the degree of the denominator. This is not the case for high-pass, bandpass and bandstop filters which have equal numerator and denominator degrees. Here the numerator and denominator polynomials have to be divided prior to the procedures described above. The result of this division is the sum of a constant, which is equivalent to the gain of the system for $\omega \to \infty$, and a rational fractional function with the degree of the numerator reduced by one.

Applying an impulse to the input of the parallel combination of first- and second-order subsystems (Fig. 1-1b) results in an impulse response which is composed of the sum of all partial responses. These are decaying exponentials and exponentially decaying sinusoids. The decay rate and the frequency of these partial responses are determined by the poles of the system. For the design of digital filters which are intended to reproduce the properties of analog filters in the time domain, partial-fraction expansion of transfer functions is therefore of special interest.

Example

We now expand (1.16) into partial fractions. In the first step, we have to divide the numerator by the denominator polynomial, which is necessary as the degrees of numerator and denominator are equal. Equation (1.20a) shows the result.

$$H(p) = 1 + \frac{4p^2 + 12p + 10}{(p+1+j)(p+1-j)(p+3)} \tag{1.20a}$$

In the second step, we perform the partial fractional expansion (1.20b).

$$H(p) = 1 + \frac{1}{p+1+j} + \frac{1}{p+1-j} + \frac{2}{p+3} \tag{1.20b}$$

In order to obtain realisable sub-systems, the pair of conjugate-complex terms is combined to a transfer function of second order (1.20c).

$$H(p) = 1 + \frac{2p+2}{p^2+2p+2} + \frac{2}{p+3} \qquad (1.20c)$$

Partial systems of first and second order are of special importance for the design of filters, as already mentioned. These are the smallest realisable units into which transfer functions can be decomposed. Chapter 8 explains in more detail the advantages of a design based on low-order partial systems. The transfer function of a general second-order filter has the following form:

$$H(p) = \frac{b_0 + b_1 p + b_2 p^2}{a_0 + a_1 p + a_2 p^2} \; . \qquad (1.21a)$$

In practice, this transfer function is often normalised such that $a_0 = 1$.

$$H(p) = \frac{b_0 + b_1 p + b_2 p^2}{1 + a_1 p + a_2 p^2} \qquad (1.21b)$$

It is sometimes advantageous to factor a constant out of the numerator polynomial, which then appears as a gain factor V in front of the transfer function.

$$H(p) = V \frac{1 + b_1 p + b_2 p^2}{1 + a_1 p + a_2 p^2} \qquad (1.21c)$$

It is easy to see that V is the gain of the system at the frequency $p = 0$.
    The transfer function of a complex pole pair of the form

$$H(p) = \frac{1}{p^2 + a_1 p + a_0}$$

can be characterised in three ways:
- by the coefficients a_1 and a_0 of the denominator polynomial
- by the real part $\mathrm{Re}\, p_\infty$ and imaginary part $\mathrm{Im}\, p_\infty$ of the pole
- by the resonant frequency ω_0 and quality factor Q

$$H(p) = \frac{1}{p^2 + (\omega_0 / Q)p + \omega_0^2} = \frac{1}{p^2 - 2\,\mathrm{Re}\, p_\infty p + (\mathrm{Re}\, p_\infty)^2 + (\mathrm{Im}\, p_\infty)^2} \qquad (1.22)$$

The quality factor Q has two illustrative interpretations:
- The quality factor is the ratio of the gain at the resonant frequency ω_0 to the gain at $\omega = 0$.

$$Q = \left| H(j\omega_0) / H(0) \right|$$

- The quality factor corresponds approximately to the reciprocal of the 3-dB bandwidth of the resonant peak of the transfer function.

$$Q \approx \omega_0 / \Delta\omega_{3dB}$$

From (1.22) we can derive a relationship between the quality factor and the location of the pole:

$$\left| \mathrm{Im}\, p_\infty / \mathrm{Re}\, p_\infty \right| = \sqrt{4Q^2 - 1} \; . \tag{1.23}$$

Equation (1.23) shows that poles of constant Q are located on straight lines in the complex p-plane as depicted in Fig. 1-2.

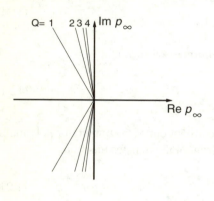

Fig. 1-2
Location of poles with constant quality factor Q in the complex p-plane

2 Analog Filters

In this chapter, we introduce the analog filters that are intended to serve as prototypes for the design of corresponding digital filters. Besides describing characteristics of Butterworth, Chebyshev, Cauer and Bessel filters, hints are given as to how the coefficients of these filters are calculated. This will enable the reader to perform calculations going beyond the numerical data given in the design tables in the Appendix.

We start by considering low-pass filters as prototypes for the digital filter design. After the introduction of the characteristics of the ideal low-pass filter, we throw some light on the filter types named above, each of which is, in a certain sense, an optimum approximation of the ideal low-pass characteristic.

Appropriate transformations allow the conversion of low-pass filters into high-pass, bandpass and bandstop filters with related characteristics. The corresponding transformation rules are also introduced in this chapter.

2.1 The Ideal Low-Pass Filter

In the context of the characteristics of the ideal low-pass filter, we first consider the distortionless system. If the form of a signal is to be preserved when passing the system, multiplication of the signal by a constant factor k (frequency-independent gain) and a frequency-independent delay t_0 are the only allowed modifications. This yields the following general relationship between input signal $x(t)$ and output signal $y(t)$ of a distortionless system in the time domain:

$$y(t) = k\,x(t - t_0) \,.$$

According to the theory of continuous-time systems as introduced in Chap. 1, the corresponding relation in the frequency domain can be expressed as

$$Y(j\omega) = X(j\omega)\,k\,e^{-j\omega t_0} \qquad \text{and}$$

$$H(j\omega) = \frac{Y(j\omega)}{X(j\omega)} = k\,e^{-j\omega t_0} \,.$$

Magnitude, phase and group delay response can be calculated using (1.12).

$$a(\omega) = -20 \times \log k$$

$$b(\omega) = \omega\, t_0$$

$$\tau_g(\omega) = t_0$$

Magnitude and group delay response are frequency independent, and the phase response is linear.

The ideal low-pass filter also features a linear phase in the passband. For $0 \le \omega < \omega_c$ the gain is constantly unity. In the stopband $\omega > \omega_c$, however, the filter cuts off completely (Fig. 2-1).

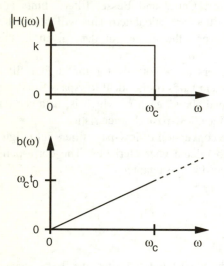

Fig. 2-1
Magnitude and phase response of the ideal low-pass filter

Such a filter cannot be realised with finite cost. Instead, we have to find rational functions that can be used to approximate the ideal frequency response according to given criteria. With increasing order N of the used polynomials, connected with increasing complexity of the implementation, the approximation to the ideal frequency response can always be improved. The starting point of many approximation methods is a low-pass response of the form

$$\left| H(j\omega) \right|^2 = \frac{1}{1 + A_N(\omega^2)} .$$
(2.1)

The function $A_N(\omega^2)$ characterises the kind of approximation. The index N, equivalent to the order of the filter, determines the degree of the approximation and thus the complexity of the implementation. A_N has the property of assuming low values between $\omega = 0$ and a certain cutoff frequency ω_c, but increasing rapidly above this frequency.

The desired magnitude characteristic of a low-pass filter is often specified in form of a tolerance scheme, which has to be satisfied by the filter (Fig. 2-2).

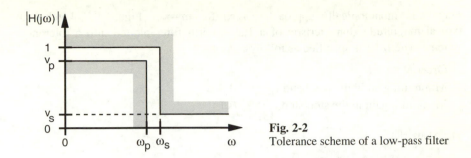

Fig. 2-2
Tolerance scheme of a low-pass filter

ω_p is the edge frequency of the passband, ω_s marks the beginning of the stopband. The region $\omega_p < \omega < \omega_s$ is the transition band between the passband and the stopband. The relative width of this region,

$$\omega_s / \omega_p,$$

corresponds to the sharpness of the cutoff of the filter which, apart from other parameters, largely determines the effort necessary for the implementation of the filter. v_p specifies the allowable passband deviation of the gain with respect to the ideal case, while v_s corresponds to the maximum allowable stopband gain. Both figures are often specified in dB. The corresponding tolerance scheme is open towards the bottom as the gain $|H(j\omega)| = 0$ approaches negative infinity in the logarithmic scale.

2.2 Butterworth Filters

With this filter type, the approximation function $A_N(\omega^2)$ assumes the following form:

$$A_N(\omega^2) = \left(\frac{\omega}{\omega_c}\right)^{2N} . \tag{2.2}$$

Insertion into (2.1) yields (2.3), which represents the squared magnitude response of the Butterworth filter.

$$|H(j\omega)|^2 = \frac{1}{1 + \left(\dfrac{\omega}{\omega_c}\right)^{2N}} \tag{2.3}$$

This filter type features an optimally flat magnitude characteristic in the passband. All derivatives of the magnitude function are zero at $\omega = 0$ which leads to the named behaviour. All zeros of the filter are infinite. The consequence is that the magnitude monotonically approaches zero for $\omega \to \infty$. Figure 2-3 sketches a typical magnitude characteristic of a Butterworth filter plotted into a tolerance scheme. The filter is specified as follows:

Order N: 4

Minimum gain in the passband v_p: 0.9

Maximum gain in the stopband v_s: 0.1

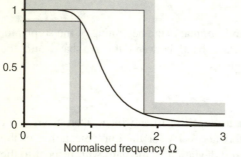

Fig. 2-3
Magnitude characteristic of a fourth-order Butterworth filter

Normalising the frequency axis simplifies the further mathematical treatment of the filter.

$$\Omega = \frac{\omega}{\omega_c} \qquad (2.4)$$

In the normalised representation of the magnitude characteristic, the cutoff frequency of the filter is located at $\Omega = 1$.

$$|H(j\Omega)|^2 = \frac{1}{1 + \Omega^{2N}} \qquad (2.5)$$

At the cutoff frequency ω_c ($\Omega = 1$), the gain amounts to $1/\sqrt{2}$, independently of the order of the filter, which corresponds in good approximation to a loss of 3 dB.

For the design of the filter and later on for the conversion to a digital filter, it is important to determine the poles and thus the transfer function $H(j\Omega)$ of the filter, for which we currently only know the magnitude (2.5). For that purpose, we have to calculate the zeros of the denominator polynomial in (2.5). In the first step, the magnitude is represented as a function of the normalised frequency P. With $P = j\Omega$ or $\Omega = -jP$, (2.5) becomes

$$|H(P)|^2 = \frac{1}{1 + (-1)^N P^{2N}} = (-1)^N \frac{1}{P^{2N} + (-1)^N} \qquad (2.6)$$

$$\text{for } P = j\Omega \ .$$

So the zeros of the polynomial

$$P^{2N} + (-1)^N = 0 \tag{2.7}$$

have to be determined. The $2N$ zeros have a magnitude of 1 and are equally spaced around the unit circle in the normalised P-plane. They can be calculated analytically. The $2N$ solutions are

$$P_{\infty v} = j e^{j\pi(2v+1)/2N} \qquad \text{with } 0 \le v \le 2N - 1$$

or, after separation into the real and imaginary parts,

$$P_{\infty v} = -\sin[\pi(2v+1)/2N] + j\cos[\pi(2v+1)/2N] \tag{2.8}$$
$$\text{with } 0 \le v \le 2N - 1 \ .$$

Figure 2-4 sketches the location of the poles of the example shown in Fig. 2-3. Half of the poles lie in the right half, the other half in the left half of the P-plane. Equation (2.6) can therefore be decomposed into two partial functions where the one contains the poles in the left, the other those in the right half-plane.

$$|H(P)|^2 = (-1)^N H_l(P) H_r(P) \tag{2.9}$$
$$\text{for } P = j\Omega$$

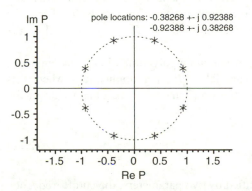

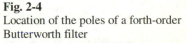

Fig. 2-4
Location of the poles of a forth-order Butterworth filter

As for each pole in the left half-plane there exists a corresponding pole located symmetrically about the origin in the right half-plane, we can write

$$(-1)^N H_r(P) = H_l(-P) \ .$$

Substitution of this equation in (2.9) yields

$$|H(P)|^2 = H_l(P) H_l(-P)$$
$$\text{for } P = j\Omega \ .$$

Replacing P by $j\Omega$ again leads to

$$|H(j\Omega)|^2 = H_1(j\Omega)\, H_1(-j\Omega) = H_1(j\Omega)\, H_1(j\Omega)^* = |H_1(j\Omega)|^2 . \qquad (2.10)$$

$H_1(j\Omega)$ is the frequency response of a stable filter with the transfer function $H_1(P)$ because all poles lie in the left half of the P-plane. As the square of the frequency response $H_1(j\Omega)$ equals the squared magnitude function (2.5), $H_1(P)$ is the desired transfer function of the Butterworth filter. $H(P) = H_1(P)$ is simply composed of the N poles of (2.5) in the left half-plane, which are defined by (2.8).

$$H(P) = \frac{1}{(P - P_{\infty 1})(P - P_{\infty 2})\, \ldots \,(P - P_{\infty N})} \qquad (2.11)$$

According to Sect. 1.3, this transfer function can be realised by a cascade of second-order sections and, in case of an odd filter order, with an additional first-order filter block.

$$H(P) = \frac{1}{1 + A_{11}P + A_{21}P^2} \, \cdots \, \frac{1}{1 + A_{1k}P + A_{2k}P^2} \, \frac{1}{1 + A_{1(k+1)}P}$$

The filter design table in the Appendix shows the coefficients A_{1i} and A_{2i} for filter orders up to $N = 12$. If N is even, $k = N/2$ second-order sections are needed to realise the filter. For odd orders, the filter is made up of one first-order and $k = (N-1)/2$ second-order sections. Up to now, we have only considered transfer functions in the normalised form. Substitution of P by p/ω_c yields the unnormalised transfer function $H(p)$.

Multiplying out the poles in (2.11), we obtain a polynomial in the denominator, which is called the Butterworth polynomial $B_n(P)$. For applications where this polynomial is needed in closed form, the coefficients can be obtained from the corresponding filter design tables.

$$H(P) = \frac{1}{B_n(P)} \qquad (2.12)$$

The Butterworth filter is fully specified by two parameters, the cutoff frequency ω_c and the filter order N. These have to be derived from characteristic data of the tolerance scheme. A closer look at Fig. 2-3 shows that the magnitude graph just fits into the tolerance scheme if it passes through the edges (v_p, Ω_p) and (v_s, Ω_s). This yields two equations that can be used to determine the two unknown filter parameters in (2.3). As a result, the required filter order is specified by

$$N = \frac{\log \dfrac{\sqrt{1/v_s^2 - 1}}{\sqrt{1/v_p^2 - 1}}}{\log \dfrac{\omega_s}{\omega_p}} .$$

For the mathematical treatment of the filter types considered here and in later sections, it proved to be advantageous to introduce a normalised gain factor,

$$v_{norm} = v / \sqrt{(1 - v^2)} \, ,$$

which will simplify many relations. This applies, for instance, to the calculation of the filter order:

$$N = \frac{\log(v_{pnorm} / v_{snorm})}{\log(\omega_s / \omega_p)} \, . \qquad (2.14a)$$

The filter order calculated using (2.14a) is not an integer and has to be rounded up in order to comply with the prescribed tolerance scheme. Thus the order is almost always a little higher than needed to fulfil the specification. This somewhat higher complexity can be used to improve the filter performance with respect to the slope, the passband ripple or the stopband attenuation.

Using (2.14b), the cutoff frequency of the low-pass filter can be determined in such a way that the passband edge is still touched. The behaviour in the stopband, however, is better than that specified.

$$\omega_c = \omega_p \, v_{pnorm}^{1/N} \qquad (2.14b)$$

2.3 Chebyshev Filters

With this filter type, we use the approximation function

$$A_N(\Omega^2) = \varepsilon^2 \, T_N^2(\Omega) \qquad (2.15)$$

to approximate the ideal low-pass filter. The corresponding squared magnitude function is of the form

$$\left| H(j\Omega) \right|^2 = \frac{1}{1 + \varepsilon^2 \, T_N^2(\Omega)} \, . \qquad (2.16)$$

$T_n(\Omega)$ is the Chebyshev polynomial of the first kind of the order n. There are various mathematical representations of this polynomial. The following is based on trigonometric and hyperbolic functions.

$$T_n(\Omega) = \begin{cases} \cos(n \arccos \Omega) & 0 \leq \Omega \leq 1 \\ \cosh(n \cosh^{-1} \Omega) & \Omega > 1 \end{cases} \qquad (2.17)$$

Applying some theorems of hyperbolic functions, the following useful representation can be derived [9].

$$T_n(\Omega) = \frac{\left(\Omega + \sqrt{\Omega^2 - 1}\right)^n + \left(\Omega + \sqrt{\Omega^2 - 1}\right)^{-n}}{2} \qquad (2.18)$$

For integer orders n, (2.18) yields polynomials with integer coefficients. The first polynomials for $n = 1 \ldots 5$ are given below.

$$T_1(\Omega) = \Omega$$

$$T_2(\Omega) = 2\Omega^2 - 1$$

$$T_3(\Omega) = 4\Omega^3 - 3\Omega$$

$$T_4(\Omega) = 8\Omega^4 - 8\Omega^2 + 1$$

$$T_5(\Omega) = 16\Omega^5 - 20\Omega^3 + 5\Omega$$

A simple recursive formula enables the computation of further Chebyshev polynomials.

$$T_{n+1}(\Omega) = 2\Omega T_n(\Omega) - T_{n-1}(\Omega) \qquad (2.19)$$

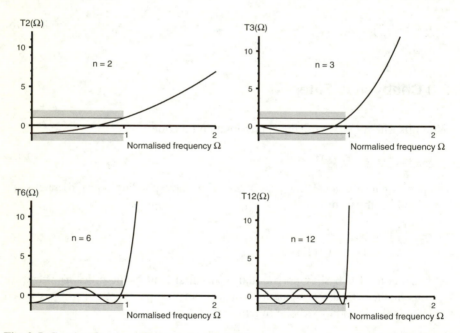

Fig. 2-5 Graphs of various Chebyshev polynomials

Figure 2-5 shows the graphs of some Chebyshev polynomials. These oscillate between +1 and −1 in the frequency range $0 \leq \Omega \leq 1$. The number of maxima and minima occurring is proportional to the degree n of the polynomial. For $\Omega > 1$ the functions monotonously approach infinity. The slope of the curves also increases proportionally to the order of the polynomial. ε is another independent design parameter which can be used to influence the ripple of the magnitude in the passband, as can be seen upon examination of (2.16). The smaller ε is, the smaller is the amplitude of the ripple in the passband. At the same time, the slope of the magnitude curve in the stopband decreases. For a given filter order, sharpness of the cutoff and flatness of the magnitude characteristic are contradictory requirements. Comparison of Fig. 2-3 and Fig. 2-6 demonstrates that the achievable sharpness of the cutoff of the Chebyshev filter is higher compared than that of the Butterworth filter, if one assumes equal filter order, passband ripple and stopband attenuation.

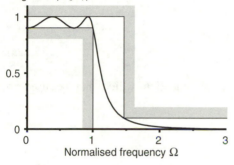

Fig. 2-6
Magnitude characteristic of a fourth-order Chebyshev filter
$v_p = 0.9$, $\varepsilon = 0.484$

In order to determine the poles of the Chebyshev filter, we start with a change to the complex frequency variable $P = j\Omega$.

$$|H(P)|^2 = \frac{1}{1 + \varepsilon^2 T_N^2(-jP)}$$

$$\text{for } P = j\Omega$$

The zeros of the denominator polynomial are the poles of the transfer function $H(P)$. So the poles can be determined by equating the denominator with zero.

$$1 + \varepsilon^2 T_N^2(-jP) = 0 \quad \text{or}$$
$$T_N(-jP) = \pm j / \varepsilon$$

By substitution of (2.17), we obtain the relationship

$$\cos[N \arccos(-jP)] = \pm j / \varepsilon \ .$$

Solving for P to obtain the poles is performed in two steps. We start with the introduction of a complex auxiliary variable $R + jI$.

$$R + jI = \arccos(-jP) \tag{2.20}$$

Thus the problem reduces, for the time being, to the solution of

$$\cos[N(R+jI)] = \pm j / \varepsilon .$$

Comparison of the real and imaginary parts on both sides of the equals sign yields two equations for determining R and I.

$$\cos NR \cosh NI = 0 \tag{2.21a}$$

$$\sin NR \sinh NI = \pm 1 / \varepsilon \tag{2.21b}$$

As $\cosh x \neq 0$ for all real x, we get an equation from (2.21a) which can be used to directly determine the real part R of the auxiliary variable.

$$\cos NR = 0$$

$$NR = \pi / 2 + v\pi$$

$$R_v = \frac{\pi}{2N} + v\frac{\pi}{N} \tag{2.22a}$$

Substitution of (2.22a) in (2.21b) yields an equation which can be used to determine the imaginary part I.

$$(-1)^v \sinh NI = \pm 1 / \varepsilon$$

$$I_v = \pm \frac{1}{N} \mathrm{arsinh}(1/\varepsilon) \tag{2.22b}$$

The next step in calculating the poles of the Chebyshev filter is to solve for P in (2.20).

$$R + jI = \arccos(-jP_{\infty v})$$

$$-jP_{\infty v} = \cos(R + jI)$$

$$P_{\infty v} = j\cos R_v \cosh I_v + \sin R_v \sinh I_v$$

Substitution of the already determined auxiliary variable (2.22a,b) leads to the desired relationship.

$$P_{\infty v} = \sin[\pi(2v+1)/2N]\sinh[\mathrm{arsinh}(1/\varepsilon)/N]$$
$$+ j\cos[\pi(2v+1)/2N]\cosh[\mathrm{arsinh}(1/\varepsilon)/N] \tag{2.23}$$
$$\text{with } 0 \leq v \leq 2N - 1$$

$2N$ solutions are obtained, which are located on an ellipse in the P-plane. Compare this to the situation concerning Butterworth filters in the last section, where the poles were found to be on the unit circle. Figure 2-7 shows a plot of the pole locations for the fourth-order Chebyshev filter whose magnitude graph has already been sketched in Fig. 2-6. Using the same arguments as in section 2.2, the transfer function of the desired filter is simply composed of the N poles in the left half-plane. This leads again to a transfer function of the form

$$H(P) = \frac{1}{(P - P_{\infty 1})(P - P_{\infty 2}) \ \cdots \ (P - P_{\infty N})} \qquad (2.24)$$

or, after separation into first- and second-order filter blocks, to a product representation of partial transfer functions.

$$H(P) = V \frac{1}{1 + A_{11}P + A_{21}P^2} \ \cdots \ \frac{1}{1 + A_{1k}P + A_{2k}P^2} \frac{1}{1 + A_{1(k+1)}P}$$

The factor V normalises the maximum gain of the low-pass filter to unity and can be calculated as

$$V = \begin{cases} 1 & \text{for } N \text{ odd} \\ 1/\sqrt{1 + \varepsilon^2} & \text{for } N \text{ even.} \end{cases}$$

For Chebyshev filters, tables are included in the Appendix, too, to show the coefficients A_{1i} and A_{2i} for filter orders up to $N = 12$. If N is even, $k = N/2$ second-order sections are needed to realise the filter. For odd orders, the filter is made up of one first-order and $k = (N-1)/2$ second-order sections. Substitution of P by p/ω_c yields the unnormalised transfer function $H(p)$.

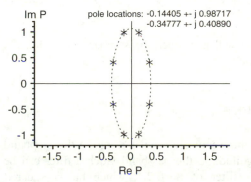

Fig. 2-7
Plot of the poles of a fourth-order Chebyshev filter (related magnitude graph is shown in Fig. 2-6)

The characteristics of a Chebyshev filter are determined by three parameters: the cutoff frequency ω_c, the parameter ε and the filter order N. According to (2.16), the parameter ε is directly related to the allowable deviation of the passband gain v_p.

$$v_p^2 = 1 / (1 + \varepsilon^2) \qquad (2.25)$$

The cutoff frequency of the filter, to which the frequency axis is normalised, is not the point of 3-dB attenuation, as in the case of the Butterworth filter, but the edge frequency, where the curve of the magnitude leaves the passband in the tolerance scheme (Fig. 2-6).

$$\omega_c = \omega_p \quad \text{or} \quad \Omega_p = 1 \tag{2.26a}$$

The relation to calculate the required order of the filter is derived from the condition that the graph just touches the stopband edge (v_s, ω_s).

$$N = \frac{\cosh^{-1}(v_{pnorm} / v_{snorm})}{\cosh^{-1}(\omega_s / \omega_p)} \tag{2.26b}$$

The similarity to the corresponding relation for the Butterworth filter is striking. The logarithm is simply replaced by the inverse hyperbolic cosine. For definite compliance with the prescribed tolerance scheme, the result of (2.26b) has to be rounded up to the next integer.

By appropriate choice of the parameter ε, we can determine how the "surplus" filter order, which is a result of rounding up, will be used to improve the filter performance beyond the specification. With

$$\varepsilon = 1 / v_{pnorm} \tag{2.26c}$$

we get a filter that has exactly the prescribed ripple in the passband, whereas the stopband behaviour is better than required. With

$$\varepsilon = \frac{1}{v_{snorm} \cosh\left(N \cosh^{-1}(\omega_s / \omega_p)\right)} , \tag{2.26d}$$

on the other hand, the magnitude plot will just touch the stopband edge, but the passband ripple is smaller than specified.

2.4 Inverse Chebyshev Filters

Occasionally, there are applications that require a regular ripple in the stopband and a monotonic course for the magnitude in the passband. Such a filter can be easily derived from the Chebyshev filter of Sect. 2.3, which has a squared magnitude function (2.16) of the form

$$|H(j\Omega)|^2 = \frac{1}{1 + \varepsilon^2 T_N^2(\Omega)} .$$

In the first step, we apply the transformation $\Omega \to 1/\Omega$ to invert the frequency axis, which yields a high-pass response.

$$|H(j\Omega)|^2 = \frac{1}{1 + \varepsilon^2 T_N^2(1/\Omega)} \tag{2.27}$$

In the second step, (2.27) is subtracted from unity which results in a low-pass magnitude function again. The ripple can now be found in the stopband.

$$|H(j\Omega)|^2 = 1 - \frac{1}{1 + \varepsilon^2 T_N^2(1/\Omega)}$$

$$|H(j\Omega)|^2 = \frac{\varepsilon^2 T_N^2(1/\Omega)}{1 + \varepsilon^2 T_N^2(1/\Omega)} \tag{2.28}$$

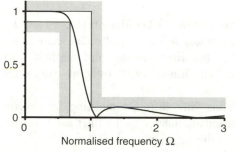

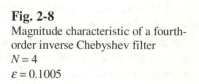

Fig. 2-8
Magnitude characteristic of a fourth-order inverse Chebyshev filter
$N = 4$
$\varepsilon = 0.1005$

Figure 2-8 shows the corresponding magnitude characteristic, which does not approach zero monotonically any more. This behaviour requires finite zeros of the transfer function, which are provided by the numerator polynomial in (2.28). These zeros can be calculated from the inverse of the zeros of the Chebyshev polynomial.

$$P_{0\mu} = \pm j \frac{1}{\cos[\pi(2\mu + 1)/2N]} \tag{2.29}$$

$$\text{with } 0 \le \mu \le M/2 - 1$$

The order of the numerator polynomial M is always even. It is related to the order N of the denominator polynomial in the following way:

$M = N$ for N even and

$M = N - 1$ for N odd.

The normalised poles of the inverse Chebyshev filter equal the reciprocal of the poles of the Chebyshev filter as introduced in Sect. 2.3. Figure 2-9 shows the location of the poles and zeros for a particular Chebyshev filter.

In the case of a cascade of first- and second-order filter blocks, we have the following product representation of partial transfer functions.

$$H(P) = \frac{1 + B_{21}P^2}{1 + A_{11}P + A_{21}P^2} \cdots \frac{1 + B_{2k}P^2}{1 + A_{1k}P + A_{2k}P^2} \frac{1}{1 + A_{1(k+1)}P} \tag{2.30}$$

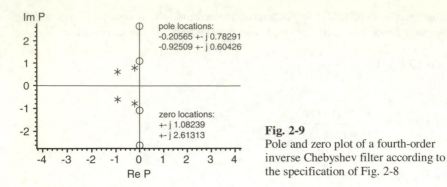

Fig. 2-9
Pole and zero plot of a fourth-order
inverse Chebyshev filter according to
the specification of Fig. 2-8

The filter design tables in the Appendix show the coefficients A_{1i}, A_{2i} and B_{2i} for filter orders up to $N = 12$. If N is even, $k = N/2$ second-order sections are needed to realise the filter. For odd orders, the filter is made up of one first-order and $k = (N-1)/2$ second-order sections. Substitution of P by p/ω_c yields the unnormalised transfer function $H(p)$.

The parameters ε and ω_c have a somewhat different meaning for inverse Chebyshev filters. ε determines the stopband attenuation of the filter and is related to the maximum gain v_s in the stopband by the equation

$$v_s^2 = \varepsilon^2 / (1 + \varepsilon^2).$$

The cutoff and normalisation frequency ω_c is now the edge frequency of the stopband.

$$\omega_c = \omega_s \quad \text{or} \quad \Omega_s = 1 \tag{2.32a}$$

For the determination of the required filter order, (2.26b) is still valid.

$$N = \frac{\cosh^{-1}(v_{pnorm} / v_{snorm})}{\cosh^{-1}(\omega_s / \omega_p)} \tag{2.32b}$$

The parameter ε specifies the maximum normalised gain v_{snorm} and thus determines the ripple in the stopband.

$$\varepsilon = v_{snorm} \tag{2.32c}$$

The presented design rules result in a frequency response which has exactly the prescribed ripple in the stopband. The rounded up filter order leads to a better behaviour in the passband then specified.

For a given tolerance scheme, Chebyshev and inverse Chebyshev filters require the same filter order. The complexity of the inverse filter, however, is higher because the zeros have to be implemented, too, but consequently we obtain a better group delay behaviour than that of the Chebyshev filter.

2.5 Elliptic Filters

We have shown that, compared to the Butterworth filter, a given tolerance scheme can be satisfied with a lower filter order if, with respect to the magnitude characteristic, a regular ripple is allowed in the passband (Chebyshev filter) or stopband (inverse Chebyshev filter). The next logical step is to simultaneously allow a ripple in the passband and stopband. Figure 2-10 shows the magnitude characteristic of a corresponding forth-order filter.

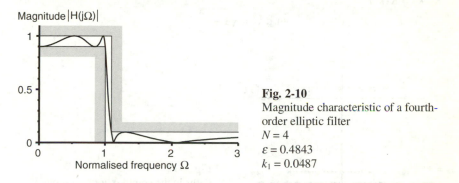

Fig. 2-10
Magnitude characteristic of a fourth-order elliptic filter
$N = 4$
$\varepsilon = 0.4843$
$k_1 = 0.0487$

The filter type with this property is called Cauer or elliptic filter because elliptic functions are used to calculate the poles and zeros. The squared magnitude function of the elliptic filter is of the form

$$|H(u)|^2 = \frac{1}{1 + \varepsilon^2 \operatorname{sn}^2(u, k_1)} .$$
(2.33)

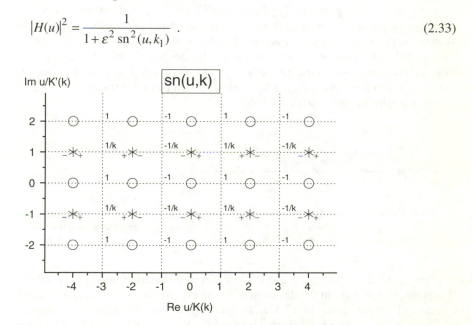

Fig. 2-11 Location of the poles and zeros of the Jacobi elliptic function

sn(u,k) is a Jacobian elliptic function [31] which in turn is the inverse function of the incomplete elliptic integral of the first kind. u is a complex variable. In the directions of both the real and the imaginary axes, this function has periodically recurrent poles and zeros in the complex u-plane (Fig. 2-11). The distances or periods in both directions are determined by the parameter k. The period $K(k)$ in the direction of the real axis is the complete elliptic integral of the first kind with the parameter k. The period in the direction of the imaginary axis is the complementary elliptic integral $K'(k)$.

$$K'(k) = K(k')$$

$$\text{with } k' = \sqrt{1 - k^2}$$

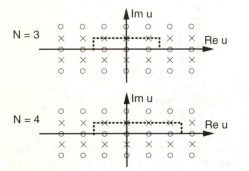

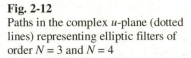

Fig. 2-12
Paths in the complex u-plane (dotted lines) representing elliptic filters of order $N = 3$ and $N = 4$

Along the dotted paths in Fig 2-11, sn(u,k) assumes real values. Complete elliptic integrals can be obtained from tables or calculated to an arbitrary precision by using appropriate series expansions. By skilful choice of a path in the u-plane through the poles and zeros, we can obtain the desired magnitude function for $H(u)$ with regular ripple in the pass and stopband. Such a path is a closed rectangle where N determines how many poles and zeros are passed through. Figure 2-12 shows examples for the orders $N = 3$ and $N = 4$.

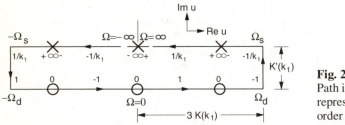

Fig. 2-13
Path in the u-plane representing a third-order elliptic filter

Figure 2-13 shows the rectangle for $N = 3$ in more detail. Starting at $\Omega = 0$ and following the rectangle in the direction of the arrow, we first pass the zeros, which correspond to the poles in (2.33). In this region, the magnitude of sn(u,k_1) oscillates between 0 and 1, which gives values between 1 and $1/(1+\varepsilon^2)$ for the squared magnitude of $H(u)$. The minimum gain in the passband can therefore be expressed as

$$v_p = 1 \Big/ \sqrt{1 + \varepsilon^2} \quad . \tag{2.34a}$$

After the zeros we pass the poles on the rectangle, which correspond to the zeros of the transfer function (2.33). The magnitude of $sn(u,k_1)$ oscillates here between ∞ and $1/k_1$, which gives values between 0 and $1/(1+\varepsilon^2/k_1^2)$ for the squared magnitude characteristic. So the maximum gain in the stopband is related to the design parameters ε and k_1 as follows:

$$v_s = 1 \Big/ \sqrt{1 + \varepsilon^2 / k_1^2} \quad . \tag{2.34b}$$

Starting from $\Omega = 0$ in the opposite direction, we obtain the same magnitude characteristic of the transfer function. This path corresponds to negative frequencies Ω.

From (2.34), we can derive relationships to determine the design parameters ε and k_1 from data of the tolerance scheme.

$$\varepsilon = \sqrt{1/v_p^2 - 1} = 1/v_{pnorm} \tag{2.35a}$$

$$k_1 = \sqrt{1/v_p^2 - 1} \Big/ \sqrt{1/v_s^2 - 1} = v_{snorm} / v_{pnorm} \tag{2.35b}$$

For the calculation of the poles and zeros, we analyse the denominator of (2.33) in more detail.

$$1 + \varepsilon^2 sn^2(u, k_1) \tag{2.36}$$

The poles of (2.36) and of $sn(u,k_1)$ are identical. The zeros fulfil the following condition:

$$1 + \varepsilon^2 sn^2(u, k_1) = 0$$

$$sn(u, k_1) = \pm j/\varepsilon$$

$$u_0(l,m) = \pm jU_0 + 2lK(k_1) + j2mK'(k_1)$$

$$\text{with } U_0 = \left| sn^{-1}(j/\varepsilon, k_1) \right| \quad .$$

Figure 2-14a shows a plot of the periodically occurring poles and zeros of the denominator term. Here we have directly interchanged the symbols for the poles and zeros (x,o), since the squared magnitude of $H(u)$ is the reciprocal of (2.36). Thus poles and zeros reverse roles. For a filter of Nth order, the following location of the poles and zeros can be read from the plot in u-plane:

$$u_{0i} = 2iK(k_1) + jK'(k_1)$$

$$u_{\infty i} = 2iK(k_1) + jU_0$$

$$\text{with } i = -(N-1)/2 \dots +(N-1)/2 \quad \text{for } N \text{ odd}$$

$$\text{and } i = -N/2 + 1 \dots +N/2 \quad \text{for } N \text{ even} \quad . \tag{2.37}$$

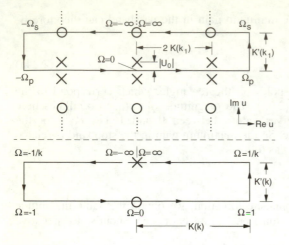

Fig. 2-14a
Poles and zeros of the squared
magnitude function $|H(u)|^2$

Fig. 2-14b
Mapping of the $j\Omega$-axis onto the
rectangle in the u-plane

For even filter orders, poles and zeros lie asymmetrically about the imaginary axis
in the u-plane as already suggested in Fig. 2-12.

In order to turn (2.33) into a magnitude response dependent on Ω, we have to
find a relationship that transforms the rectangle in the u-plane into the $j\Omega$-axis in
the P-plane. Jacobian elliptic functions are also involved in this problem.

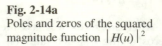

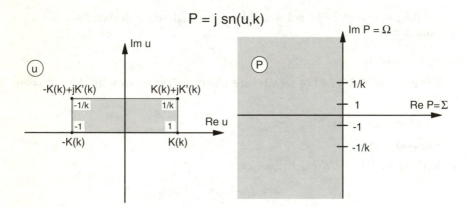

Fig. 2-15 Conformal mapping of a rectangle in the u-plane into the left half of the P-plane

The conformal mapping

$$P = j \, sn(u, k) \tag{2.38}$$

has exactly the desired property. Figure 2-15 shows that a rectangle with the
corners $-K$, $+K$, $K+jK'$ and $-K+jK'$ is transformed into the entire left half of the
P-plane. The edge of the rectangle turns into the $j\Omega$-axis.

$$\Omega = sn(u, k) \qquad \text{for } u \text{ on the rectangle}$$

For an arbitrary rectangle with the width $2B$ and the height H we have to change the scaling in (2.38) which results in

$$P = \mathrm{jsn}(u\,K(k)\,/\,B,k) \ . \tag{2.39a}$$

At the same time, the condition

$$\frac{K'(k)}{K(k)} = H\,/\,B \tag{2.39b}$$

has to be fulfilled. Applying (2.39) to the rectangle in Fig. 2-14a, which includes the poles and zeros to be transformed, yields the following relationships, which are found by comparison of Fig. 2-14a and Fig. 2.14b:

$$B = N\,K(k_1) \qquad \text{and}$$
$$H = K'(k_1) \ .$$

Hence, in our special case, the equations for the transformation from the u-plane into the P-plane read as follows:

$$P = \mathrm{jsn}\left(u\,\frac{K(k)}{NK'(k_1)},k \right) \tag{2.40a}$$

with $$\frac{K'(k)}{K(k)} = \frac{K'(k_1)}{NK(k_1)} \ . \tag{2.40b}$$

The parameter k which we have not specified in detail up to now has a clear meaning, too, which is illustrated by Fig. 2-14. The bottom right corner of the rectangle corresponds to the passband edge, which assumes the value 1 on the $j\Omega$-axis. The upper right corner corresponds to the stopband edge of the filter, which, as a consequence of the transformation, assumes the value $1/k$. This means that

$$k = \Omega_d\,/\,\Omega_s \ . \tag{2.41}$$

The parameter k is the quotient of the passband and stopband edge frequencies and represents the slope of the filter. Equation (2.40b) thus relates the characteristic parameters filter order (N), stopband attenuation and passband ripple (by choice of k_1) as well as the slope (by choice of k). If two parameters are given, the third one can be calculated. If k and k_1 are prescribed in a tolerance scheme, (2.40b) provides a relation to determine the required filter order.

$$N = \frac{K'(k_1)\,/\,K(k_1)}{K'(k)\,/\,K(k)} \tag{2.42}$$

Since complete elliptic integrals cannot be expressed by elementary functions, (2.42) is a somewhat unwieldy relation, which makes it necessary to use tables or computer support to determine the required filter order. A good approximation of (2.42) can be derived, because the conditions $k \approx 1$ and $k_1 \ll 1$ are well fulfilled in practical filter design problems. In these two extreme cases, complete elliptic integrals of the first kind can be approximated in the following way [31]:

$$K(k) = \pi / 2 \qquad\qquad\qquad \text{for } k \ll 1 \quad \text{ and}$$

$$K(k) = \ln(4 / k') = \ln\left(4 / \sqrt{1 - k^2} \right) \qquad \text{for } k \approx 1 \quad .$$

Thus (2.42) simplifies to

$$N \approx \frac{2}{\pi^2} \times \ln\frac{4}{k_1} \times \ln\frac{8}{1-k} \;. \qquad\qquad\qquad (2.43)$$

After rounding N up to the next integer, k and k_1 can be optimised with the help of (2.43). The "surplus" filter order can be used to increase the slope or the stopband attenuation or to decrease the passband ripple beyond the original specification.

With a parameter set k, k_1 and N that is harmonised in such a way that (2.42) is fulfilled, we can start to transform the poles and zeros from the u-plane (2.37) into the P-plane. Using the transformation (2.40) yields

$$P_{0i} = j\,\mathrm{sn}(2iK(k) / N + jK'(k), k) \quad \text{and} \qquad\qquad (2.44a)$$

$$P_{\infty i} = j\,\mathrm{sn}(2iK(k) / N + jU_0 K'(k) / K'(k_1), k) \qquad\qquad (2.44b)$$

$$\text{with } i = -(N-1)/2 \,...\, +(N-1)/2 \quad \text{for } N \text{ odd}$$
$$\text{and } i = -N/2+1 \,...\, +N/2 \qquad \text{for } N \text{ even} \quad .$$

By application of appropriate theorems of elliptic functions, we finally obtain real relations between the real and imaginary parts in the u- and P-plane.

$$P_r = \frac{-\mathrm{cn}(u_r,k)\,\mathrm{dn}(u_r,k)\,\mathrm{sn}(u_i,k')\,\mathrm{cn}(u_i,k')}{1 - \mathrm{dn}^2(u_r,k)\,\mathrm{sn}^2(u_i,k')} \quad \text{and} \qquad (2.45a)$$

$$P_i = \frac{\mathrm{sn}(u_r,k)\,\mathrm{dn}(u_i,k')}{1 - \mathrm{dn}^2(u_r,k)\,\mathrm{sn}^2(u_i,k')} \qquad\qquad (2.45b)$$

$$\text{with } \quad \mathrm{cn}^2(u,k) = 1 - \mathrm{sn}^2(u,k),$$
$$\mathrm{dn}^2(u,k) = 1 - k^2 \,\mathrm{sn}^2(u,k)$$
$$\text{and } \quad k' = \sqrt{1 - k^2}$$

The Pascal routine Cauer in the Appendix saves us the complicated mathematics and calculates the poles and zeros according to the specified parameters based on (2.44) and (2.45). It is true that some filter coefficient tables are included in the Appendix, but these can cover only a very limited range of the possibilities of elliptic filters because, compared to the Chebyshev filter, another design parameter (k_1) is available which offers an additional degree of freedom. Figure 2-16 sketches the poles and zeros of the example introduced in Fig. 2-10.

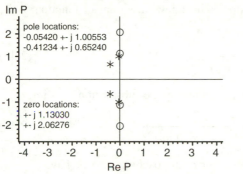

Fig. 2-16
Pole and zero plot of a fourth-order elliptic filter according to the specification of Fig. 2-10

For a cascade implementation of partial filters of first and second order, we have the following transfer function:

$$H(P) = V \frac{1 + B_{21}P^2}{1 + A_{11}P + A_{21}P^2} \cdots \frac{1 + B_{2k}P^2}{1 + A_{1k}P + A_{2k}P^2} \frac{1}{1 + A_{1(k+1)}P} \quad .$$

The factor V normalises the maximum gain of the low-pass filter to unity. V is given by

$$V = \begin{cases} 1 & \text{for } N \text{ odd} \\ 1/\sqrt{1+\varepsilon^2} & \text{for } N \text{ even} \end{cases} \quad .$$

The degree of the numerator polynomial M is always even. It is related to the degree N of the denominator polynomial in the following way:

$M = N$ for N even   and

$M = N - 1$ for N odd   .

2.6 Bessel Filters

Bessel low-pass filters are used if the form of a given input signal is to be preserved after filtering. Impulse deformations are mainly due to group delay distortions which occur especially in the vicinity of the cutoff frequency of the filter. Bessel filters have poles with a very low quality factor close to the aperiodic borderline case, which avoids any overshot in the output signal if, for instance, a step function is applied. The group delay characteristic is optimally flat, comparable to the magnitude response of the Butterworth filter. Thus Bessel filters can be specifically used to delay signals. The passband of the filter has to be chosen according to the bandwidth of the signal to be delayed. The larger the product of the bandwidth of the signal and the delay time is, the higher is the filter order needed.

We begin the following considerations with a low-pass transfer function of the form

$$H(p) = \frac{1}{1 + a_1 p + a_2 p^2 + a_3 p^3 + \ldots a_N p^N} \quad . \tag{2.46}$$

We obtain the corresponding frequency response by the substitution $p \to j\omega$.

$$H(j\omega) = \frac{1}{1 + ja_1\omega - a_2\omega^2 - ja_3\omega^3 + \ldots + a_N(j\omega)^N} \tag{2.47}$$

The coefficient a_1 has a special meaning. For ω approaching zero we have

$$H(j\omega) \approx \frac{1}{1 + ja_1\omega} \quad .$$

Calculation of the phase response yields for low frequencies

$$b(\omega) \approx \arctan\frac{a_1\omega}{1} = \arctan(a_1\omega) \quad .$$

The corresponding group delay characteristic of the filter takes the form

$$\tau_g(\omega) = \frac{d}{d\omega}\arctan(a_1\omega) = \frac{a_1}{1 + (a_1\omega)^2} \quad .$$

It appears that the coefficient a_1 equals the group delay τ_0 for ω approaching zero.

$$a_1 = \tau_g(0) = \tau_0$$

Starting from (2.46), we arrive at a normalised representation of the transfer function if we use the definitions $P = p\,\tau_0$ and $A_i = a_i / \tau_0^i$.

$$H(P) = \frac{1}{1 + P + A_2 P^2 + A_3 P^3 + \ldots + A_N P^N} \tag{2.48}$$

Using the definition $\Omega = \omega\,\tau_0$, the corresponding normalised frequency response reads

$$H(j\Omega) = \frac{1}{1 + j\Omega - A_2\Omega^2 - jA_3\Omega^3 + \ldots + A_N(j\Omega)^N} \quad . \tag{2.49}$$

Using (1.12b), the resulting phase response is of the following form:

$$b(\Omega) = \arctan \frac{\Omega - A_3\Omega^3 + A_5\Omega^5 - \dots}{1 - A_2\Omega^2 + A_4\Omega^4 - \dots} \ . \tag{2.50}$$

We shall next determine the coefficients A_i in such a way that we obtain a phase response that comes as close as possible to the ideal case

$$b(\omega) = \omega\,\tau_0 \qquad \text{or} \qquad b(\Omega) = \Omega \ .$$

τ_0 is the desired constant delay time we are aiming at. The A_i should therefore satisfy the following relationship over a large range of frequencies:

$$\Omega \approx \arctan \frac{\Omega - A_3\Omega^3 + A_5\Omega^5 - \dots}{1 - A_2\Omega^2 + A_4\Omega^4 - \dots} \tag{2.51a}$$

or

$$\tan\Omega \approx \frac{\Omega - A_3\Omega^3 + A_5\Omega^5 - \dots}{1 - A_2\Omega^2 + A_4\Omega^4 - \dots} \ . \tag{2.51b}$$

(2.51b) suggests choosing the A_i as the coefficients of truncated series expansions of the sine in the numerator and of the cosine in the denominator. These are well known and read

$$A_i = 1\,/\,i! \ .$$

(2.51b) would then take the form

$$\tan\Omega \approx \frac{\Omega - \frac{1}{6}\Omega^3 + \frac{1}{120}\Omega^5 - \dots}{1 - \frac{1}{2}\Omega^2 + \frac{1}{24}\Omega^4 - \dots} \ . \tag{2.52}$$

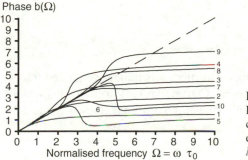

Phase b(Ω)

Normalised frequency $\Omega = \omega\,\tau_0$

Fig. 2-17
Phase response of low-pass filters with coefficients based on truncated series expansions of sine and cosine, $N = 1 \dots 10$

This choice of low-pass coefficients does not yield satisfactory results, however. The linear phase response is only reached in a very small range at low frequencies (Fig. 2-17) because the sine and cosine representations used here are series expansions about the point $\Omega = 0$. For higher frequencies, we obtain very different graphs of the phase response, depending on the filter order. The same is true for the magnitude characteristic, as shown in Fig. 2-18. The chosen approach to approximate the tangent by the quotient of truncated series expansions of sine and cosine is apparently not suitable to solve the present problem. For degrees of approximation $N > 4$ the poles and zeros do not occur at real values of Ω, as would be expected from the tangent function. Furthermore, stability is not guaranteed with these coefficients, as the corresponding transfer functions have poles in the right half of the P-plane.

Magnitude $|H(j\Omega)|$

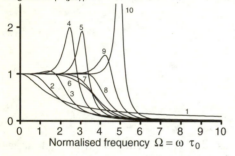

Normalised frequency $\Omega = \omega\, \tau_0$

Fig. 2-18
Magnitude response of low-pass filters with coefficients based on the truncated series expansion of sine and cosine, $N = 1 \dots 10$

Usable sets of coefficients are obtained, however, if the rational fractional function (2.52) is expanded into a continued fraction. Continued-fraction expansion is a mathematical method that is used, for instance, for the synthesis of passive two-terminal networks from given rational fractional impedance or admittance functions [48]. Applying this method to our problem, we have the following representation of the tangent function:

$$\tan \Omega \approx \cfrac{1}{\cfrac{1}{\Omega} - \cfrac{1}{\cfrac{3}{\Omega} - \cfrac{1}{\cfrac{5}{\Omega} - \cfrac{1}{\cfrac{7}{\Omega} - \cdots}}}} \quad .$$

According to the desired degree of approximation, this infinite continued fraction is truncated after the Nth partial quotient. Conversion back to rational fractional functions after truncation yields the following approximations for $N = 1$, 2 and 3:

$$N = 1 \qquad \tan \Omega \approx \Omega \,,$$

$$N = 2 \qquad \tan \Omega \approx \frac{3\Omega}{3 - \Omega^2} \text{ and}$$

$$N = 3 \qquad \tan \Omega \approx \frac{15\Omega - \Omega^3}{15 - 6\Omega^2} \,. \tag{2.53}$$

This truncated continued fraction shows the tendency to preserve important properties of the tangent function:

- All poles and zeros are real.
- Poles and zeros alternate.
- The coefficients lead to stable low-pass filters for all orders.

Figure 2-19 demonstrates the much better behaviour with respect to the phase response compared to Fig. 2-17. By differentiation we obtain the corresponding group delay characteristic, which is constant over a wide range of frequencies (Fig. 2-20). The magnitude (Fig. 2-21) shows a monotonically decreasing course, similar to the Butterworth filter but with a lower slope.

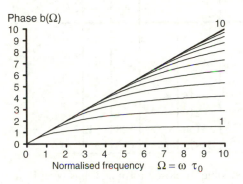

Phase b(Ω)

Normalised frequency $\Omega = \omega\, \tau_0$

Fig. 2-19
Phase response of low-pass filters with coefficients based on the truncated continued fraction of the tangent, $N = 1 \dots 10$

Inserting the coefficients of the tangent approximation (2.53) into (2.48) we obtain the following transfer functions:

$$N = 1 \qquad H(P) = \frac{1}{1+P}$$

$$N = 2 \qquad H(P) = \frac{3}{3+3P+P^2} \qquad\qquad (2.54)$$

$$N = 3 \qquad H(P) = \frac{15}{15+15P+6P^2+P^3} \; .$$

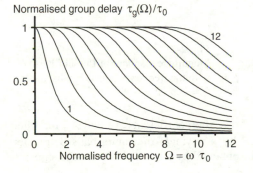

Normalised group delay $\tau_g(\Omega)/\tau_0$

Normalised frequency $\Omega = \omega\, \tau_0$

Fig. 2-20
Group delay characteristics of low-pass filters with coefficients based on the truncated continued-fraction expansion of the tangent, $N = 1 \dots 12$

In Table 2-1, the coefficients are summarised up to filter orders of $N = 8$. These need not be determined with the effort of evaluating the continued fraction expansions but can be calculated using relations of Bessel polynomials [58]. For inclusion in filter design programs, we provide the subroutine **Bessel** in the Appendix, which is an implementation of (2.55). The Bessel filter coefficients can be expressed as

$$a_v = \frac{(2n-v)!}{(n-v)!\,v!\,2^{n-v}} \,. \tag{2.55}$$

Table 2-1 Coefficients of Bessel filters

n	a_0	a_1	a_2	a_3	a_4	a_5	a_6	a_7	a_8
1	1	1							
2	3	3	1						
3	15	15	6	1					
4	105	105	45	10	1				
5	945	945	420	105	15	1			
6	10395	10395	4725	1260	210	21	1		
7	135135	135135	62370	17325	3150	378	28	1	
8	2027025	2027025	945945	270270	51975	6930	630	36	1

The calculation of the poles is not possible in closed form as it has been in the case of the filter types treated up to now. The only way is to calculate the coefficients as described above and to numerically determine the roots of the denominator polynomial in (2.48). For a cascade implementation of partial filters of first and second order, we have the transfer function

$$H(P) = \frac{1 + B_{21}P^2}{1 + A_{11}P + A_{21}P^2} \ \cdots \ \frac{1 + B_{2k}P^2}{1 + A_{1k}P + A_{2k}P^2}\,\frac{1}{1 + A_{1(k+1)}P} \,. \tag{2.56}$$

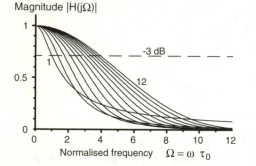

Fig. 2-21
Magnitude response of Bessel filters with the frequency axis normalised to τ_0, $N = 1 \ldots 12$

The design table in the Appendix shows the coefficients A_{1i} and A_{2i} for filter orders up to $N = 12$. Substitution of P by $p\,\tau_0$ yields the unnormalised transfer function $H(p)$ whose coefficients can be obtained by the following relationships:

$$a_{1i} = A_{1i}\,\tau_0 \qquad \text{and} \qquad a_{2i} = A_{2i}\,\tau_0^2 \,. \tag{2.57}$$

In our considerations thus far, the time delay τ_0 was the design parameter which determined, together with the filter order, the characteristics of the Bessel filter. It is of interest, of course, to know the corresponding 3-dB cutoff frequency as well. For $N > 3$, the following equation to determine the cutoff frequency Ω_{3dB} can only be solved numerically.

$$\left|H(j\Omega)\right|^2 = \frac{1}{(1 - A_2\Omega^2 + A_4\Omega^4 - \dots)^2 + (\Omega - A_3\Omega^3 + A_5\Omega^5 - \dots)^2} = \frac{1}{2}$$

This equation has N solutions, one of which is real and positive. This solution corresponds to the desired cutoff frequency. Table 2-2 shows the results up to the order $N = 12$.

N	$\Omega_{3dB}(N)$
1	1.000
2	1.361
3	1.756
4	2.114
5	2.427
6	2.703
7	2.952
8	3.180
9	3.392
10	3.591
11	3.780
12	3.959

Table 2-2
Normalised 3-dB cutoff frequencies

The determination of the 3-dB cutoff frequency provides an interesting relationship, which is valid for Bessel filters in general. The normalised and unnormalised cutoff frequencies are related as

$$\Omega_{3dB}(N) = \omega_{3dB}\,\tau_0 \,. \tag{2.58}$$

From (2.58) it follows that, for a given filter order N, the product of the group delay and the cutoff frequency is a constant. The higher the cutoff frequency is, the lower the attainable group delay, and vice versa. Equation (2.58) can also be used to estimate the filter order required to delay a signal of bandwidth $\omega_B \approx \omega_{3dB}$ at an amount of τ_0 with low distortion.

Finally, (2.58) opens up the possibility of designing Bessel filters with a prescribed 3-dB cutoff frequency. By substitution in (2.57), we obtain relationships determining the corresponding unnormalised filter coefficients:

$$a_{1i} = A_{1i}\,\Omega_{3dB} / \omega_{3dB} \qquad \text{and} \qquad a_{2i} = A_{2i}\,\Omega_{3dB}^2 / \omega_{3dB}^2 \,. \tag{2.59a}$$

From (2.59a) we can derive a set of normalised coefficients A'_{1i} and A'_{2i} such that the transfer function (2.56) is no longer normalised to the time delay τ_0, but to the 3-dB cutoff frequency ω_{3dB}.

$$A'_{1i} = A_{1i}\,\Omega_{3dB} \qquad\qquad A'_{2i} = A_{2i}\,\Omega^2_{3dB} \qquad\qquad (2.59b)$$

These coefficients can also be found in a table in the Appendix for filter orders up to 12. When using these coefficients, the normalised frequency P has to be replaced by p/ω_{3dB} in (2.56) in order to obtain the unnormalised transfer function. Figure 2-22 shows the corresponding logarithmic magnitude responses.

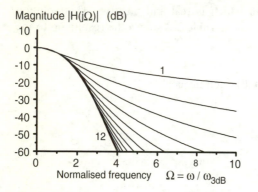

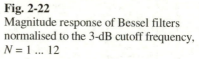

Fig. 2-22
Magnitude response of Bessel filters normalised to the 3-dB cutoff frequency, $N = 1 \ldots 12$

It is interesting to note that, up to the cutoff frequency, the magnitude response is almost independent of the chosen filter order. According to [58], the attenuation characteristic can be approximated by a square function in this region.

$$a(\omega\,/\,\omega_{3dB}) \approx 3 \times (\omega\,/\,\omega_{3dB})^2 \qquad \text{(dB)}$$

Similarly, the width of the transition band between passband and stopband is scarcely influenced by the filter order. However, for higher frequencies the filter order significantly influences the slope of the graph, which shows an asymptotic behaviour with a decay of $N\times 6$ dB/octave.

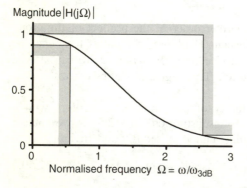

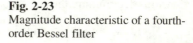

Fig. 2-23
Magnitude characteristic of a fourth-order Bessel filter

For comparison with Butterworth, Chebyshev and elliptic filters, we show the graph of the magnitude characteristic of a fourth-order Bessel filter plotted into a tolerance scheme (Fig. 2-23). The course is monotonic, similar to the behaviour of the Butterworth filter. The width of the transition band is larger, however. This is the price that has to be paid to enforce linearity of the phase response.

2.7 Filter Transformations

In the previous sections, we only considered low-pass filters with different characteristics in the passband and stopband. The coefficient tables in the Appendix also refer to low-pass filters only. The preferred analysis of this filter type is justified by the fact that the transition to other filter types such as high-pass, bandpass and bandstop filters is possible through relatively simple transformations of the frequency variable p. In the first step, the tolerance scheme of the filter to be designed has to be converted to the specification of the prototype low-pass filter. Afterwards the approximated prototype filter is converted back to the desired filter type using appropriate frequency transformations. In this section, we introduce the transformations for high-pass, bandpass and bandstop filters. Passband ripple and stopband attenuation are directly adopted into the tolerance scheme of the prototype low-pass filter (Fig. 2-24). The slope ω_s/ω_p of the prototype filter is derived from the edge frequencies of the tolerance scheme of the filter to be designed.

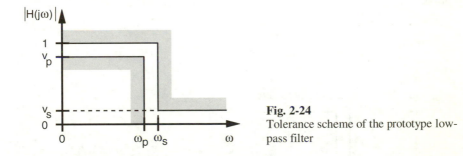

Fig. 2-24
Tolerance scheme of the prototype low-pass filter

The transformations introduced in the following sections are applied to transfer functions of the prototype low-pass filter, which are normalised with respect to the frequency axis. As a result, we again obtain a normalised transfer function, whose reference frequency depends on the type of transformation. For the high-pass transformation, the reference frequencies of high-pass and prototype low-pass filter are identical. For the bandpass and bandstop transformations, the frequency variable of the transfer function is normalised to the mid-band frequency of the passband or stopband respectively.

Figure 2-25 shows the magnitude characteristic of a fourth-order elliptic filter. This filter will be used in the following sections to demonstrate the effect of the various transformations on the frequency response.

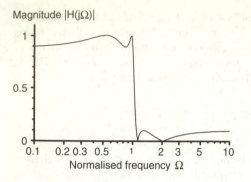

Magnitude $|H(j\Omega)|$

Normalised frequency Ω

Fig. 2-25
Magnitude characteristic of a fourth-
order elliptic filter
Passband ripple: 1 dB
Stopband attenuation: 20 dB

2.7.1 Low-Pass High-Pass Transformation

In this case, the slope of the low-pass prototype filter is derived from the edge frequencies of the high-pass tolerance scheme in Fig. 2-26 by the following relationship:

$$\omega_s / \omega_p = \omega_2 / \omega_1 . \tag{2.60}$$

After the design of the prototype filter, the high-pass transfer function complying with the tolerance scheme in Fig. 2-26 is obtained by the following transformation:

$$P \rightarrow 1/P . \tag{2.61}$$

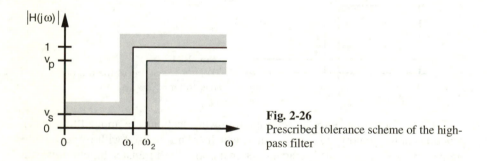

$|H(j\omega)|$

Fig. 2-26
Prescribed tolerance scheme of the high-
pass filter

In the transfer function of the prototype low-pass filter, the normalised frequency variable P has simply to be replaced by its reciprocal $1/P$. It is easy to see that this transformation reverses the frequency axis, which turns a low-pass into a high-pass filter. Figure 2-27 shows the magnitude characteristic of the high-pass filter which results from Fig. 2-25 by the transformation (2.61).

For Chebyshev and elliptic filters, the passband edge is the frequency to which the transfer functions of the prototype low-pass filter and the desired high-pass filter are normalised. The 3-dB cutoff frequency is the reference for Butterworth filters and it is also preserved after the transformation. If the Butterworth filter is not specified by the cutoff frequency but by the edge frequencies of a tolerance scheme, the coefficients of the prototype low-pass filter have to be transformed in such a way that the transfer function is normalised to the passband edge frequency. This modification of the coefficients, with the help of (2.62a), guarantees that the given tolerance scheme is satisfied.

$$B_2' = B_2\, f^2 \qquad A_2' = A_2\, f^2 \qquad A_1' = A_1\, f \qquad\qquad (2.62\text{a})$$

with f defined as

$$f = \frac{\omega_p}{\omega_{3\text{dB}}} = \left(\frac{1}{v_p^2} - 1\right)^{1/(2N)} \qquad\qquad (2.62\text{b})$$

Magnitude $|H(j\Omega)|$

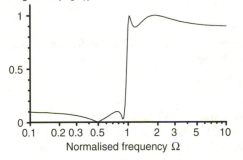

Normalised frequency Ω

Fig. 2-27
Magnitude characteristic of a fourth-order high-pass filter:
Passband ripple: 1 dB
Stopband attenuation: 20 db
(resulting from Fig. 2-25 by transformation of the frequency variable)

2.7.2 Low-Pass Bandpass Transformation

Equation (2.63) is used to calculate the slope of the prototype low-pass filter from the edge frequencies of the tolerance scheme of the desired bandpass filter (Fig. 2-28).

$$\omega_s\,/\,\omega_p = (\omega_4 - \omega_1)\,/\,(\omega_3 - \omega_2) \qquad\qquad (2.63)$$

For the specification of the tolerance scheme, it has to be observed that the applied frequency transformation leads to symmetrical bandpass filters which feature equal relative width of both transition bands. Hence we have

$$\omega_1\,\omega_4 = \omega_2\,\omega_3. \qquad\qquad (2.64)$$

Three edge frequencies can be chosen arbitrarily. The fourth one is determined by (2.64).

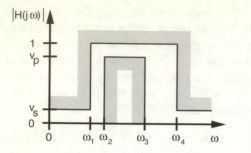

Fig. 2-28
Tolerance scheme of the bandpass filter

After the design of the prototype filter, the bandpass transfer function complying with the tolerance scheme in Fig. 2-28 is obtained by the following transformation:

$$P \quad \rightarrow \quad (P + 1/P)/B \ . \tag{2.65a}$$

B is the bandwidth of the filter normalised to the mid-band frequency.

$$B = (\omega_3 - \omega_2)/\omega_m \tag{2.65b}$$

ω_m is the mid-band frequency of the filter, which is related in the following way to the edge frequencies of the tolerance scheme:

$$\omega_m = \sqrt{\omega_1 \omega_4} = \sqrt{\omega_2 \omega_3} \ . \tag{2.65c}$$

The bandpass transfer function that we obtain by (2.65a) is normalised to this frequency. For Butterworth filters, we have to apply (2.62) to the prototype low-pass filter prior to the transformation, in order to comply with the given tolerance scheme. If we omit this step, the bandwidth B would be the difference between the two 3-dB points, and not between the passband edges.

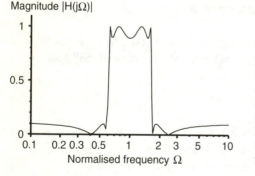

Fig. 2-29
Magnitude characteristic of an eighth-order bandpass filter:
Passband ripple: 1 dB
Stopband attenuation: 20 db
Bandwidth $B = 1$
(created by LP to BP transformation from Fig. 2-25)

Transformation (2.65) doubles the filter order. This means that the bandpass filter is of twice the order of the prototype low-pass filter. If we replace P by $j\Omega$ in (2.65), it is easy to see that the transformation turns a low-pass filter into a bandpass filter:

$$\Omega \quad \rightarrow \quad \Omega' = (\Omega - 1/\Omega)/B \;.$$

If Ω passes through the range $0 \leq \Omega \leq 1$, the low-pass characteristic is passed in the reverse direction from the stopband to the passband ($\Omega' = -\infty \ldots 0$). If Ω passes through the range $1 \leq \Omega \leq \infty$, we start in the passband and end up in the stopband ($\Omega' = 0 \ldots \infty$). Both ranges together yield a bandpass behaviour. Figure 2-29 shows the magnitude characteristic of the eighth-order bandpass filter, which results from the low-pass prototype in Fig. 2-25.

2.7.3 Low-Pass Bandstop Transformation

Equation (2.66) is used to calculate the slope of the prototype low-pass filter from the edge frequencies of the tolerance scheme of the desired bandstop filter (Fig. 2-30).

$$\omega_s / \omega_p = (\omega_4 - \omega_1)/(\omega_3 - \omega_2) \tag{2.66}$$

The same symmetry property as expressed by (2.64), which imposes constraints on the choice of the edge frequencies, applies to the bandstop filter. After the design of the prototype filter, the bandstop transfer function complying with the tolerance scheme in Fig. 2-30 is obtained by the transformation

$$P \quad \rightarrow \quad B/(P+1/P) \;. \tag{2.67a}$$

B is the bandwidth of the filter normalised to the mid-band frequency.

$$B = (\omega_4 - \omega_1)/\omega_m \tag{2.67b}$$

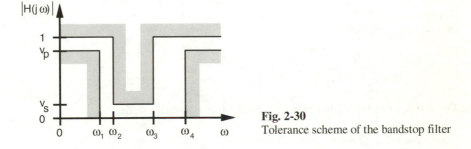

Fig. 2-30
Tolerance scheme of the bandstop filter

ω_m is the mid-band frequency of the filter, which is related to the edge frequencies of the tolerance scheme in the following way:

$$\omega_m = \sqrt{\omega_1 \omega_4} = \sqrt{\omega_2 \omega_3} \;. \tag{2.67c}$$

All the comments that we made in the previous section concerning the normalisation of the transfer function, the special handling of Butterworth filters, and the doubling of the filter order, also apply to the bandstop filter discussed here. With the same argumentation as presented there, it is easy to understand that the low-pass characteristic is passed in the first place from the passband to the stopband and afterwards in the opposite direction, which yields a bandstop characteristic. Figure 2-31 shows the magnitude characteristic of the eighth-order bandstop filter that results from the low-pass prototype in Fig. 2-25.

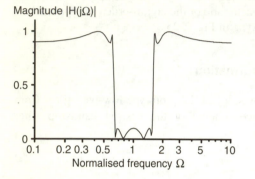

Magnitude |H(jΩ)|

Normalised frequency Ω

Fig. 2-31
Magnitude characteristic of an eighth-order bandstop filter:
Passband ripple: 1 dB
Stopband attenuation: 20 db
Bandwidth B = 1
(created by LP to BS transformation from Fig. 2-25)

2.7.4 Filter Transformations in Practice

The transformation relationships (2.61), (2.65) and (2.67) look quite simple at first glance. In practical application it turns out, however, that the computation of the coefficients of the transformed filter blocks may be rather complex in some cases. This is especially true for the low-pass bandpass and low-pass bandstop transformations. The formulas given in this section are derived in such a way that the coefficients can be easily calculated even with a pocket calculator.

All first- and second-order low-pass filter blocks that may occur with Butterworth, Chebyshev and elliptic filters have transfer functions of the following general form:

$$H(P) = \frac{1 + B_2 P^2}{1 + A_1 P + A_2 P^2} \quad .$$

Three types of transfer functions can be distinguished in practice:

Type 1: single real pole

$$B_2 = 0 \qquad A_2 = 0 \qquad A_1 \neq 0$$

Type 2: conjugate-complex pole pair

$$B_2 = 0 \qquad A_2 \neq 0 \qquad A_1 \neq 0$$

Type 3: conjugate-complex pole pair with imaginary zero pair

$$B_2 \neq 0 \qquad A_2 \neq 0 \qquad A_1 \neq 0 \quad .$$

The transformed blocks are of the form

$$H(P) = V \frac{B_{10} + B_{11}P + B_{12}P^2}{1 + A_{11}P + A_{12}P^2} \left[\frac{B_{20} + B_{21}P + B_{22}P^2}{1 + A_{21}P + A_{22}P^2} \right].$$

Because of the doubling of the filter order, the low-pass bandpass and low-pass bandstop transformations result in two filter blocks in both cases. One first- or second-order filter is the result of the low-pass high-pass transformation. In the following, we present the computational rules for the three transformation types.

Low-Pass High-Pass Transformation

Type 1

$B_{12} = 0$	$B_{11} = 1$	$B_{10} = 0$
$A_{12} = 0$	$A_{11} = 1/A_1$	$V = 1/A_1$

Type 2

$B_{12} = 1$	$B_{11} = 0$	$B_{10} = 0$
$A_{12} = 1/A_2$	$A_{11} = A_1/A_2$	$V = 1/A_2$

Type 3

$B_{12} = 1/B_2$	$B_{11} = 0$	$B_{10} = 1$
$A_{12} = 1/A_2$	$A_{11} = A_1/A_2$	$V = B_2/A_2$

Low-Pass Bandpass Transformation

Type 1

$B_{12} = 0$	$B_{11} = 1$	$B_{10} = 0$
$A_{12} = 1$	$A_{11} = B/A_1$	$V = B/A_1$

Type 2

$$r_1 = -A_1 B/(4 A_2)$$

$$i_1 = B \sqrt{4 A_2 - A_1^2}/(4 A2)$$

$$s = r_1^2 - i_1^2 - 1$$

$$t = 2 r_1 i_1$$

$$r_2 = \sqrt{0.5\left(\sqrt{s^2 + t^2} + s\right)}$$

$$i_2 = \sqrt{0.5\left(\sqrt{s^2 + t^2} - s\right)}$$

$$r_{p1} = r_1 + r_2$$

$$r_{p2} = r_1 - r_2$$

$$a_{p1} = (r_1 + r_2)^2 + (i_1 - i_2)^2$$

$$a_{p2} = (r_1 - r_2)^2 + (i_1 + i_2)^2$$

$$B_{12} = 0 \qquad\qquad B_{11} = 1 \qquad\qquad B_{10} = 0$$
$$A_{12} = 1/a_{p1} \qquad A_{11} = -2\,r_{p1}/a_{p1} \qquad V = B_2/A_2$$

$$B_{22} = 0 \qquad\qquad B_{21} = 1 \qquad\qquad B_{20} = 0$$
$$A_{22} = 1/a_{p2} \qquad A_{21} = -2\,r_{p2}/a_{p2}$$

Type 3

The denominator coefficients A_{ij} are calculated as in the case of type 2. The numerator coefficients B_{ij} and the gain V are obtained as follows:

$$i_1 = 0.5\,B/\sqrt{B_2} \qquad\qquad i_2 = \sqrt{1 + i_1^2}$$

$$B_{12} = 1/(i_1 - i_2)^2 \qquad B_{11} = 0 \qquad\qquad B_{10} = 1$$
$$A_{12} \text{ as for type 2} \qquad A_{11} \text{ as for type 2} \qquad V = B_2/A_2$$

$$B_{22} = 1/(i_1 + i_2)^2 \qquad B_{21} = 0 \qquad\qquad B_{20} = 1$$
$$A_{22} \text{ as for type 2} \qquad A_{21} \text{ as for type 2}$$

Low-Pass Bandstop Transformation

Type 1

$$B_{12} = 1 \qquad\qquad B_{11} = 0 \qquad\qquad B_{10} = 1$$
$$A_{12} = 1 \qquad\qquad A_{11} = B\,A_1 \qquad\qquad V = 1$$

Type 2

$$r_1 = -A_1\,B/4 \qquad\qquad i_1 = B\sqrt{4\,A_2 - A_1^2}\big/4$$
$$s = r_1^2 - i_1^2 - 1 \qquad\qquad t = 2\,r_1\,i_1$$
$$r_2 = \sqrt{0.5\left(\sqrt{s^2 + t^2} + s\right)} \qquad i_2 = \sqrt{0.5\left(\sqrt{s^2 + t^2} - s\right)}$$
$$r_{p1} = r_1 + r_2 \qquad\qquad r_{p2} = r_1 - r_2$$
$$a_{p1} = (r_1 + r_2)^2 + (i_1 - i_2)^2 \qquad a_{p2} = (r_1 - r_2)^2 + (i_1 + i_2)^2$$

$$B_{12} = 1 \qquad\qquad B_{11} = 0 \qquad\qquad B_{10} = 1$$
$$A_{12} = 1/a_{p1} \qquad A_{11} = -2\,r_{p1}/a_{p1} \qquad V = 1$$

$B_{22} = 1$	$B_{21} = 0$	$B_{20} = 1$
$A_{22} = 1/a_{p2}$	$A_{21} = -2\, r_{p2}/a_{p2}$	

Type 3

The denominator coefficients A_{ij} are calculated as in the case of type 2. The numerator coefficients B_{ij} and the gain V are obtained as follows:

$$i_1 = 0.5\,B\sqrt{B_2} \qquad\qquad i_2 = \sqrt{1 + i_1^2}$$

$B_{12} = 1/(i_1 - i_2)^2$	$B_{11} = 0$	$B_{10} = 1$
A_{12} as for type 2	A_{11} as for type 2	$V = 1$
$B_{22} = 1/(i_1 + i_2)^2$	$B_{21} = 0$	$B_{20} = 1$
A_{22} as for type 2	A_{21} as for type 2	

3 Discrete-Time Systems

In Chap. 1, we introduced the relations between the input and output signals of continuous-time LTI systems. In the time domain, the output signal $y(t)$ is calculated by convolution of the input signal $x(t)$ with the impulse response $h(t)$ of the system (1.1). In the frequency domain, the spectrum of the output signal $Y(j\omega)$ is simply obtained by multiplication of the spectrum of the input signal $X(j\omega)$ by the transfer function $H(j\omega)$ (1.10). In this chapter we derive the corresponding relations for discrete-time systems, which look quite similar.

Let us first consider the context of discrete-time systems. Input and output signals are no longer physical quantities, but sequences of dimensionless numbers. Having in mind a physical meaning behind a discrete-time signal, the physical quantity can always be obtained by multiplication by a reference value of this quantity. This is exactly what digital-to-analog converters do. In the following, the input sequence of the system is expressed by $x(n)$, and the output sequence by $y(n)$ where n points to the nth element of the sequence. If the sequence is related to a real-time signal or process, an association with the actual time axis can be made by writing $x(nT)$ or $y(nT)$ where T is the quantisation period of the time axis. A discrete-time system is the implementation of a mathematical algorithm, which transforms the input sequence $x(n)$ into the output sequence $y(n)$. The algorithm may be realised in dedicated hardware or in software running on Digital Signal Processors (DSPs) or on general purpose processors. The processing of signal sequences may take place in real-time, which requires synchronisation with other processes like A/D and D/A conversion happening at the same time. On the other hand, signal data may be processed off-line without any relationship to the time axis. An example is processing of data that are available on disk and that are written back to disk after processing.

The terms "discrete-time" and "digital" are often used synonymously. But strictly speaking, discrete-time means quantisation of the time axis only, whereas digital is related to a real technical implementation in which signal levels have to be quantised too, as they can only be represented by a finite number of bits. The effects resulting from the limited resolution of signal levels are discussed in detail in Chap. 8.

In the discrete-time domain, there are also transfer systems that have the properties of linearity and time-invariance. Instead of time-invariance, however, we prefer to use the term "shift-invariance", since the processing of data may take place without any relation to the real time axis, as stated above. Thus we speak of linear shift-invariant (LSI) systems.

Linearity means that two superimposed signals pass through a system without mutual interference. Let $y_1(n)$ be the response of the system to $x_1(n)$ and $y_2(n)$ the response to $x_2(n)$. Applying a linear combination of both individual signals $x(n) = a\,x_1(n) + b\,x_2(n)$ to the input results in $y(n) = a\,y_1(n) + b\,y_2(n)$ at the output.

The property of shift-invariance can be described as follows: Let $y(n)$ be the reaction of the system to the input sequence $x(n)$. If one applies the same sequence to the input but delayed for n_0 cycles of the algorithm or n_0 clock cycles of the hardware, then the response to $x(n-n_0)$ is simply $y(n-n_0)$.

If both properties apply, the characteristics of the system can be fully described by the reaction to simple test signals such as pulses, steps and harmonic functions.

3.1 Discrete-Time Convolution

The unit-sample sequence $\delta(n)$ adopts the role of the delta function as a test function for the behaviour of the system in the discrete time domain. The unit-sample sequence is defined as

$$\delta(n) = \begin{cases} 1 & \text{for } n = 0 \\ 0 & \text{for } n \neq 0 \ . \end{cases} \tag{3.1}$$

The response of the system to $\delta(n)$ is the unit-sample response $h(n)$. Provided that the system is linear and shift-invariant, we can derive a general relationship between the input and output signals. We start by considering the reaction of the system to a single sample out of the whole of the input sequence. The n_0th pulse of the input sequence $x(n)$ can be written in the form

$$x(n_0)\,\delta(n - n_0) \ .$$

This single pulse leads to the output signal

$$x(n_0)\,h(n - n_0) \ .$$

To obtain this result, we have made use of the properties of linearity and shift-invariance. If we now superimpose the responses of the individual input samples by summation over all possible n_0, we obtain the response of the system to the whole input sequence $x(n)$, again making use of the linear property.

$$\sum_{n_0=-\infty}^{+\infty} x(n_0)\,h(n - n_0) = y(n) \tag{3.2}$$

According to (3.2) we can calculate the output sequence $y(n)$ as a response to any arbitrary input sequence $x(n)$ if the unit-sample response $h(n)$ is known.

Relation (3.2) is called discrete convolution. It is the analogous relation to (1.1). In the discrete-time case, a summation takes the place of the integral. $x(n)$ and $h(n)$ can reverse roles in (3.2), which yields the following two representations:

$$y(n) = \sum_{k=-\infty}^{+\infty} x(k)\, h(n-k) = \sum_{k=-\infty}^{+\infty} h(k)\, x(n-k) \ . \tag{3.3}$$

The symbol for the convolution is an asterisk (*). So (3.3) can be written in the abbreviated form

$$y(n) = x(n) * h(n) = h(n) * x(n) \ .$$

The following example is to illustrate the method of discrete convolution.

Example

A unit-step sequence, which is defined as

$$x(n) = 1 \quad \text{for } n \geq 0 \qquad \text{and} \qquad x(n) = 0 \quad \text{for } n < 0$$

is applied to the input of the system under consideration. The unit-sample response of this system is specified as follows:

$$h(n) = \begin{cases} 10 & \text{for } n = 0 \\ 4 & \text{for } n = 1 \\ 2 & \text{for } n = 2 \\ 1 & \text{for } n = 3 \\ 0 & \text{otherwise} \ . \end{cases}$$

In order to perform the convolution according to (3.3), we write down the sequences $h(n)$ and $x(n)$ in the form of a table. The following example shows the calculation of the output sample $y(n=2)$.

k	-3	-2	-1	0	1	2	3	4	5	6	7
x(k)	0	0	0	1	1	1	1	1	1	1	1
h(2-k)	0	0	1	2	4	10	0	0	0	0	0
y(2)	0 + 0 + 0 + 2 + 4 + 10 + 0 + 0 + 0 + 0 + 0 = **16**										

The values of $h(n)$ and $x(n)$ that appear in the table one below the other have to be multiplied. Afterwards all products are summed to get the output sample $y(2)$. To calculate other output samples, the position of the unit-sample response, which appears in reversed order in the 3rd line, has to be shifted accordingly. For other values of n, the convolution yields the following output samples:

n	-2	-1	0	1	2	3	4	5	6	7	8	9 .....
x(n)	0	0	1	1	1	1	1	1	1	1	1	1 .....
y(n)	0	0	10	14	16	17	17	17	17	17	17	17 .....

As a reaction to the unit-step sequence, we observe a slow rise up to the final value at the output, which is a clear low-pass behaviour.

The response to the unit-step sequence fully characterises the systems in the time domain as well. The unit-step function is defined as

$$u(n) = \begin{cases} 1 & \text{for } n \geq 0 \\ 0 & \text{for } n < 0 \ . \end{cases} \tag{3.4}$$

Substitution of (3.4) in the convolution sum (3.3) gives the following relationship between unit-sample response $h(n)$ and unit-step response $a(n)$ of the system:

$$a(n) = \sum_{m=-\infty}^{+\infty} u(n-m)\, h(m)$$

$$a(n) = \sum_{m=-\infty}^{n} h(m) \ . \tag{3.5a}$$

Conversely, the unit-sample response can be expressed as the difference of two consecutive samples of the unit-step response.

$$h(n) = a(n) - a(n-1) \tag{3.5b}$$

Digital filters, in principle, perform discrete convolutions according to (3.3). Numerous algorithms and filter structures are available for the practical implementation of the convolution which differ in complexity and performance. For the best choice, the respective application has to be considered. In Chap. 5, we introduce the most important filter structures and highlight the characteristics of the various realisations.

3.2 Relations in the Frequency Domain

In order to describe the properties of the discrete-time system in the frequency domain, we choose a complex harmonic oscillation as a test function, as we did previously in the continuous-time case. The discrete-time harmonic sequence is obtained by the substitution $t = nT$ in the continuous-time form:

$$x(t) = e^{j\omega_0 t}$$

$$x(n) = e^{j\omega_0 nT} \ .$$ (3.6)

T is the period at which the complex exponential is sampled. The reciprocal of T is commonly called sampling frequency f_s.

$$f_s = 1/T$$

The corresponding angular frequency is defined as

$$\omega_s = 2\pi f_s = 2\pi/T \ .$$

In the case of digital real-time applications, where signal processing is synchronised with other processes in the continuous-time domain, such as A/D and D/A conversion, frequency f or ω and period T have the usual meaning of "cycles per time unit" and "period of sampling". But if the samples $x(n)$ are available in stored form, on a hard disk for instance, it is no longer possible to conclude how these samples have been generated. A 200 Hz sinusoid sampled at 1 kHz yields the same sequence as a 1000 Hz sinusoid sampled at 5 kHz. Hence it seems reasonable to use the normalised frequency ωT which specifies how the angle progresses between two successive samples of the harmonic oscillation. Using normalised frequencies allows the treatment of discrete-time systems independent of the real time axis.

No special symbol has been introduced in the literature for the frequency normalised to the sampling period. Some authors use ω for this purpose, as well, and indicate a different meaning in the course of the text. Using this style, the samples of an harmonic exponential oscillation can be expressed as

$$x(n) = e^{j\omega n} \ .$$

Throughout this text, the combination ωT will be used exclusively. We leave it to the reader to assume either processes in real time, with the corresponding meaning of ω and T, or signal processing without any relation to the real time axis, where the interpretation of ωT as normalised frequency is more appropriate.

Convolution of (3.6) with the unit-sample response yields the response of the system to the complex exponential.

$$y(n) = \sum_{m=-\infty}^{+\infty} e^{j\omega_0 T(n-m)} h(m)$$

$$y(n) = e^{j\omega_0 Tn} \sum_{m=-\infty}^{+\infty} e^{-j\omega_0 Tm} h(m)$$ (3.7)

The summation term in (3.7) results in a complex variable $H(e^{j\omega_0 T})$, which is a function of the normalised frequency $\omega_0 T$. The relation

$$H\left(e^{j\omega_0 T}\right) = \sum_{m=-\infty}^{+\infty} e^{-j\omega_0 Tm}\, h(m)$$

is commonly referred to as the Fourier transform. $H(e^{j\omega_0 T})$ is the Fourier transform of the sequence $h(m)$. Thus (3.7) becomes

$$y(n) = e^{j\omega_0 Tn}\, H(e^{j\omega_0 T}).$$

$H(e^{j\omega_0 T})$ modifies the complex exponential with respect to amplitude and phase and therefore represents the gain and phase shift of the system specified by its unit-sample response $h(m)$. The Fourier transform (FT) of the unit-sample response evaluated at arbitrary frequencies ω is thus the frequency response of the system.

$$H\left(e^{j\omega T}\right) = \sum_{n=-\infty}^{+\infty} e^{-j\omega Tn}\, h(n) \tag{3.8}$$

According to (3.8), the frequency response of discrete-time systems is a function of the term $e^{j\omega T}$. As a consequence, the frequency response is periodic with the period $2\pi/T = \omega_s$. This result is closely related to the fact that the sequences

$$x(n) = e^{j(\omega + k\omega_s)nT} = e^{j(\omega nT + kn2\pi)}$$

are identical for all integer values of k. Since the input sequences are identical, the reactions $y(n)$ of the system will also be identical for all k. Consequently, the frequency response of the system assumes the same complex values at frequencies $\omega + k\,\omega_s$, which means periodicity with period ω_s.

Application of the FT to (3.3) yields the starting point for the derivation of a general relationship between discrete-time input and output signals in the frequency domain.

$$Y\left(e^{j\omega T}\right) = \sum_{n=-\infty}^{+\infty} e^{-j\omega Tn}\, y(n)$$

$$Y\left(e^{j\omega T}\right) = \sum_{n=-\infty}^{+\infty} e^{-j\omega Tn} \sum_{k=-\infty}^{+\infty} x(k)\, h(n-k)$$

Rearrangement gives

$$Y\left(e^{j\omega T}\right) = \sum_{k=-\infty}^{+\infty} x(k)\, e^{-j\omega Tk} \sum_{n=-\infty}^{+\infty} h(n-k)\, e^{-j\omega T(n-k)}.$$

With $m = n-k$ it follows that

$$Y\left(e^{j\omega T}\right) = \sum_{k=-\infty}^{+\infty} x(k)\, e^{-j\omega Tk} \sum_{m=-\infty}^{+\infty} h(m)\, e^{-j\omega Tm}$$

$$Y\left(e^{j\omega T}\right) = X\left(e^{j\omega T}\right) H\left(e^{j\omega T}\right) . \tag{3.9}$$

Thus the FT of the output signal is the product of the FT of the input signal and the frequency response. This relationship is analogous to the one for continuous-time systems.

As already mentioned, the Fourier transform of a sequence $f(n)$ can be written in the form

$$F\left(e^{j\omega T}\right) = \sum_{n=-\infty}^{+\infty} e^{-j\omega Tn}\, f(n) . \tag{3.10a}$$

$F(e^{j\omega T})$ has the meaning of the spectrum of the number sequence $f(n)$. Conversely, if the spectrum is known, the sequence $f(n)$ can be calculated. For the derivation of the inverse relationship, we start by multiplying both sides of (3.10a) by $e^{j\omega kT}$ and integrating over one period of the periodic spectrum.

$$\int_{-\omega_s/2}^{+\omega_s/2} F(e^{j\omega T})e^{j\omega kT}\, d\omega = \int_{-\omega_s/2}^{+\omega_s/2} \sum_{n=-\infty}^{+\infty} f(n)e^{-j\omega nT} e^{j\omega kT}\, d\omega$$

$$\int_{-\omega_s/2}^{+\omega_s/2} F(e^{j\omega T})e^{j\omega kT}\, d\omega = \sum_{n=-\infty}^{+\infty} f(n) \int_{-\omega_s/2}^{+\omega_s/2} e^{j\omega(k-n)T}\, d\omega$$

The integral on the right side is nonzero for $n = k$ only, since the integral is otherwise evaluated over whole periods of the complex exponential.

$$\int_{-\omega_s/2}^{+\omega_s/2} F(e^{j\omega T})e^{j\omega kT}\, d\omega = f(k) \int_{-\omega_s/2}^{+\omega_s/2} 1 \times d\omega = f(k)\,\omega_s$$

$$f(k) = \frac{1}{\omega_s} \int_{-\omega_s/2}^{+\omega_s/2} F(e^{j\omega T})e^{j\omega kT}\, d\omega = \frac{1}{2\pi} \int_{-\pi}^{+\pi} F(e^{j\omega T})e^{j\omega kT}\, d\omega T \tag{3.10b}$$

Relation (3.10b) constitutes the inverse Fourier transform of a discrete-time signal. If the sequence $f(n)$ is real, the spectrum $F(e^{j\omega T})$ has to fulfil a certain symmetry condition:

$$F(e^{j\omega T}) = F^*(e^{-j\omega T}) . \tag{3.11}$$

Only if (3.11) is met, will the integral (3.10b) yield real values for $f(k)$.

3.3 The z-Transform

3.3.1 Definition of the z-Transform

In the context of the FT of discrete-time signals, we come across the same problems that we have already encountered with continuous-time signals. Assuming idealised signals such as the unit-step sequence or a sinusoid sequence extending from $n = -\infty$ to $n = +\infty$, the Fourier sum (3.10a) does not converge. On the other hand, these functions are often used in theoretical considerations. If the FT is applied to calculate the spectrum of such discrete-time signals, the only way out is to treat the occurring singularities using the concept of distributions. The latter allows the characterisation of the energy, which is concentrated on one frequency or distributed over several or an infinite number of frequencies, by delta functions. We already used the same approach in the context of continuous-time signals. In many applications, however, calculations are preferably performed in the frequency domain, to make use of the much simpler multiplication, instead of the equivalent convolution in the time domain. In these cases, a similar approach to the Laplace transform is available in the discrete-time domain.

First of all, we only consider sequences whose elements are zero for $n < 0$ (right-sided transform). The samples to be transformed are multiplied by a damping term of the form $e^{-\sigma n T}$. σ is a real parameter that can be freely chosen such that the Fourier sum converges in any case.

$$F(e^{j\omega T}, \sigma) = \sum_{n=0}^{+\infty} f(n) e^{-\sigma n T} e^{-j\omega n T}$$

Combining the exponentials yields

$$F(e^{j\omega T}, \sigma) = \sum_{n=0}^{+\infty} f(n) e^{-(\sigma + j\omega) n T} \ .$$

The term $e^{(\sigma + j\omega)T}$ is commonly represented by the letter z.

$$F(z) = \sum_{n=0}^{+\infty} f(n) z^{-n} \qquad (3.12)$$

Equation (3.12) constitutes the so-called right-sided z-transform (ZT). With σ approaching zero, equivalent to replacing z by $e^{j\omega T}$, the z-transform turns into the Fourier transform. This transition requires caution, however: it is only possible without problems if the Fourier sum of the sequence $f(n)$ converges for all frequencies ω!

3.3.2 Properties of the z-Transform

The properties of the z-transform are discussed in this section only in so far as they are needed for the understanding of the following chapters. Emphasis is put on the convolution and shift theorems whose application is illustrated in a number of examples. Detailed analysis of the z-transform, also with methods of the theory of complex functions, can be found in [24, 46].

The Convolution Theorem

Let $c(n)$ be the result of the convolution of $a(n)$ and $b(n)$.

$$c(n) = \sum_{k=-\infty}^{+\infty} a(k)b(n-k)$$

According to (3.12) the z-transform of $c(n)$ can be calculated as

$$C(z) = \sum_{n=0}^{\infty} \left(\sum_{k=-\infty}^{+\infty} a(k)b(n-k) \right) z^{-n} \ .$$

Interchanging the order of summation yields

$$C(z) = \sum_{k=-\infty}^{+\infty} a(k) \sum_{n=0}^{\infty} b(n-k)z^{-n} \ .$$

The substitution $m = n-k$ results in

$$C(z) = \sum_{k=-\infty}^{+\infty} a(k) \sum_{m=-k}^{+\infty} b(m)z^{-(m+k)} \ .$$

As we only consider right-sided sequences, there are no contributions to the sums over m and k for $m < 0$ and $k < 0$.

$$C(z) = \sum_{k=0}^{\infty} a(k) \left(\sum_{m=0}^{\infty} b(m)z^{-m} \right) z^{-k}$$

If we choose σ, the real part of z, such that the two sums converge, we can express $C(z)$ as the product of two sums:

$$C(z) = \sum_{k=0}^{\infty} a(k)z^{-k} \sum_{m=0}^{\infty} a(m)z^{-m}$$

and hence

$$C(z) = A(z) \times B(z) \ .$$

It turns out that the property of the Fourier transform, that convolution in the time domain is equivalent to multiplication in the frequency domain, is also valid in the context of the z-transform. Applied to discrete-time systems, this means that the product of the z-transform of the input $X(z)$ and of the unit-sample response of the system $H(z)$ yields the z-transform of the output $Y(z)$.

$$Y(z) = X(z) \times H(z) \tag{3.13}$$

$H(z)$ is called the system or transfer function. If the Fourier sum of $h(n)$ converges, as stated above, z may be replaced by $e^{j\omega T}$ to get the frequency response of the system. Figure 3-1 summarises again the relationships between the input and output signals of discrete-time systems.

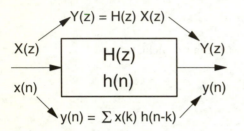

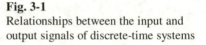

Fig. 3-1
Relationships between the input and output signals of discrete-time systems

The Shift Theorem

Let $F(z)$ be the z-transform of the sequence $f(n)$ and $F_d(z)$ the z-transform of the delayed sequence $f_d(n) = f(n-n_0)$ with $n_0 > 0$.

$$F_d(z) = \sum_{n=0}^{\infty} f(n-n_0)z^{-n}$$

We shall show how $F(z)$ and $F_d(z)$ are related. Substitution of $n-n_0$ by m yields

$$F_d(z) = \sum_{m=-n_0}^{\infty} f(m)z^{-(m+n_0)}$$

$$F_d(z) = z^{-n_0} \sum_{m=-n_0}^{\infty} f(m)z^{-m} \ .$$

Since $f(m)$ is zero for $m < 0$, there is no contribution to the sum for negative m. Hence we can write

$$F_d(z) = z^{-n_0} \sum_{m=0}^{\infty} f(m)z^{-m} = z^{-n_0} F(z) \ . \tag{3.14}$$

The delay of a sequence by n_0 sample periods in the time domain leads to a multiplication by z^{-n_0} in the frequency domain. Multiplication by the term z^{-1}, in particular, means a delay by one sample period. The term z^{-1} is therefore called the unit-delay operator. The unit delay is of great importance as it is a basic building block of digital signal processing algorithms, so it will often be found in the following chapters. Table 3-1 summarises the most important rules of the z-transform.

Time domain	Frequency domain
$f(n-n_0)$	$F(z)z^{-n_0}$
$f(n)z_0^n$	$F(z/z_0)$
$n f(n)$	$-z F'(z)$
$f_1(n) * f_2(n)$	$F_1(z) \times F_2(z)$

Table 3-1
A selection of the most important rules of the z-transform

Inverse z-Transform

For completeness, we introduce in this section the inverse z-transform, which can be mathematically expressed as

$$f(n) = \oint_C F(z)\, z^{n-1}\, dz \ . \tag{3.15}$$

The integration path C is a contour in the convergence region of $F(z)$, which encircles the origin and is passed through counterclockwise. In the case of the right-sided z-transform considered here, C encloses all poles of $F(z)$. The integration according to (3.15) needs not be carried out explicitly very often. Extensive tables are available from which the most important z-transform pairs needed in practice can be obtained. Most of the problems in the context of first- and second-order systems may be solved with the transform pairs shown in Table 3-2.

3.3.3 Examples

z-Transform of the Unit-Sample Sequence $\delta(n)$

$$f(n) = \delta(n)$$

$$F(z) = \sum_{n=0}^{\infty} \delta(n) z^{-n}$$

According to the definition of $\delta(n)$, there is only one contribution to the above sum for $n = 0$. For all other values of n, the unit-sample sequence is zero.

$$F(z) = 1 \times z^{-0} = 1$$

Time domain	Frequency domain
$\delta(n)$	1
$u(n)$	$\dfrac{z}{z-1}$
$n\,u(n)$	$\dfrac{z}{(z-1)^2}$
$z_0^n\,u(n)$	$\dfrac{z}{z-z_0}$
$(1-z_0^n)\,u(n)$	$\dfrac{z(1-z_0)}{(z-1)(z-z_0)}$
$\cos(\omega_0 n)\,u(n)$	$\dfrac{z^2-2z\cos\omega_0}{z^2-2z\cos\omega_0+1}$
$\sin(\omega_0 n)\,u(n)$	$\dfrac{z\sin\omega_0}{z^2-2z\cos\omega_0+1}$
$z_0^n\cos(\omega_0 n)\,u(n)$	$\dfrac{z^2-zz_0\cos\omega_0}{z^2-2zz_0\cos\omega_0+z_0^2}$
$z_0^n\sin(\omega_0 n)\,u(n)$	$\dfrac{zz_0\sin\omega_0}{z^2-2zz_0\cos\omega_0+z_0^2}$

Table 3-2
z-transforms of selected signal sequences

z-Transform of the Unit-Step Sequence *u(n)*

All samples of the unit-step sequence assume the value 1 for $n \geq 0$. Hence

$$F(z) = \sum_{n=0}^{\infty} u(n)\, z^{-n} = \sum_{n=0}^{\infty} z^{-n} \ .$$

This geometric progression converges for $|z^{-1}| < 1$, which is equivalent to choosing $\sigma > 0$, and yields

$$F(z) = \frac{1}{1-z^{-1}} = \frac{z}{z-1} \ . \tag{3.16}$$

Decaying Exponential

$$f(n) = e^{-\alpha nT}\, u(n) \qquad \text{with } \alpha > 0$$

$$F(z) = \sum_{n=0}^{\infty} e^{-\alpha nT}\, z^{-n} = \sum_{n=0}^{\infty} (e^{-\alpha T}\, z^{-1})^n$$

Again we have to deal with the sum of an infinite geometric progression. Convergence is assured providing that $|e^{-\alpha T} z^{-1}| < 1$ or $\sigma > -\alpha$. The infinite summation yields

$$F(z) = \frac{1}{1 - e^{-\alpha T} z^{-1}} = \frac{z}{z - e^{-\alpha T}} \ . \tag{3.17}$$

An Example to Illustrate the Convolution Theorem

Let a discrete-time system be characterised by an exponentially decaying unit-sample response of the form $h(n) = e^{-\alpha n T} u(n)$. As an input signal we apply the unit-step sequence $u(n)$. The output sequence $y(n)$ may be obtained in the time domain by convolution of $h(n)$ and $u(n)$.

$$y(n) = \sum_{k=-\infty}^{+\infty} u(k) h(n - k)$$

But in this example, we want to make use of the convolution theorem and perform our calculations in the frequency domain, where the convolution is replaced by the multiplication of the *z*-transforms.

$$Y(z) = X(z) \times H(z)$$

With the results of the previous examples, we can directly write down the *z*-transform of the output signal which is simply the product of (3.16) and (3.17).

$$Y(z) = \frac{z}{z-1} \frac{z}{z - e^{-\alpha T}} = \frac{z^2}{(z-1)(z - e^{-\alpha T})}$$

Partial fraction expansion yields

$$Y(z) = \frac{1}{1 - e^{-\alpha T}} \frac{z}{z-1} - \frac{e^{-\alpha T}}{1 - e^{-\alpha T}} \frac{z}{z - e^{-\alpha T}} \ .$$

Again using the results from these examples, it is easy to return to the time domain.

$$y(n) = \frac{1}{1 - e^{-\alpha T}} u(n) - \frac{e^{-\alpha T}}{1 - e^{-\alpha T}} e^{-\alpha n T} u(n)$$

$$y(n) = \frac{1}{1 - e^{-\alpha T}} \left(1 - e^{-\alpha(n+1)T}\right) u(n)$$

Figure 3-2 shows the input signal, the unit-sample response and the output signal of this example.

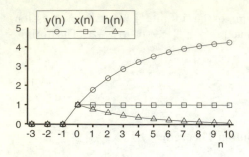

Fig. 3-2
Unit-step response of a filter with the unit-sample response $h(n) = \mathrm{e}^{-\alpha nT} u(n)$, $\alpha T = 0.25$

Calculation of the Frequency Response and Magnitude

Let us consider the transfer function of the system in the last example.

$$H(z) = \frac{z}{z - \mathrm{e}^{-\alpha T}} = \frac{1}{1 - \mathrm{e}^{-\alpha T} z^{-1}}$$

The corresponding frequency response is obtained by the transition $z \to \mathrm{e}^{j\omega T}$.

$$H(\mathrm{e}^{j\omega T}) = \frac{1}{1 - \mathrm{e}^{-\alpha T} \mathrm{e}^{-j\omega T}} \tag{3.18}$$

We multiply (3.18) by $H(\mathrm{e}^{-j\omega T})$ to get the squared magnitude of the transfer function.

$$\left| H\left(\mathrm{e}^{j\omega T}\right) \right|^2 = \frac{1}{1 - \mathrm{e}^{-\alpha T} \mathrm{e}^{-j\omega T}} \; \frac{1}{1 - \mathrm{e}^{-\alpha T} \mathrm{e}^{+j\omega T}}$$

$$\left| H\left(\mathrm{e}^{j\omega T}\right) \right|^2 = \frac{1}{1 + \mathrm{e}^{-2\alpha T} - \mathrm{e}^{-\alpha T}\left(\mathrm{e}^{+j\omega T} + \mathrm{e}^{-j\omega T}\right)}$$

$$\left| H\left(\mathrm{e}^{j\omega T}\right) \right|^2 = \frac{1}{1 + \mathrm{e}^{-2\alpha T} - 2\,\mathrm{e}^{-\alpha T} \cos \omega T}$$

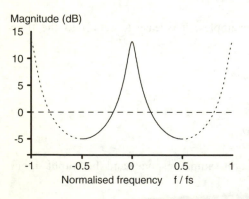

Fig. 3-3
Logarithmic magnitude of a first order low-pass filter
$h(n) = \mathrm{e}^{-\alpha Tn} u(n)$, $\alpha T = 0.25$

Figure 3-3 shows the logarithmic magnitude in dB, which is periodic with period f_s. The magnitude shows a low-pass characteristic, as could be expected from the slow rise of the reaction of the system to the unit-step sequence.

3.4 Stability Criterion in the Time Domain

There are a number of definitions of stability. One approach is the definition of stability by Lyapunov [37], who considers the behaviour of the system starting from an initial state with vanishing input signal. We come back to this definition later in Chap. 8. In this discussion we assume the so-called BIBO-stability (Bounded Input – Bounded Output), which requires that a limited input signal always results in a limited output signal. The requirement can be formulated mathematically as follows:

$$|y(n)| = \left| \sum_{k=-\infty}^{+\infty} x(n-k)h(k) \right| < B \ . \tag{3.19}$$

We are looking for constraints imposed on $h(k)$ which guarantee that the magnitude of the convolution sum stays below a certain upper boundary B whenever the input sequence is bounded. By application of the triangle inequality to the magnitude of the convolution sum, we obtain an expression which is guaranteed to be equal or larger than the term in (3.19). If this expression is bounded, then (3.19) also is fulfilled.

$$\sum_{k=-\infty}^{+\infty} |x(n-k)| \, |h(k)| < B$$

Replacing the magnitude of the input signal $|x(n)|$ by its maximum x_{max} again yields an expression whose magnitude is equal to or larger than the previous one.

$$\sum_{k=-\infty}^{+\infty} x_{max} |h(k)| = x_{max} \sum_{k=-\infty}^{+\infty} |h(k)| < B \tag{3.20}$$

In order to fulfil (3.20), it is required that the absolute sum of the unit-sample response is bounded.

$$\sum_{k=-\infty}^{+\infty} |h(k)| < \infty \tag{3.21}$$

The inequality (3.21) is the desired constraint on the unit-sample response guaranteeing BIBO-stability. If $h(k)$ is absolutely summable, then the system is stable in the discussed sense.

The inequality (3.21) is a criterion in the time domain. As the behaviour of systems is mostly characterised by the transfer function in the frequency domain, it would be desirable to have available a stability criterion in the frequency domain, as well. We come back to that problem later in Chap. 5, where we investigate the structures of digital filters and their transfer functions.

3.5 Further Properties of the Unit-Sample Response $h(n)$

The absolute sum of the unit-sample response has an important meaning with respect to the stability of the system, as shown in the previous section. Occasionally, the sum of the sequence and the sum of the squared sequence are of interest, too.

Let us start with considering (3.8) for the special case $\omega = 0$. It turns out that the sum of the unit-sample response corresponds to the gain of the system at the frequency $\omega = 0$.

$$\sum_{n=-\infty}^{+\infty} h(n) = H(0) \tag{3.22a}$$

For $\omega = \pi/T$, the exponential in (3.8) alternates between +1 and −1. Hence we have

$$\sum_{n=-\infty}^{+\infty} h(n)(-1)^n = H(\pi/T) \ . \tag{3.22b}$$

The equations (3.22a,b) can be used to normalise the unit-sample response $h(n)$ to an arbitrary system gain with respect to $\omega = 0$ or $\omega = \pi/T$.

The starting point of the following calculations is relation (3.8) again.

$$H\left(e^{j\omega T}\right) = \sum_{n=-\infty}^{+\infty} e^{-j\omega Tn} \, h(n) \tag{3.23}$$

The conjugate-complex frequency response can be written as

$$H^*\left(e^{j\omega T}\right) = \sum_{m=-\infty}^{+\infty} e^{+j\omega Tm} \, h(m) \ . \tag{3.24}$$

Integration of the product of (3.23) and (3.24) over a whole period $0 \ldots 2\pi/T$ of the periodic frequency response yields the following relationship. We have directly taken into account that the product of H and H^* corresponds to the squared magnitude of the frequency response.

$$\int_0^{2\pi/T} \left|H(e^{j\omega T})\right|^2 d\omega = \int_0^{2\pi/T} \sum_{n=-\infty}^{+\infty} h(n) e^{-j\omega Tn} \sum_{m=-\infty}^{+\infty} h(m) e^{+j\omega Tm} \, d\omega \qquad (3.25a)$$

$$\int_0^{2\pi/T} \left|H(e^{j\omega T})\right|^2 d\omega = \sum_{n=-\infty}^{+\infty} h(n) \sum_{m=-\infty}^{+\infty} h(m) \int_0^{2\pi/T} e^{j\omega T(m-n)} \, d\omega \qquad (3.25b)$$

When interchanging summation and integration in (3.25a), we tacitly assumed that the system under consideration is stable. The integral on the right side of (3.35b) is only nonzero for $m = n$ because the integration otherwise extends over whole periods of the periodic complex exponential.

$$\int_0^{2\pi/T} \left|H(e^{j\omega T})\right|^2 d\omega = \sum_{n=-\infty}^{+\infty} h(n) \sum_{m=-\infty}^{+\infty} h(m) \; 2\pi\delta(n-m)/T$$

$$= 2\pi/T \sum_{n=-\infty}^{+\infty} h^2(n)$$

$$\sum_{n=-\infty}^{+\infty} h^2(n) = \frac{T}{2\pi} \int_0^{2\pi/T} \left|H(e^{j\omega T})\right|^2 d\omega = \frac{T}{\pi} \int_0^{\pi/T} \left|H(e^{j\omega T})\right|^2 d\omega \qquad (3.26)$$

The sum over the squared unit-sample response provides information about the mean-square gain of the system. Equation (3.26) is known in the literature as one of the forms of Parseval's theorem.

The sum of a squared sequence is commonly referred to as the energy of the sequence. Equation (3.26) opens up the possibility of calculating the energy of the signal in the frequency domain by integrating the squared magnitude of the spectrum.

Example

Let the gain of a low-pass filter be unity up to a certain cutoff frequency ω_c. Beyond this frequency, up to $\omega = \pi/T$, the gain is zero. Evaluation of (3.26) with these data yields the following result:

$$\sum_{n=-\infty}^{+\infty} h^2(n) \approx \frac{T}{\pi} \int_0^{\omega_c} 1 \, d\omega = \frac{\omega_c T}{\pi} = 2 f_c T$$

Thus, the energy of the unit-sample response is proportional to the cutoff frequency of the low-pass filter if the gain in the passband is kept constant.

4 Sampling Theorem

4.1 Introduction

The sampling theorem, popularised by Shannon in 1948 [54], is one of the most important laws of signal theory. It makes the connection between the domain of the continuous-time physical signals and the abstract digital world, where signals are represented by sequences of numbers and the behaviour of electrical, mechanical or hydraulic systems is imitated by mathematical algorithms. The sampling theorem is a prerequisite for the pulse code modulation (PCM) which forms the basis for time multiplex techniques and digital measurement and control, as well as for the possibility of high quality transmission, processing and storage of signals in digital form.

The sampling theorem states that it is, under certain circumstances, sufficient to know samples of a continuous-time signal only at discrete times in order to reconstruct the original continuous-time signal completely.

The following considerations will highlight some important relationships between continuous-time and discrete-time signals.

4.2 Sampling

After periodic sampling of a continuous-time signal $f_a(t)$, we only know the values $f(n) = f_a(nT)$ of the original signal at equidistant instants nT. We want to show how the missing information about the course of the signal between the samples manifests itself in the frequency domain.

The time function $f_a(t)$ and spectrum $F_a(j\omega)$ of analogue signals are related by the transform pair of the Fourier transform. Provided that we are only interested in the signal values at the instants $t = nT$, the inverse transform (1.11b) takes on the form

$$f(n) = f_a(nT) = \frac{1}{2\pi} \int_{-\infty}^{+\infty} F_a(j\omega) e^{j\omega Tn} \, d\omega \ . \tag{4.1}$$

Decomposition of integral (4.1) into intervals of $2\pi / T$ and summation over all partial integrals yields

$$f(n) = \frac{1}{2\pi} \sum_{k=-\infty}^{+\infty} \int_{-\pi/T}^{+\pi/T} F_a(j\omega + jk2\pi/T) e^{j(\omega Tn + 2\pi kn)} \, d\omega \ .$$

Using the relation $\omega_s = 2\pi/T$, and taking into account that k and n are integers, it follows

$$f(n) = \frac{1}{2\pi} \sum_{k=-\infty}^{+\infty} \int_{-\omega_s/2}^{+\omega_s/2} F_a(j\omega + jk\omega_s) e^{j\omega Tn} \, d\omega \ .$$

By interchanging integration and summation we obtain

$$f(n) = \frac{1}{2\pi} \int_{-\omega_s/2}^{+\omega_s/2} e^{j\omega Tn} \sum_{k=-\infty}^{+\infty} F_a[j(\omega + k\omega_s)] \, d\omega \ . \tag{4.2a}$$

The sum under the integral in (4.2a) corresponds to the periodically continued spectrum of the original analog signal. For the calculation of the samples from the spectrum of the analog signal, we have, according to (4.2a), to calculate the inverse Fourier transform over one period of the priodically continued spectrum. Figure 4-1 shows a corresponding example. It is obvious from (4.2a) that the bold detail in the interval $\omega_s/2 \leq \omega \leq +\omega_s/2$ of the periodically continued spectrum contains the complete information of the samples $f(n)$.

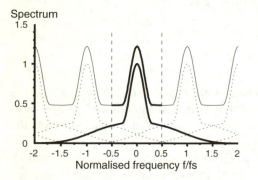

Spectrum

Normalised frequency f/fs

Fig. 4-1
Periodical continuation of the
spectrum (Shannon condition not met)

Comparison of (4.2a) with (3.10b) yields a relation between the Fourier transform of $f(n)$ and the spectrum of the original analog signal $f_a(t)$.

$$\frac{1}{2\pi} \sum_{k=-\infty}^{+\infty} F_a[j(\omega + k\omega_s)] = \frac{1}{\omega_s} F(e^{j\omega T})$$

Thus sampling of the analog signal $f_a(t)$ with the period T leads to the following equivalent relations in the time and frequency domains:

time domain: $f(n) = f_a(nT)$

frequency domain: $F\left(e^{j\omega T}\right) = \dfrac{1}{T} \displaystyle\sum_{k=-\infty}^{+\infty} F_a[j(\omega + k\omega_s)]$.   (4.3)

The Fourier transform of the sequence $f(n)$ is, apart from the constant factor $1/T$, identical to the periodically continued spectrum of the analog signal $f_a(t)$. By insertion of (3.10a) in (4.3), we obtain the reverse relation to (4.2a), which provides information about the analog spectrum derived from the samples $f(n)$.

$$\sum_{k=-\infty}^{+\infty} F_a[j(\omega + k\omega_s)] = T \sum_{n=-\infty}^{+\infty} f(n)\,e^{-j\omega Tn} \qquad\qquad (4.2b)$$

From (4.2b) it can be seen that the loss of information about the course of the analog signal $f_a(t)$ between the samples $f_a(nT)$ has the consequence that only the periodically continued spectrum is available. The original spectrum of the analog signal, in general, can not be recovered any more. The reason is that we cannot identify how the bold part of the periodical spectrum in Fig. 4-1 is composed of the periodically continued parts of the original spectrum. There is an inifinite number of possibilities for decomposing this spectrum into partial spectra. The equivalent statement in the time domain is that there is an infinite number of possible analog functions $f_a(t)$ to interpolate the given samples $f_a(nT)$.

4.3 Band-Limited Signals

If only samples of the analog signal $f_a(t)$ are available, the original spectrum $F_a(j\omega)$ cannot usually be recovered. The only exception is the special situation where the analog spectrum $F_a(j\omega)$ is limited to the frequency range $0 \le |\omega| < \omega_s/2$. No overlapping or "aliasing" takes place if the spectrum is periodically continued (Fig. 4-2). The original spectrum and hence the original analog signal can be recovered by simply removing the periodically continued parts of the spectrum of the discrete-time signal, which is accomplished by low-pass filtering at a cutoff frequency of $\omega_s/2$. This low-pass filtering interpolates the discrete-time samples and yields the original analog signal. Hence no information is lost if a signal is sampled at a frequency which is at least double the highest frequency contained in the spectrum of the signal. This is the essential message of the sampling theorem.

The mathematical relations between the samples $f(n)$ and the spectrum of the analog signal $F_a(j\omega)$ become considerably simpler if the condition of the sampling theorem is fulfilled. We can omit the periodical continuation in (4.2a) because there is no aliasing, and the spectrum in the interval $-\omega_s/2 < \omega < +\omega_s/2$ is identical to the original spectrum of the analog signal.

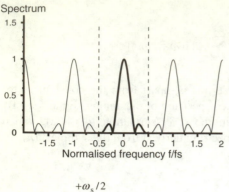

Spectrum
Normalised frequency f/fs

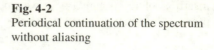

Fig. 4-2
Periodical continuation of the spectrum
without aliasing

$$f(n) = \frac{1}{2\pi} \int\limits_{-\omega_s/2}^{+\omega_s/2} F_a(j\omega) e^{j\omega Tn} \, d\omega \tag{4.4a}$$

The missing aliasing also simplifies the inverse relation (4.2b). The spectrum $F_a(j\omega)$ can be directly calculated from the samples $f(n)$.

$$F_a(j\omega) = \begin{cases} T \sum\limits_{n=-\infty}^{+\infty} f(n) e^{-j\omega Tn} = T\,F(e^{j\omega T}) & \text{for } |\omega| < \pi/T \\ 0 & \text{for } |\omega| \geq \pi/T \end{cases} \tag{4.4b}$$

In (4.4b) the spectrum of the analog signal $F_a(j\omega)$ is formally calculated as the inverse Fourier transform of the samples $f(n)$, setting all spectral components outside the range $-\omega_s/2 < \omega < +\omega_s/2$ to zero. The latter corresponds to an ideal low-pass filtering at a cutoff frequency of $\omega_s/2$. In Chap. 2.1 we have already seen that such an ideal filter is not realisable. Nevertheless we will apply the inverse Fourier transform (1.11b)

$$f_a(t) = \frac{1}{2\pi} \int\limits_{-\infty}^{+\infty} e^{j\omega t} \, F_a(j\omega) \, d\omega$$

to the spectrum according to (4.4b), which will provide us with a relation between the interpolated analog signal $f_a(t)$ and the samples $f(n)$. We will thus gain insight into the details of the interpolation mechanism for the case of ideal low-pass filtering. Insertion of (4.4b) into the inverse Fourier integral leads to

$$f_a(t) = \frac{1}{2\pi} \int\limits_{-\omega_s/2}^{+\omega_s/2} e^{j\omega t} \, T \sum\limits_{n=-\infty}^{+\infty} f(n) e^{-j\omega Tn} \, d\omega \; .$$

This integral covers only the range $-\omega_s/2 < \omega < +\omega_s/2$, which corresponds to ideal low-pass filtering. Interchanging summation and integration yields

$$f_a(t) = \frac{T}{2\pi} \sum_{n=-\infty}^{+\infty} f(n) \int_{-\omega_s/2}^{+\omega_s/2} e^{j\omega(t-nT)} \, d\omega \ .$$

Carrying out the integration results in

$$f_a(t) = \frac{1}{\omega_s} \sum_{k=-\infty}^{+\infty} f(n) \frac{e^{j\omega_s(t-nT)/2} - e^{-j\omega_s(t-nT)/2}}{j(t-nT)}$$

$$f_a(t) = \sum_{k=-\infty}^{+\infty} f(n) \frac{\sin \omega_s(t-nT)/2}{\omega_s(t-nT)/2} \ .$$

Introducing the sinc function that is defined as

$$\text{sinc}(x) = \frac{\sin x}{x} \ ,$$

we finally obtain

$$f_a(t) = \sum_{k=-\infty}^{+\infty} f(n)\,\text{sinc}[\omega_s(t-nT)/2] = \sum_{k=-\infty}^{+\infty} f(n)\,\text{sinc}[\pi(t/T - n)] \ . \tag{4.5}$$

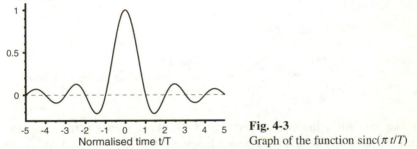

sin(π t/T)/(π t/T)

Fig. 4-3
Graph of the function sinc($\pi t/T$)

Normalised time t/T

Relation (4.5) shows that the interpolation is performed through sinc functions. These have the special property that they become unity at $t = 0$ and vanish at all other sampling instants $t = nT$ ($n \neq 0$) (Fig. 4-3). If we assign to each sampling instant a sinc function with an amplitude according to the value of the respective sample, we obtain the desired interpolated signal by summation of all shifted and weighted sinc functions (Fig. 4-4). Because of the previously mentioned property, the sinc functions do not interfere so that the interpolated signal assumes the same values at the sampling instants as the samples that have been interpolated. If the original analog signal $f_a(t)$ was band-limited to half the sampling frequency, interpolation of the samples $f_a(nT)$ through sinc functions yields a perfect reconstruction of the original signal.

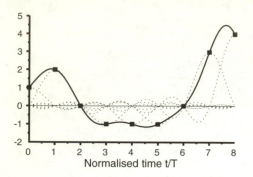

Fig. 4-4
Interpolation of samples through the sinc function

4.4 Practical Aspects of Signal Conversion

4.4.1 Analog-to-Digital Conversion

The process of A/D-conversion consists in practice of three essential steps, as shown in Fig. 4-5:

- low-pass filtering
- sample and hold
- the actual analog-to-digital conversion

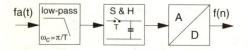 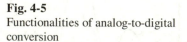

Fig. 4-5
Functionalities of analog-to-digital conversion

The low-pass filter limits the spectrum of the input signal $f_a(t)$ to half the sampling frequency. This avoids any overlapping of spectral components if the spectrum is periodically continued by the subsequent sampling. Half the sampling frequency is often referred to as Nyquist frequency. In the following, we will use the term baseband for the frequency range up to the Nyquist frequency, which corresponds to the frequency band usable for user information if the signal is digitised at a given sample rate. Low-pass filtering has to be performed carefully. Alias spectral components that are mirrored into the baseband cannot be removed later. It is impossible to distinguish after sampling which part of the spectrum is the original one and which part stems from the periodical continuation. Without having the original spectrum available, it is also impossible to restore the signal later in the time domain.

The usable range of the baseband is further reduced by the transition band of the low-pass filter, which cannot be made arbitrarily small in real implementations. The effort that has to be spent on filtering depends on the efficiency with which the baseband is to be used and on quality requirements with respect to the conservation of the original form of the signal. For the latter aspect, the group delay distortions in the vicinity of the cutoff frequency are an important factor.

Samplers that only sample a given signal at double the highest frequency contained in the signal are called Nyquist samplers. In Chap. 9 we will describe the concept of oversampling, which uses sampling rates much higher than needed to fulfil the sampling theorem. Conversion techniques based on this concept allow us to shift the expenditure for filtering and interpolation into the digital domain, to reduce the complexity of A/D and D/A converters and to improve the overall performance of signal processing by linear-phase filtering, which is not possible in the continuous-time domain.

In practice, sampling of signals is mostly performed using "sample and hold" circuits. The result is a stepped signal $f_s(t)$ which keeps the signal value at the sampling instant $f_a(nT)$ over a whole sampling period T (Fig. 4-6). This is necessary because most of the A/D converter types require the sampled value for a while to determine the corresponding numerical value in digital form. Some types of converters need multiple access to the sample value. If the value changes during the conversion process, the result would be incorrect.

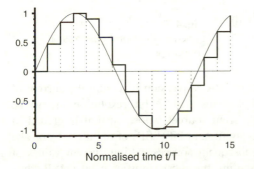

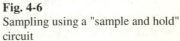

Fig. 4-6
Sampling using a "sample and hold" circuit

The optimal choice of an A/D converter architecture depends on the respective applications. Factors such as accuracy, resolution, the nature of the analog input signal, conversion speed, and environmental conditions have to be considered in the process of selection. Most of the converters on the market are based on one of the following principles or on variations of these:

- flash
- sub-ranging
- successive approximation
- dual slope
- voltage-to-frequency conversion (VFC)
- delta-sigma

Common to all architectures is the function to convert the rounded or truncated ratio of the analog input voltage to a certain reference voltage into a numerical value represented by a number of bits. Delta-sigma converters are often used in combination with the concept of oversampling. This principle of operation will be explained in more detail in Chap. 9.

4.4.2 Digital-to-Analog Conversion

We have shown by (4.4b) that the reconstruction of the analog signal $f_a(t)$ is conceptionally achieved by low-pass filtering of the spectrum $F(e^{j\omega T})$ of the corresponding discrete-time sequence $f(n)$. This procedure interpolates the course of the signal between the digital samples. What does it mean in a real physical implementation, however, to interpolate a sequence of numbers by means of an analog low-pass filter as suggested by (4.4b)? We need an intermediate step, in which the digital sample $f(n)$ is represented by some analog equivalent which can then be filtered using real analog hardware. This is the actual function of the D/A converter, which generates at its output a sequence of electrical pulses, at a rate of $1/T$, whose amplitude is proportional to the value of the corresponding digital samples $f(n)$. Figure 4-7 shows the essential steps of digital-to-analog conversion, which include the actual D/A conversion, deglitching and low-pass filtering.

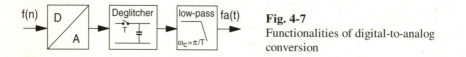

Fig. 4-7
Functionalities of digital-to-analog conversion

The most frequently used principle of D/A conversion is to sum the current of switched current sources. If digital samples with N-bit resolution are to be converted, the converter provides N current sources whose current is graded in binary exponential steps. Let the least significant bit (LSB) of the digital word be represented by a current of I_0. The subsequent sources exhibit current values of $2\,I_0$, $4\,I_0$, $8\,I_0$, etc. The source representing the most significant bit (MSB) thus exhibits a value of $2^{N-1}\,I_0$. Given a digital word to be converted, the current sources are switched on or off depending on the value of the corresponding bits. The current thus accumulated is then converted to a proportional voltage, which appears at the output of the converter. The advantage of using current sources in contrast to voltage sources is the higher achievable conversion speed.

In practical implementations, it can not be guaranteed that all current sources are switched at the same instant in the transition period from one digital input value to the next one. The result is a period in which the analog output value is uncertain. A so-called deglitcher is used to sample the output of the converter at an instant when it has reached a stable voltage and to provide an impulse at this voltage with a well-defined length. The function is thus similar to the sample and hold circuit used in A/D conversion. It is common to choose the length of the pulses equal to the sampling period T, which results in a stepped signal. The influence that the choice of the pulse length has on the conversion process will be discussed in more detail below in the mathematical analysis of interpolation by low-pass filtering.

The low-pass filter removes the periodically continued parts of the spectrum and interpolates the digital samples. Depending on the application, more or less effort is spent to come as close as possible to the behaviour of the ideal low-pass filter. If the best possible preservation of the form of a signal is an issue, oversampling techniques which allow linear-phase filtering in the digital domain are superior.

4.5 Mathematical Analysis of the D/A Conversion Process

As described above, the input signal to the interpolation low-pass filter is a sequence of pulses at a rate of $1/T$. These pulses have a width of τ and an amplitude according to the value of the corresponding digital sample $f(n)$. In most practical applications, τ is chosen equal to the sampling period T, which results in a stepped signal similar to the "sample and hold" signal shown in Fig. 4-6. In the following, however, we want to discuss the general case of an arbitrary impulse width $\tau \le T$. The sequence of pulses $f_s(t)$, weighted according to the value of the related digital samples $f(n)$, can be mathematically expressed as

$$f_s(t) = \sum_{n=-\infty}^{+\infty} f(n)\ \mathrm{rect}\left(\frac{t - nT}{\tau}\right), \tag{4.6}$$

where $\mathrm{rect}(x)$ is the unit rectangle as defined in Fig. 4-8. From that definition, it is easy to understand that $\mathrm{rect}[(t-t_0)/\tau]$ is the more general case of a rectangle occurring at t_0 and having a width of τ.

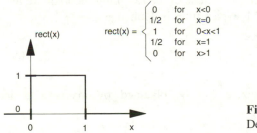

$$\mathrm{rect}(x) = \begin{cases} 0 & \text{for} & x<0 \\ 1/2 & \text{for} & x=0 \\ 1 & \text{for} & 0<x<1 \\ 1/2 & \text{for} & x=1 \\ 0 & \text{for} & x>1 \end{cases}$$

Fig. 4-8
Definition of the unit rectangle $\mathrm{rect}(x)$

With $f_s(t)$ we have found a continuous-time representation of the discrete-time sequence $f(n)$ that we want to interpolate. Applying the Fourier transform to (4.6), we obtain the spectrum of $f_s(t)$.

$$F_s(j\omega) = \int\limits_{-\infty}^{+\infty} \sum_{n=-\infty}^{+\infty} f(n)\, \text{rect}\left(\frac{t-nT}{\tau}\right) e^{-j\omega t}\, dt$$

Interchanging summation and integration yields

$$F_s(j\omega) = \sum_{n=-\infty}^{+\infty} f(n) \int\limits_{nT}^{nT+\tau} e^{-j\omega t}\, dt \ .$$

By carrying out the integration, we obtain

$$F_s(j\omega) = \frac{e^{-j\omega\tau}-1}{-j\omega} \sum_{n=-\infty}^{+\infty} f(n)\, e^{-j\omega Tn}$$

$$F_s(j\omega) = \frac{\tau}{T} \frac{e^{-j\omega\tau}-1}{-j\omega\tau}\, T \sum_{n=-\infty}^{+\infty} f(n)\, e^{-j\omega Tn} \ . \tag{4.7}$$

Substitution of (4.2b) in (4.7) yields

$$F_s(j\omega) = \frac{\tau}{T} \frac{e^{-j\omega\tau}-1}{-j\omega\tau} \sum_{k=-\infty}^{+\infty} F_a(j\omega + jk\omega_s) \ .$$

Finally, multiplying numerator and denominator of the fraction by $e^{j\omega\tau/2}$ results in

$$F_s(j\omega) = \frac{\tau}{T} \frac{\sin \omega\tau/2}{\omega\tau/2}\, e^{-j\omega\tau/2} \sum_{k=-\infty}^{+\infty} F_a(j\omega + jk\omega_s) \ . \tag{4.8}$$

Apart from the three terms preceding the sum, which we will discuss later, the spectrum of $f_s(t)$ equals the periodically continued spectrum of the original analog signal $f_a(t)$ that was sampled. According to Sect. 4.3 the analog signal can be recovered by low-pass filtering if the conditions of the sampling theorem are met. The frequency response of an ideal low-pass filter is defined as

$$H_{LP}(j\omega) = \begin{cases} 1 & \text{for } |\omega| \le \omega_c \\ 0 & \text{for } |\omega| > \omega_c \ . \end{cases} \tag{4.9}$$

The corresponding impulse response can be obtained by inverse Fourier transform:

$$h_{LP}(t) = \frac{1}{2\pi} \int\limits_{-\infty}^{+\infty} H_{LP}(j\omega)\, e^{j\omega t}\, d\omega = \frac{1}{2\pi} \int\limits_{-\omega_c}^{+\omega_c} e^{j\omega t}\, d\omega$$

$$h_{LP}(t) = \frac{1}{2\pi} \frac{e^{j\omega_c t} - e^{-j\omega_c t}}{jt} = \frac{\sin \omega_c t}{\pi t} \ .$$

The cutoff frequency of the interpolation filter is chosen as half the sampling frequency $\omega_c = \pi / T$.

$$h_{LP}(t) = \frac{\sin \pi t / T}{\pi t} = \frac{1}{T} \frac{\sin \pi t / T}{\pi t / T} = \frac{1}{T} \operatorname{sinc}(\pi t / T) \qquad (4.10)$$

Thus ideal low-pass filtering according to (4.9) corresponds to convolution with the sinc function in the time domain. In Sect. 4.3 we have already shown that digital samples are interpolated by sinc functions if ideal filtering is assumed to remove all spectral components outside the baseband. If we represent the sequence of samples $f(n)$ by a continuous-time function such as $f_s(t)$ (4.6), the interpolation of the samples is then accomplished by convolution of $f_s(t)$ with the sinc function (4.10).

As mentioned earlier, the ideal low-pass filter is not realisible, and (4.10) also shows that the impulse response is not causal. Real filters have to be used, with all their associated imperfections with respect to the magnitude and group delay characteristics. It is evident that interpolation will not be done by means of sinc functions in practical implementations. If the preservation of the form of a signal is of importance, we should consider reconstructing the signal exactly, at least at the sampling instants. This can be accomplished by means of interpolating impulse responses with properties similar to the sinc function (all values but one on a grid with period T are zero). An example of such an impulse response is shown in Fig. 4-9.

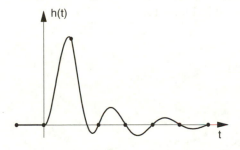

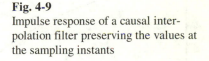

Fig. 4-9
Impulse response of a causal interpolation filter preserving the values at the sampling instants

Filters with this property fulfil the so-called first Nyquist condition, which plays an important role in digital data transmission. The overall impulse response of a transmission section which is determined by the pulse shaper on the transmitter side, by the characteristics of the transmission medium and by the frequency response of the equaliser at the receiver side, should approximate as close as possible a course that fulfils the Nyquist condition. This minimizes interference between neighbouring bits (intersymbol interference, ISI) and thus reduces the risk of wrong decisions concerning the recovered bit values at the receiver side.

Such interference-free low-pass filters exhibit an interesting behaviour in the frequency domain, as sketched in Fig. 4-10: the slope in the transition band is symmetrical about the frequency $f_s/2$. If periodically continued, the overall frequency response sums up to a constant.

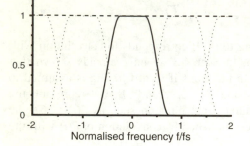

Fig. 4-10
Example of a frequency response that fulfils the 1st Nyquist condition

To complete the picture of signal restoration in the D/A conversion, we have to consider the three terms preceding the sum in (4.8).

$e^{-j\omega\tau/2}$ simply represents a delay at an amount of half the pulse width τ, which is more or less caused by the sluggish behaviour of the interpolating low-pass filter.

τ/T is a gain factor which attenuates the output signal of the filter according to the duty cycle of the pulse sequence. This is a result of the averaging behaviour of the low-pass filter, which spreads the energy of the pulses in time to yield a continuous analog signal.

The third term, a sinc function, requires a closer look, because it introduces a frequency-dependent distortion, as depicted in Fig. 4-11.

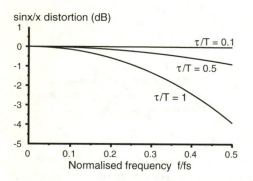

Fig. 4-11
Frequency response of the sinx/x-distortion (τ/T =1, 0.5 and 0.1)

The frequency response of this distortion, which is sometimes referred to as sinx/x-distortion, is, according to (4.8):

$$H_d(j\omega) = \frac{\sin(\omega\tau/2)}{\omega\tau/2} \ .$$

At first glance, the best solution seems to be the choice of narrow pulses. With decreasing τ, the sinc function approaches unity, which avoids the need for any further correction of the frequency response. The disadvantage of this approach is that the gain factor τ / T becomes smaller and smaller, which must be compensated for by additional amplification of the signal. Assuming a certain amount of noise in the environment of the D/A conversion circuitry, this means a deterioration in the achievable signal-to-noise ratio. In practical implementations, it is harder to reduce the environmental noise by 20 dB, for instance, than to incorporate a correction into the frequency response of the interpolation filter that amounts to about 4 dB at half the sampling frequency if τ is chosen equal to T (Fig. 4-11). So most of the D/A converters on the market deliver a stepped signal with $\tau = T$, which requires the previously mentioned compensation of the $\sin x/x$-distortion.

5 Filter Structures

5.1 Graphical Representation of Discrete-Time Networks

Digital filter algorithms are usually represented in form of block diagrams and flowgraphs. Both representations are based on the following three basic elements:
- Addition of two signal sequences
- Multiplication of a sequence by a constant
- Delay of a sequence by one sample period T

The last-mentioned function corresponds to the operation of a D flip-flop in digital systems. Triggered by the rising or falling edge of the applied clock signal, the information at the input is stored and passed to the output. This leads to the situation that we find a sample value at the output of the delay element that was applied to the input during the previous clock period. Shift-register structures, in which sample sequences can be stored and shifted, are realised by serial arrangement of the named delay elements.

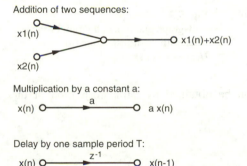

Addition of two sequences:

x1(n)

x2(n)

x1(n)+x2(n)

Multiplication by a constant a:

x(n)   a   a x(n)

Delay by one sample period T:

x(n) z^{-1} x(n-1)

Fig. 5-1
Elements of the flowgraph

Flowgraphs are comparable to the graphs used for the analysis of electrical networks. The topology of the network is described by means of nodes and branches. Figure 5-1 shows the elements of a flowgraph. The delay element is denoted with the unit-delay operator z^{-1}. Multiplication by a constant a is represented by a branch denoted with a. The addition of two symbols has no special symbol. Signals running up to a common node are added at this node.

Block diagrams depict the complexity of a digital filter. The amount of memory and the number of mathematical operations such as additions and multiplications required to realise the filter become visible. Figure 5-2 shows the symbols used.

Addition of two sequences:

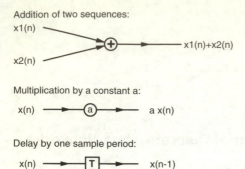

Multiplication by a constant a:

x(n) ———▶—(a)——— a x(n)

Delay by one sample period:

x(n) ———▶—[T]—▶—— x(n-1)

Fig. 5-2
Elements of the block diagram

As an introductory example, we consider a digital filter that calculates the average of two successive samples. Figure 5-3 shows the corresponding flowgraph and block diagram. The algorithm of the filter can be expressed as follows:

$$y(n) = 0.5 \left[x(n) + x(n-1) \right] \; .$$

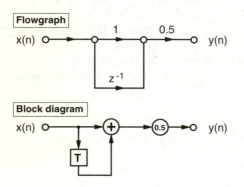

Fig. 5-3
Flowgraph and block diagram of the filter with the transfer function
$H(z) = 0.5 \, (1 + z^{-1})$

The transfer function of this filter can be obtained by z-transformation of the algorithm.

$$Y(z) = 0.5 \left[X(z) + X(z) \, z^{-1} \right]$$

$$Y(z) = 0.5 \left(1 + z^{-1} \right) X(z)$$

This results in the transfer function

$$H(z) = \frac{Y(z)}{X(z)}$$

$$H(z) = 0.5 \left(1 + z^{-1} \right) = 0.5 \, \frac{z+1}{z} \; .$$

The inverse transform of $H(z)$ into the time domain yields the unit-sample response of the filter. With the results of section 3.3 we obtain:

$$H(z) = 0.5 \times 1 + 0.5 \times 1 \times z^{-1}$$

$$h(n) = 0.5\,\delta(n) + 0.5\,\delta(n-1) \;.$$

The filter reacts to the unit-sample sequence with two unit-samples of half the amplitude (Fig. 5-4). Inspection of the flowgraph leads to the same result.

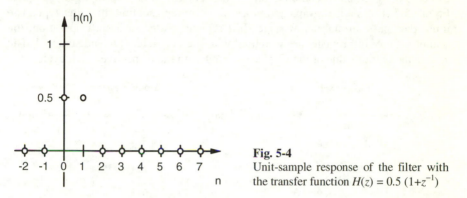

Fig. 5-4
Unit-sample response of the filter with the transfer function $H(z) = 0.5\,(1+z^{-1})$

For the hardware realisation of this filter, a memory is required that keeps the current input sample for calculating the average with the following input sample. Furthermore we need an adder. The multiplication by 0.5 can be simply realised by a shift operation in binary arithmetic.

Instead of developing special hardware, digital filters are often realised in the form of mathematical algorithms implemented on microcomputers or signal processors. These solutions are more flexible and are especially appropriate for applications in which the filter parameters have to be modified often, or during operation. Block diagrams can be used as the basis for the development of the corresponding program code. The following Pascal pseudo-code is implementing the algorithm of our example.

```
{ X = current input sample
  X1= memory to keep the current input sample
  Y = currently calculated output sample
  ********************************************************************** }
X1:=0; { initialisation of the memory }
Loop:
X:="Input sample";
{ In real-time applications, the "Input sample" is provided by
an A/D converter or input port of the hardware. Otherwise it is
taken from a data array or hard disk. }
Y:=0.5*(X+X1);
X1:=X; { keep input sample }
"Output sample":=Y;
{ The result is stored back to memory or output via a port or
a D/A converter. }
Goto Loop;
```

5.2 FIR Filters

5.2.1 The Basic Structure of FIR Filters

In the case of FIR (Finite Impulse Response) filters, the output sequence is calculated by a linear combination of the current and M past input samples. Figure 5-5 shows a flowgraph and block diagram representing this filter type. The input data pass through an M-stage shift register. After each shift operation, the outputs of the shift register are weighted with the coefficients b_r and summed. The sum is the output value of the filter associated with the respective clock cycle.

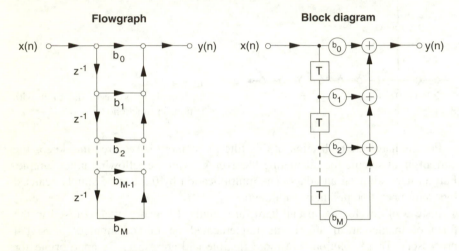

Fig. 5-5 Flowgraph and block diagram of the FIR filter

The weighting coefficients B_r, which determine the characteristics of the filter, have a clear meaning. Let us consider the unit-sample response of the filter, which we obtain by applying a unit-sample sequence $\delta(n)$ to the input. This sequence has only one nonzero sample with the value 1. This unit sample propagates step-by-step through the shift register. It is evident that the coefficients B_r will appear at the output one after the other starting with b_0. This means that the coefficients B_r are the samples of the unit-sample response $h(r)$ of the filter. From the flowgraph we can derive the following algorithm:

$$y(n) = b_0 x(n) + b_1 x(n-1) + \ \dots \ + b_M x(n-M)$$

or expressed in compact form

$$y(n) = \sum_{r=0}^{M} x(n-r)\, b_r \ . \tag{5.1}$$

Relation (5.1) is the convolution sum that we derived in section 3.1, because b_r can be replaced by $h(r)$ as in the previous discussion.

$$y(n) = \sum_{r=0}^{M} x(n-r)\,h(r)$$

The FIR filter, therefore, has a structure that realises the convolution sum in direct form. The length of the shift register determines the length of the unit-sample response, which is finite and gives the filter its name. This is the reason why the summation in (5.1) extends over a finite number of elements which is in contrast to the general convolution sum (3.3). The transfer function of the FIR filter is obtained by z-transformation of the unit-sample response $h(n)$.

$$H(z) = \sum_{r=0}^{M} h(r)\,z^{-r}$$

Replacing $h(r)$ by b_r yields a relation between the transfer function and the filter coefficients.

$$H(z) = \sum_{r=0}^{M} b_r\,z^{-r} \tag{5.2}$$

The transition from z to $e^{j\omega T}$ yields the frequency response of the filter.

$$H\!\left(e^{j\omega T}\right) = \sum_{r=0}^{M} b_r\,e^{-j\omega rT} \tag{5.3}$$

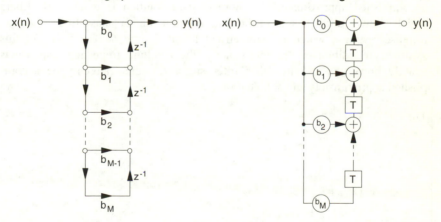

Fig. 5-6 Flowgraph and block diagram of the transposed FIR filter

An alternative filter structure can be found by transposition of the flowgraph. There are a number of transform rules that can be applied to flowgraphs without changing the transfer function. A transposed filter is obtained by reversing the direction of signal flow in all branches and interchanging the input and output. Figure 5-6 shows the flowgraph and block diagram of the transposed structure, as derived from the corresponding representations in Fig. 5-5. While the original structure was characterised by a continuous shift register, the transposed structure features single memory cells separated by adders.

The complexity of FIR filters may be quite high in standard filter applications, because many multiplications and additions are required to reproduce the complete impulse response in the time domain. The complexity is determined by the number of coefficients and consequently by the length of the impulse response to be realised. The kind of filter characteristic and the accuracy of the approximation of the frequency response highly determine the number of coefficients. Filters with a hundred coefficients and more can be often found in practice. However, the FIR filter has some interesting properties that make it attractive in realisation and application:

1. The FIR filter has a highly regular structure which is advantageous for the implementation.
2. The FIR filter is definitely stable since the stability criterion (3.21) is always fulfilled.
3. The filter always quiets down if the input signal vanishes. Uncontrolled oscillations, which may occur with recursive systems due to numerical problems, are excluded.
4. There are filter applications where the marked phase distortions occurring in the vicinity of the cutoff frequency are not tolerable, and hence a linear phase filter is required. An alternative possibility to solve this problem is to linearise the phase response by means of complex all-pass networks which can only be accomplished approximately. A more elegant solution is to use FIR filters which allow a direct design in the time domain. Linear-phase filters have an impulse response which is symmetrical about a time t_0 (Fig. 5-7). As this symmetry condition can be met exactly, the resulting frequency response is exactly linear-phase. With other filter structures, this can only be accomplished approximately and at high cost.

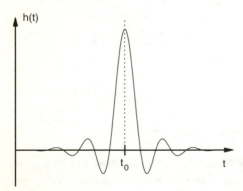

Fig. 5-7
Example of a linear-phase impulse response

5.2.2 Poles and Zeros of the FIR Filter

An alternative representation of (5.2) can be found by factoring out the term z^{-M}.

$$H(z) = \frac{\sum_{r=0}^{M} b_r z^{M-r}}{z^M} \tag{5.4}$$

All exponents in the sum then become positive. In the numerator we get a polynomial in z of degree M, resulting in M zeros of the transfer function. The denominator of (5.4) represents an Mth-order pole at $z = 0$.

$$H(z) = b_0 \frac{(z - z_{01})(z - z_{02}) \dots (z - z_{0M})}{z^M}$$

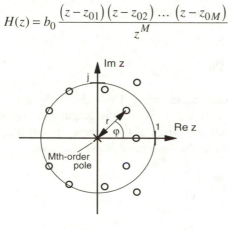

Fig. 5-8
Possible location of poles and zeros of an Mth-order FIR filter

Figure 5-8 shows the possible locations of poles and zeros in the z-plane. The Mth-order pole in the origin is independent of the realised transfer function. It only depends on the filter order and provides the following contribution to the overall magnitude, phase and group delay response.

$$H(\omega) = 1$$

$$b(\omega) = \omega \, M \, T$$

$$\tau_g(\omega) = M \, T$$

The magnitude is constant, and the phase characteristic linear. The group delay is frequency independent and amounts to M sample periods. The zeros may be located within, on or outside the unit circle. Since the coefficients b_r are real, the zeros must be real or occur in complex-conjugate pairs. Let us consider now what a zero of the form

$$H(z) = z - z_0$$

contributes to the overall frequency response of the filter. The location of the zero is specified by the distance from the origin r_0 and the angle φ_0 with respect to the positive real axis (Fig. 5-8).

$$H(z) = z - r_0\, e^{j\varphi_0} = z\left(1 - r_0\, e^{j\varphi_0}\, z^{-1}\right)$$

Replacing z by $e^{j\omega T}$ yields the frequency response of the zero.

$$H\!\left(e^{j\omega T}\right) = e^{j\omega T}\left(1 - r_0\, e^{j\varphi_0}\, e^{-j\omega T}\right) = e^{j\omega T}\left(1 - r_0\, e^{-j(\omega T - \varphi_0)}\right)$$

$$H\!\left(e^{j\omega T}\right) = e^{j\omega T}\left(1 - r_0 \cos(\omega T - \varphi_0) + j r_0 \sin(\omega T - \varphi_0)\right) \qquad (5.5)$$

The amplitude response is the magnitude of the complex expression (5.5).

$$\left|H\!\left(e^{j\omega T}\right)\right|^2 = \left(1 - r_0 \cos(\omega T - \varphi_0)\right)^2 + \left(r_0 \sin(\omega T - \varphi_0)\right)^2$$

$$\left|H\!\left(e^{j\omega T}\right)\right|^2 = 1 + r_0^2 - 2 r_0 \cos(\omega T - \varphi_0) \qquad (5.6)$$

The minimum of the magnitude occurs at $\omega T = \varphi_0$. The gain is $1+r_0$ in the maximum and $1-r_0$ in the minimum. Figure 5-9a shows the magnitude response for various distances r_0 of the zeros from the origin of the z-plane. By calculation of the negative argument of (5.5) we obtain the associated phase response.

$$b(\omega) = -\omega T - \arctan\!\left(\frac{r_0 \sin(\omega T - \varphi_0)}{1 - r_0 \cos(\omega T - \varphi_0)}\right) \qquad (5.7)$$

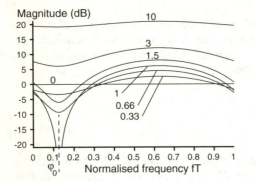

Fig. 5-9a
Magnitude of a zero with r_0 as parameter

The group delay of the zero is calculated as the first derivative of the phase with respect to ω.

$$\tau_g(\omega) = -T\,\frac{1 - r_0 \cos(\omega T - \varphi_0)}{1 + r_0^2 - 2 r_0 \cos(\omega T - \varphi_0)} \qquad (5.8)$$

Figure 5-9b shows the corresponding graphs of the group delay for various r_0.

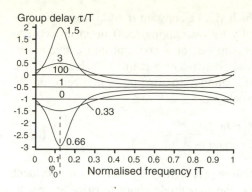
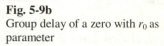

Fig. 5-9b
Group delay of a zero with r_0 as parameter

Relation (5.6) and Fig. 5-9a show that zeros close to or on the unit circle yield high attenuation around the frequency φ_0/T, while zeros close to the origin achieve low attenuation with more or less variation of the magnitude. Thus it can be expected that, as a result of FIR filter design, zeros in the stopband will be located close to the unit circle while the passband features zeros close to or within the origin. Figure 5-10 shows an example of the frequency response of a filter with twelve zeros equally spaced in angle. The maximum gain is normalised to unity. Six zeros are located in the frequency range $-0.25 < fT < 0.25$ at a radius of $r_0 = 0.4$ (low attenuation). The remaining zeros have been placed onto the left half of the unit circle (high attenuation). As a whole, the described zero configuration results in a low-pass behaviour.

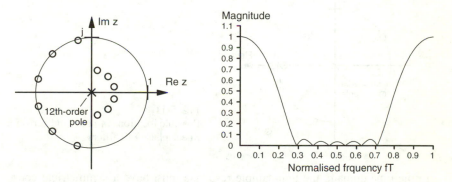

Fig. 5-10 Pole/zero chart and frequency response of a twelfth-order FIR low-pass filter

5.2.3 Linear-Phase Filters

As already mentioned, a linear-phase or constant group delay can be easily realised by means of FIR filters. Let us first consider the consequences of a linear-phase response in the frequency domain. Figure 5-9b gives a hint on how to start the discussion. Zeros with reciprocal distance from the origin in the z-plane have complementary group-delay graphs which yield a constant if added together. This can be easily established mathematically by rearranging (5.8) in an appropriate manner such that the group delay is composed of a constant and a frequency-dependent part. The sign of the frequency-dependent part in (5.9) changes if the radius r_0 is replaced by its reciprocal. Equation (5.9) also shows that zeros on the unit circle ($r_0 = 1$) have constant group delay.

$$\tau_g(\omega) = -T/2 - T/2 \frac{1/r_0 - r_0}{1/r_0 + r_0 - 2\cos(\omega T - \varphi_0)} \tag{5.9}$$

Reciprocal zeros have another interesting property. From Fig. 5-9a it is evident that, apart from a constant gain factor, the magnitude characteristics are equal. This can be easily shown mathematically by rearrangement of (5.6).

$$\left| H\left(e^{j\omega T}\right) \right|^2 = r_0 \left(1/r_0 + r_0 - 2\cos(\omega T - \varphi_0) \right) \tag{5.10}$$

The frequency-dependent term in parentheses remains unchanged if r_0 is replaced by its reciprocal. Apart from the already mentioned condition that zeros must be real or complex-conjugate, we have the following additional requirement for linear-phase filters: zeros must lie on the unit circle or occur in reciprocal pairs (Fig. 5-11).

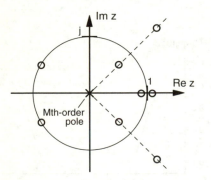

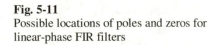

Fig. 5-11
Possible locations of poles and zeros for linear-phase FIR filters

In the time domain, the unit-sample response must have a symmetrical course as sketched in Fig. 5-7. In the following we will show that this symmetry condition really results in a linear-phase filter. An Mth-order FIR filter has $M+1$ coefficients and the following frequency response:

$$H\left(e^{j\omega T}\right) = \sum_{r=0}^{M} b_r\, e^{-j\omega rT} \quad .$$

Out of this sum we extract a delay term corresponding to half the length of the unit-sample response.

$$H\left(e^{j\omega T}\right) = e^{-j\omega TM/2} \sum_{r=0}^{M} b_r\, e^{-j\omega T(r-M/2)} \qquad (5.11a)$$

The symmetry of the impulse response and consequently of the filter coefficients results in the following relationship:

$$b_r = b_{M-r}\,. \qquad (5.12)$$

Substitution of (5.12) in (5.11a) yields

$$H\left(e^{j\omega T}\right) = e^{-j\omega TM/2} \sum_{r=0}^{M} b_{M-r}\, e^{-j\omega T(r-M/2)}\,.$$

The variable substitution $k = M-r$ results in the following form:

$$H\left(e^{j\omega T}\right) = e^{-j\omega TM/2} \sum_{k=0}^{M} b_k\, e^{j\omega T(k-M/2)}\,. \qquad (5.11b)$$

The sums in (5.11a) and (5.11b) are complex-conjugate. If both equations are added and the result is divided by two, we get the real sum

$$H\left(e^{j\omega T}\right) = e^{-j\omega TM/2} \sum_{k=0}^{M} b_k \cos\omega T(k - M/2)\,. \qquad (5.13)$$

The exponential function in front of the sum represents a linear-phase transfer function that merely causes a signal time delay of $MT/2$. It has no influence on the magnitude. The sum is purely real and represents a zero-phase filter.

$$H_0(\omega) = \sum_{k=0}^{M} b_k \cos\omega T(k - M/2) \qquad (5.14)$$

The symmetry relation (5.12) leads to a representation of the frequency response of the following form.

If M is even (odd number of coefficients), we obtain:

$$H_0(\omega) = \sum_{r=0}^{M/2} B_r \cos r\omega T$$

with $B_0 = b_{M/2}$ \hfill (5.15a)

and $B_r = 2\, b_{M/2-r} = 2\, b_{M/2+r}$ for $r = 1 \dots M/2$.

An odd filter order M (even number of coefficients) accordingly leads to

$$H_0(\omega) = \sum_{r=1}^{(M+1)/2} B_r \cos(r - 1/2)\omega T \tag{5.15b}$$

with $B_r = 2\, b_{(M+1)/2-r} = 2\, b_{(M+1)/2+r-1}$ for $r = 1 \dots (M+1)/2$.

Because of the symmetry condition, an Mth-order FIR filter is fully defined by $M/2+1$ arbitrary coefficients if M is even or $(M+1)/2$ coefficients if M is odd.

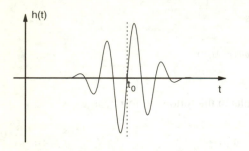

Fig. 5-12
Example of an antisymmetrical linear-phase impulse response

Finally, it must be mentioned that an antisymmetrical impulse response, as depicted in Fig. 5-12, results in a linear-phase frequency response, too. With this kind of symmetry, the phase response has an additional constant component of $\pi/2$. This property allows the realisation of orthogonal filters, differentiators and Hilbert transformers, which all require this constant phase shift of $\pi/2$.

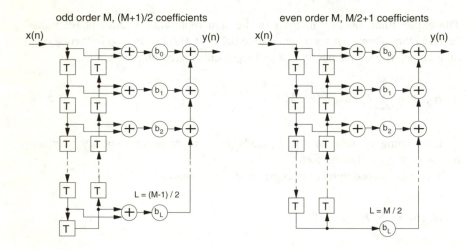

Fig. 5-13 Optimised structures for linear-phase FIR filters

Because of the symmetry of the filter coefficients, the complexity of the filter realisation can be reduced considerably. All coefficients (for odd filter order) or all but one (for even filter order) occur in pairs so that the number of multiplications can be halved as shown in the block diagrams in Fig. 5-13. From both structures we can also derive the corresponding transposed forms.

5.3 IIR Filters

5.3.1 An Introductory Example

Impulse responses with infinite length can be realised with a limited number of coefficients by also using delayed values of the output to calculate the current output sample. Remember that the FIR filter only uses the current and delayed input values. A simple first-order low-pass filter with exponentially decaying unit-sample response may serve as an example:

$$h(n) = e^{-\alpha n} u(n) = \left(e^{-\alpha}\right)^n u(n) \ .$$

z-transformation of the unit-sample response yields the transfer function of the filter. From Table 4-2 we see that

$$H(z) = \frac{z}{z - e^{-\alpha}} = \frac{1}{1 - e^{-\alpha} z^{-1}} \ .$$

The transfer function is the quotient of the z-transformed input and output sequences.

$$H(z) = \frac{Y(z)}{X(z)} = \frac{1}{1 - e^{-\alpha} z^{-1}}$$

$$Y(z)\left(1 - e^{-\alpha} z^{-1}\right) = X(z)$$

$$Y(z) - e^{-\alpha} Y(z) z^{-1} = X(z)$$

Transformation back into the time domain using the rules derived in Chap. 3 yields an algorithm that relates input sequence $x(n)$ and output sequence $y(n)$.

$$y(n) - e^{-\alpha} y(n-1) = x(n)$$

$$y(n) = x(n) + e^{-\alpha} y(n-1)$$

The resulting filter merely requires one multiplier and one adder. The calculation of the current output value is based on the current input value and the previous output value. Figure 5-14 shows the flow graph of this filter. The use of delayed output values gives the filter a feedback or recursive structure.

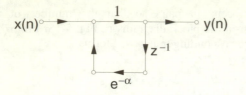

Fig. 5-14
Flow graph of an IIR filter with the
transfer function $H(z)=z/(z-e^{-\alpha})$

5.3.2 Direct Form Filters

In a general recursive filter realisation, the current and M delayed values of the input are multiplied by the coefficients b_k, N delayed values of the output are multiplied by the coefficients a_k, and all the resulting products are added. Sorting input and output values to both sides of the equals sign yields the following sum representation (5.16):

$$\sum_{r=0}^{M} b_r x(n-r) = \sum_{i=0}^{N} a_i y(n-i) \ . \tag{5.16}$$

This relation, commonly referred to as difference equation, is analogous to the differential equation (1.13), which describes the input/output behaviour of continuous-time systems in the time domain. It is usual to normalise (5.16) such that $a_0 = 1$. This leads to the following filter algorithm:

$$y(n) = \sum_{r=0}^{M} b_r x(n-r) - \sum_{i=1}^{N} a_i y(n-i) \ . \tag{5.17}$$

Figure 5-15a shows a block diagram representing (5.17). In contrast to the FIR filter, a second shift register appears that is used to delay the output sequence $y(n)$. The structure on the left of Fig. 5-15a with the coefficients b_r is the well-known FIR filter structure, which represents the non-recursive part of the filter. The structure on the right is the recursive part of the filter, whose characteristics are determined by the coefficients a_i .

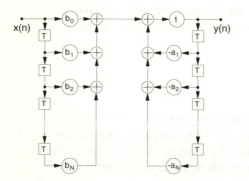

Fig. 5-15a
Block diagram of an IIR filter

Both filter parts can be moved together without changing the characteristics of the filter. In the middle of the structure we find a row of adders in which all delayed and weighted input and output values are summed (Fig. 5-15b). This structure is called direct form I.

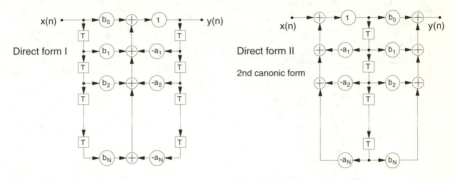

Fig. 5-15b Direct form I   **Fig. 5-15c** Direct form II

The recursive and non-recursive parts in Fig. 5-15a can be considered as two independent filters arranged in series. The order of these filters can be reversed without changing the overall function. This rearrangement leads to two parallel shift registers that can be combined into one because both are fed by the same signal in this constellation. The resulting structure, as depicted in Fig. 5-15c, manages with the minimum possible amount of memory. This structure is called direct form II. Because of the minimum memory requirement, it is also referred to as canonic form.

Transposed structures also exist for IIR filters. These are derived by reversing all arrows and exchanging input and output in the corresponding flowgraphs. Figure 5-15d shows the transposed direct form II, which also features the minimum memory requirement. This structure is therefore canonic, too. With the transposed direct form I as depicted in Fig. 5-15e, we have introduced all possible direct forms.

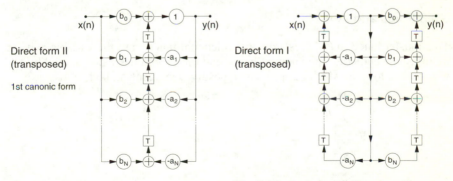

Fig. 5-15d Transposed direct form II   **Fig. 5-15e** Transposed direct form I

The transfer function of the IIR filter is obtained by z-transformation of (5.16). On this occasion we make use of the shift theorem that we have introduced in Sect. 3.3.2.

$$\sum_{r=0}^{M} b_r X(z) z^{-r} = \sum_{i=0}^{N} a_i Y(z) z^{-i}$$

$$X(z) \sum_{r=0}^{M} b_r z^{-r} = Y(z) \sum_{i=0}^{N} a_i z^{-i}$$

$$H(z) = \frac{Y(z)}{X(z)} = \frac{\displaystyle\sum_{r=0}^{M} b_r z^{-r}}{\displaystyle\sum_{i=0}^{N} a_i z^{-i}} \tag{5.18}$$

Replacing z by $e^{j\omega T}$ yields the frequency response of the filter.

$$H\left(e^{j\omega T}\right) = \frac{\displaystyle\sum_{r=0}^{M} b_r\, e^{-j\omega r T}}{\displaystyle\sum_{i=0}^{N} a_i\, e^{-j\omega i T}} \tag{5.19}$$

The filter structures depicted in Fig. 5-15a–e show that the filter coefficients are identical to the coefficients of the numerator and denominator polynomials of the transfer function. Since these filters are directly derived from the transfer function, they are also referred to as direct-form filters. These have the advantage that they manage with the lowest possible number of multipliers, which is the same as the number of coefficients of the transfer function. Moreover, choosing a canonic structure leads to a filter that realises a given transfer function with the lowest possible complexity in terms of memory and arithmetical operations.

5.3.3 Poles and Zeros

The transfer function (5.18) is a rational fractional function of the variable z. If we multiply the numerator polynomial by z^M and the denominator polynomial by z^N, we obtain a form in which z appears with positive exponents in both sums.

$$H(z) = z^{N-M}\, \frac{\displaystyle\sum_{r=0}^{M} b_r z^{M-r}}{\displaystyle\sum_{i=0}^{N} a_i z^{N-i}}$$

According to their degree, numerator and denominator polynomials have M or N zeros respectively. With respect to the transfer function, this means M zeros, whose location is determined by the coefficients b_r, and N poles, whose location is determined by the coefficients a_i. At $z = 0$ we find an $|N\!-\!M|$ th-order pole or zero depending on the difference between the degrees of numerator and denominator.

In Sect. 5.2.2 we already calculated magnitude, phase and group delay of the zeros. The corresponding relations for the poles can be directly derived from these results. The magnitude of the poles is reciprocal to the magnitude of the zeros. Logarithmic magnitude, phase and group delay have inverse sign with respect to the corresponding expressions of the zeros.

$$|H(\omega)|^2 = \frac{1}{1 + r_\infty^2 - 2r_\infty \cos(\omega T - \varphi_\infty)} \tag{5.20}$$

$$b(\omega) = \omega T + \arctan\left(\frac{r_\infty \sin(\omega T - \varphi_\infty)}{1 + r_\infty \cos(\omega T - \varphi_\infty)} \right) \tag{5.21}$$

$$\tau_g(\omega) = T \frac{1 - r_\infty \cos(\omega T - \varphi_\infty)}{1 + r_\infty^2 - 2r_\infty \cos(\omega T - \varphi_\infty)} \tag{5.22}$$

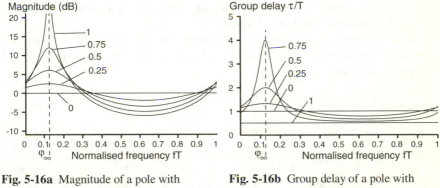

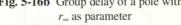

Fig. 5-16a Magnitude of a pole with r_∞ as parameter

Fig. 5-16b Group delay of a pole with r_∞ as parameter

Figure 5-16a shows the graph of the magnitude of a pole according to (5.20). This graph exhibits a more or less marked resonance peak depending on the distance of the poles from the unit circle. The closer the poles come to the unit circle, the higher the quality factor. Figure 5-16b shows the group delay behaviour according to (5.22), which assumes the highest values around the resonance frequency.

In case of selective filters, the poles are distributed over the passband, since these can realise the much higher gain needed in the passband in contrast to the stopband. Zeros, if present, are positioned in the stopband, where they support the named behaviour by providing additional attenuation. Figure 5-17 shows the

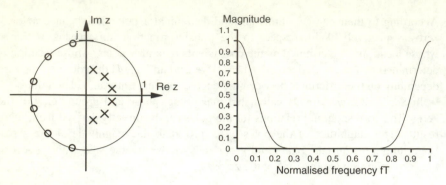

Fig. 5-17 Pole/zero plot and frequency response of a sixth-order IIR low-pass filter

example of a sixth-order filter which consists of six poles at $r_\infty = 0.5$ and six zeros on the unit circle, all equally spaced in angle.

Besides the rational fractional representation of the transfer function (5.18), which is closely related to the difference equation of the system, there are two further forms that are of importance in practice: the pole/zero and the partial-fraction representations. In section 1.3 we already explained in the context of continuous-time systems that the pole/zero representation is useful if the system is to be cut up into cascadable partial systems. In contrast, the partial-fraction representation is helpful for the parallel arrangement of partial systems. The same is valid for the transfer functions of discrete-time systems.

In order to obtain the pole/zero representation, the numerator and denominator polynomials are factorised into zero terms, which leads to the following form of the transfer function:

$$H(z) = z^{N-M} \frac{(z - z_{01})(z - z_{02}) \ldots (z - z_{0M})}{(z - z_{\infty 1})(z - z_{\infty 2}) \ldots (z - z_{\infty N})} \; . \tag{5.23}$$

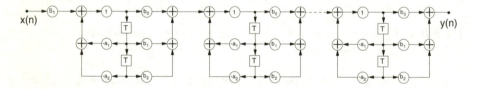

Fig. 5-18 A cascade of second-order filter sections

For reasons we will deal with in more detail in Chap. 8, it is desirable to cut up a filter into smaller subsystems with as low an order as possible. For real poles and zeros, first-order sections are the appropriate solution. Complex-conjugate pole and zero pairs are combined into second-order sections which results in real filter coefficients. Figure 5-18 shows a corresponding cascade arrangement of second-order filter sections. The scaling coefficient s_1 at the input of the cascade is chosen in such a way that no overflow occurs in the first filter section. Such scaling factors also need to be inserted between the following sections. These, however, can be directly reckoned in the nonrecursive coefficients b_0, b_1 and b_2. On the one hand, overflow has to be avoided by these scaling factors, on the other hand the available dynamic range of the signal path has to be used in an optimised way to obtain the best possible signal-to-noise ratio. Together with the pairing of poles and zeros and the sequence of subsystems, these scaling coefficients are degrees of freedom, which have no direct influence on the frequency response but which decisively determine the noise behaviour and thus the quality of the filter.

In contrast to the representations of magnitude and group delay for the zeros, Fig. 5-16a–b only shows graphs for $r < 1$. It is theoretically possible to show graphs for $r \geq 1$, too. These would not be of practical value, however, because poles on or outside the unit circle are not stable, as will be shown in the next section.

5.4 Further Properties of Transfer Functions

5.4.1 Stability Considerations in the Frequency Domain

According to the considerations in Sect. 3.4, BIBO stability is guaranteed if the unit-sample response is absolutely summable.

$$\sum_{k=-\infty}^{+\infty} |h(k)| < \infty \tag{5.24}$$

FIR filters are always stable, since their unit-sample responses consist of a finite number of samples. Which condition has to be fulfilled by a rational fractional transfer function $H(z)$ to guarantee BIBO stability? An analysis of the unit-sample response $h(n)$, which we obtain by inverse z-transform of $H(z)$ into the time domain, will answer this question. Partial-fraction expansion of $H(z)$ leads to an expression that allows a direct transformation into the time domain.

$$H(z) = \sum_{r=0}^{N} A_r \frac{z}{z - z_{\infty r}}$$

We obtain by inverse z-transform

$$h(n) = \sum_{r=0}^{N} A_r \, z_{\infty r}^n \, u(n) \ .$$

Each summand that contributes to the impulse response has to fulfil condition (5.24). This means that the absolute sum of each term has to stay below a certain boundary S to assure stability.

$$\sum_{n=0}^{\infty} \left| A_r \, z_{\infty r}^n \right| < S$$

$$|A_r| \sum_{n=0}^{\infty} \left| z_{\infty r}^n \right| < S \tag{5.25}$$

Expression (5.25) is the sum of an infinite geometric progression which converges if $|z_{\infty r}| < 1$. A plot of the possible locations in the complex z-plane shows that stable poles must lie within the unit circle (Fig. 5-19). Since the coefficients of the denominator polynomial of $H(z)$ are real, the poles have to be real or occur in complex-conjugate pairs.

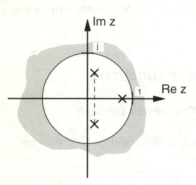

Fig. 5-19
Stability region of poles in the z-domain

There is an analogous condition in the theory of continuous-time systems. In the case of stable filters, the poles must lie within the left half of the p-plane with p being the complex frequency variable in the continuous-time domain. If we aim at imitating the behaviour of an analog filter with a digital filter, this is commonly accomplished by transformations that map the p-plane into the z-plane. In order to assure that a stable analog filter leads to a stable digital filter, the part of the p-plane that contains the poles of the analog filter has to be mapped into the interior of the unit circle in the z-plane. For a more general approach, the whole left half of the p-plane should be mapped into the unit circle of the z-plane.

Because of the condition that the poles of a stable discrete-time system have to lie within the unit circle, the coefficients a_i of the denominator polynomial cannot assume any arbitrary value. For elementary first- and second-order filter sections, the admissible range of values can be easily specified. A first-order pole is expressed as follows:

$$\frac{1}{1 + a_1 z^{-1}} = \frac{z}{z + a_1} \; .$$

A stable system has to fulfil the condition

$$|a_1| < 1 \; . \tag{5.26}$$

In the case of second-order filter sections, the coefficients a_1 and a_2 determine the location of the poles.

$$\frac{1}{1 + a_1 z^{-1} + a_2 z^{-2}} = \frac{z^2}{z^2 + a_1 z + a_2}$$

The triangle in Fig. 5-20 shows the possible combinations of coefficients in the a_1/a_2-plane that result in stable filters. A parabola subdivides the triangle into two regions. Coefficient pairs in the dark upper area lead to complex-conjugate pole pairs while coefficients in the lower part lead to two real poles.

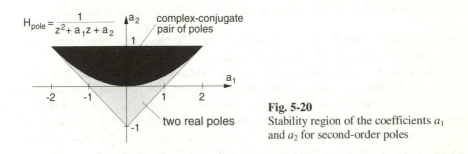

Fig. 5-20

Stability region of the coefficients a_1 and a_2 for second-order poles

In practice, there are further limitations concerning the choice of filter coefficients that we will consider in detail in Chap. 8. In the edge area of the triangle particularly, there is a potential danger of numerical instabilities. Another natural limitation follows from the fact that the coefficients can only be represented with limited precision. The possible combinations of the coefficients a_1 and a_2 form a grid in the stability triangle.

5.4.2 Minimum-Phase Filters

Minimum-phase filters realise a given magnitude with minimum delay. The minimum-phase condition is fulfilled if all zeros of a discrete-time system lie within or on the unit circle.

According to the last section, the poles of a stable system must lie within the unit circle. Moreover, we have shown in Sect. 5.2.2 that zeros with reciprocal distance r_0 and $1/r_0$ from the origin have, apart from a constant gain factor, identical magnitude. There are seemingly two equivalent alternatives when designing a filter with prescribed magnitude. A look at the group delay graphs (Fig. 5-9b) shows striking differences, however. For zeros within the unit circle ($r_0 < 1$), the group delay tends towards negative values. Zeros within the unit circle are thus able to more or less compensate for the delay of the poles which tend to positive values (Fig. 5-16b). Zeros outside the unit circle add even more delay. Filters with zeros inside the unit circle thus realise a given magnitude with the lowest possible delay, which is equivalent to the lowest possible phase variation $b(\omega)$.

5.5 State-Space Structures

Direct-form filters, as treated in the previous sections, are often used in practice because these structures realise a given transfer function with minimum cost. This filter type has the disadvantage that there are almost no possibilities to optimise all the effects that are caused by the finite precision of the coefficient and signal representation. These are:

- Sensitivity of the frequency response with respect to coefficient inaccuracies.
- Quantisation noise that results from the limitation of the word length after mathematical operations such as multiplication or floating point addition.
- Unstable behaviour, i.e. overflow oscillations and quantisation limit cycles.

Higher flexibility and far better possibilities to influence the named effects are offered by the so-called state-space structures.

In the case of state-space structures, a direct delayless path between input and output, as characterised by the coefficient b_0 of the transfer function, is realised in the form of a bypass. If the orders of the denominator and numerator polynomials are equal in a given application, the numerator polynomial first has to be divided by the denominator polynomial. This results in a constant and a new rational fractional transfer function that has no constant term in the numerator.

$$\frac{b_0 + b_1 z^{-1} + b_2 z^{-2}}{1 + a_1 z^{-1} + a_2 z^{-2}} = b_0 + \frac{(b_1 - b_0 a_1)z^{-1} + (b_2 - b_0 a_2)z^{-2}}{1 + a_1 z^{-1} + a_2 z^{-2}}$$

This constant b_0 determines the gain of the direct path between input and output of the filter.

After splitting off the direct path, the second-order direct-form filter can be easily redrawn into the form of Fig. 5-21.

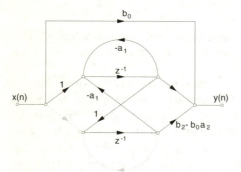

Fig. 5-21
Redrawn second-order direct form filter

If this figure is completed by the grey paths, we obtain the general form of a state-space structure (Fig. 5-22) which features feedback paths from each memory output $z_i(n)$ to each memory input $z_i(n+1)$. The coefficients in these feedback paths are a_{ij}. The input signal is multiplied by the coefficients b_i and added to the feedback signals at the input of the memories. The output signal of the filter is obtained by a linear combination of the memory outputs. The coefficients of this this linear combination are c_i. The described behaviour can be mathematically expressed as

$$z_1(n+1) = a_{11} z_1(n) + a_{12} z_2(n) + b_1 x(n) \qquad (5.27a)$$

$$z_2(n+1) = a_{21} z_1(n) + a_{22} z_2(n) + b_2 x(n) \qquad (5.27b)$$

$$y(n) = c_1 z_1(n) + c_2 z_2(n) + d x(n) . \qquad (5.27c)$$

While the rational fractional transfer function of a general direct-form filter has five coefficients which fully characterise the transfer behaviour, the corresponding state-space structure has nine coefficients. This means higher implementation complexity, but offers many degrees of freedom to optimise the filter performance. Let us now relate the coefficients of the transfer function,

$$H(z) = \beta_0 + \frac{\beta_1 z + \beta_2}{z^2 + \alpha_1 z + \alpha_2} = \beta_0 + \frac{\beta_1 z^{-1} + \beta_2 z^{-2}}{1 + \alpha_1 z^{-1} + \alpha_2 z^{-2}} , \qquad (5.28)$$

to the coefficients of the state-space representation (5.27).

$$\beta_0 = d$$
$$\beta_1 = c_1 b_1 + c_2 b_2$$
$$\beta_2 = c_1 b_2 a_{12} + c_2 b_1 a_{21} - c_1 b_1 a_{22} - c_2 b_2 a_{11}$$
$$\alpha_1 = -(a_{11} + a_{22})$$
$$\alpha_2 = a_{11} a_{22} - a_{12} a_{21}$$

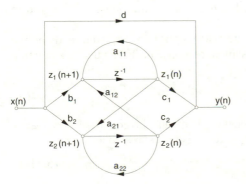

Fig. 5-22
Structure of a second-order state-space filter

Relation (5.27) can be advantageously written in vector form.

$$z(n+1) = A z(n) + b x(n) \qquad (5.29a)$$

$$y(n) = c^T z(n) + d x(n) \qquad (5.29b)$$

The vectors and matrices printed in bold have the following meaning:

$$z(n) = \begin{pmatrix} z_1(n) \\ z_2(n) \end{pmatrix} \qquad b = \begin{pmatrix} b_1 \\ b_2 \end{pmatrix} \qquad c = \begin{pmatrix} c_1 \\ c_2 \end{pmatrix} \text{ or } c^T = (c_1 \quad c_2)$$

$$A = \begin{pmatrix} a_{11} & a_{12} \\ a_{21} & a_{22} \end{pmatrix} .$$

In this vector representation, the state-space structure can be easily generalised towards higher filter orders. For a general Nth-order filter, z, b and c are vectors of dimension N, and A is a $N \times N$ matrix. With these definitions, (5.29) is valid for any arbitrary filter order. Figure 5-23 shows a corresponding general block diagram.

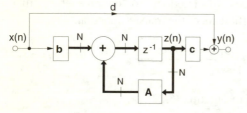
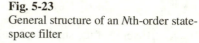

Fig. 5-23
General structure of an Nth-order state-space filter

As a consequence of the $N \times N$ matrix, the number of coefficients increases with the square of the filter order. A direct-form filter requires $2N+1$ multipliers, while a state-space filter is fully specified by $(N+1)^2$ coefficients. This is why higher-order state-space filters are not used very often in practice. Their use is limited, in general, to cascade realisations of second-order filter blocks. Compared to the straightforward implementation of an Nth-order state-space structure, however, the cascade realisation yields only sub-optimal results with regard to the optimisation of the filter performance, but is a good compromise with regard to implementation complexity.

From the vector form (5.29) of the state-space algorithm, we can directly derive the transfer function $H(z)$ and the unit-sample response $h(n)$. The transfer function can be expressed as

$$H(z) = c^T (zI - A)^{-1} b + d . \tag{5.30}$$

I is the identity matrix. The exponent -1 means inversion of the matrix expression in parenthesis. The unit-sample response is expressed as

$$h(n) = 0 \qquad\qquad \text{for } n < 0 \tag{5.31a}$$

$$h(n) = d \qquad\qquad \text{for } n = 0 \tag{5.31b}$$

$$h(n) = c^T A^{n-1} b \qquad \text{for } n > 0 . \tag{5.31c}$$

It can be shown mathematically that (5.31c) converges for $n \to \infty$ only if the magnitudes of the eigenvalues of the matrix A are less then unity. This condition is equivalent to the stability criterion with respect to the location of the poles in the z-plane, because the eigenvalues of A equal the poles of the transfer function (5.30).

We already mentioned that the state-space structure can realise a given transfer function in many ways, since this structure has more coefficients available than are needed to specify the transfer behaviour. We want to show now that a given filter structure can be easily transformed into an equivalent one if we apply simple vector and matrix transformations to the transfer function (5.30). We start by expanding (5.30) on both sides of the matrix expression in parenthesis by the term $T\,T^{-1}$.

$$H(z) = c^{\mathrm{T}} T T^{-1} (zI - A)^{-1} T T^{-1} b + d$$

T is an arbitrary nonsingular matrix. Since the product $T\,T^{-1}$ yields the identity matrix, the frequency response is not changed by this manipulation. The following assignment results in a new set of vectors and matrices:

$$A' = T^{-1}AT$$
$$b' = T^{-1}b \tag{5.32}$$
$$c'^{T} = c^{T}T \ .$$

It is easy to see that we obtain a filter with totally different coefficients but exactly the same transfer function as the original one.

Example

When introducing the state-space structure, we showed that the direct-form filter is a special case of this structure. Furthermore we became acquainted with four different variants of the direct form which have identical transfer behaviour. Let us consider direct form II and its transposed structure as an example. Figure 5-24 shows the block diagram of a second-order direct form II filter in two ways: standard direct form and equivalent state-space representation.

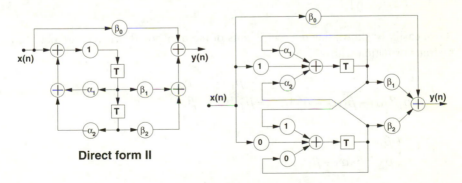

Fig. 5-24 Direct form II in standard and state-space representation

The following state-space coefficients can be read from Fig. 5-24:

$$b = \begin{pmatrix} 1 \\ 0 \end{pmatrix} \qquad c^{\mathrm{T}} = \begin{pmatrix} \beta_1 & \beta_2 \end{pmatrix}$$

$$A = \begin{pmatrix} -\alpha_1 & -\alpha_2 \\ 1 & 0 \end{pmatrix}.$$

The block diagram of the corresponding transposed structure is shown in Fig. 5-25.

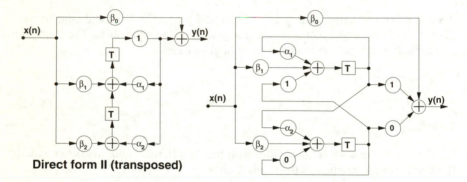

Direct form II (transposed)

Fig. 5-25 Transposed direct form II in standard and state-space representation

Figure 5-25 yields the following state-space coefficients:

$$b' = \begin{pmatrix} \beta_1 \\ \beta_2 \end{pmatrix} \qquad c'^{\mathrm{T}} = \begin{pmatrix} 1 & 0 \end{pmatrix}$$

$$A' = \begin{pmatrix} -\alpha_1 & 1 \\ -\alpha_2 & 0 \end{pmatrix}.$$

It can easily be shown that the coefficients of the transposed structure are obtained by transformation with

$$T = \frac{1}{\beta_1\beta_2\alpha_1 - \beta_1^2\alpha_2 - \beta_2^2} \begin{pmatrix} \beta_2\alpha_1 - \beta_1\alpha_2 & -\beta_2 \\ -\beta_2 & \beta_1 \end{pmatrix}$$

$$T^{-1} = \begin{pmatrix} \beta_1 & \beta_2 \\ \beta_2 & \beta_2\alpha_1 - \beta_1\alpha_2 \end{pmatrix}$$

from the coefficients of the original direct form II.

5.6 The Normal Form

The filter structure that we deal with in this section is also referred to as "coupled-loop" structure according to Gold and Rader in the literature.

A well-tried means to reduce the coefficient sensitivity of higher-order rational fractional transfer functions is the decomposition into first- and second-order partial transfer functions. It is obvious to improve the performance of the filter by further decomposition of the second-order sections. A second-order filter that realises a complex-conjugate pole pair cannot be easily subdivided into first-order partial systems because the resulting filter coefficients would be complex (5.33).

$$\frac{1}{z^2 + a_1 z + a_2} = \frac{1}{z - (\alpha + j\beta)} \frac{1}{z - (\alpha - j\beta)}$$

$$= \frac{z^{-1}}{1 - (\alpha + j\beta)z^{-1}} \frac{z^{-1}}{1 - (\alpha - j\beta)z^{-1}} \qquad (5.33)$$

with $a_1 = -2\alpha$ and $a_2 = \alpha^2 + \beta^2$

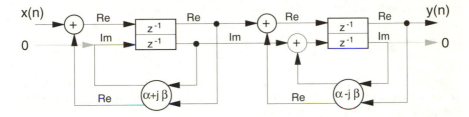

Fig. 5-26 Decomposition of a second-order pole into first-order sections

Figure 5-26 shows the block diagram of a second-order pole realised as a cascade of two first-order complex-conjugate poles. The input and output sequences of this filter must still be real. Within the structure, however, we find complex signals as a consequence of the complex coefficients. In Fig. 5-26, these complex signal paths are depicted as two parallel lines, one representing the real part, the other the imaginary part, of the respective signal. Restructuring and simplification yields the structure shown in Fig. 5-27. On this occasion we have taken into account that some of the imaginary paths are not needed because the input and output signals of the overall second-order block are real. This structure requires four coefficients to realise the complex-conjugate pole pair instead of two in the case of the direct form. The coefficients equal the real part α and the imaginary part β of the pole pair. These are, derived from (5.33), related to the coefficients of the transfer function a_1 and a_2 in the following way:

$$\alpha = -a_1/2 \qquad \beta = \sqrt{a_2 - a_1^2/4} \qquad a_1 = -2\alpha \qquad a_2 = \alpha^2 + \beta^2 \ .$$

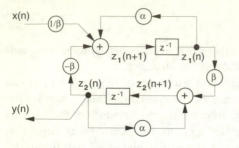

Fig. 5-27
The "coupled-loop" structure according to Rader and Gold

The normal form can be easily converted into the state-space structure according to Fig. 5-24, which yields the following state-space coefficients:

$$b = \begin{pmatrix} 1 \\ 0 \end{pmatrix} \qquad c^T = \begin{pmatrix} 0 & 1/\beta \end{pmatrix}$$

$$A = \begin{pmatrix} \alpha & -\beta \\ \beta & \alpha \end{pmatrix} .$$

Introduction of a polynomial into the numerator of (5.33) leads to a more general second-order transfer function.

$$H(z) = \frac{b_1 z + b_2}{z^2 + a_1 z + a_2}$$

State-space vector b and system matrix A are not affected by this amendment. The vector c^T takes into account the non-recursive filter part which is represented by the numerator polynomial.

$$c^T = \begin{pmatrix} b_1 & b_1 \alpha / \beta + b_2 / \beta \end{pmatrix}$$

The normal filter, too, can be derived from the direct form II by simple application of a transformation T.

$$T = \begin{pmatrix} 1 & \alpha/\beta \\ 0 & 1/\beta \end{pmatrix} \qquad T^{-1} = \begin{pmatrix} 1 & -\alpha \\ 0 & \beta \end{pmatrix}$$

5.7 Digital Ladder Filters

The filter structures, treated in this chapter up to now, are more or less direct implementations of the rational fractional transfer function in z. This is obvious especially in the case of the direct form since the coefficients of the transfer function are identical with the filter coefficients. With increasing filter order, however, the direct realisation of the transfer function becomes more and more critical with respect to noise performance, stability and coefficient sensitivity. We already showed two possibilities to avoid these problems:

- The decomposition of the transfer function into lower order partial systems, preferably of first and second order.
- The transition to state-space structures which require a higher implementation complexity but offer more flexibility with respect to the control of the named drawbacks.

The filter structures that we introduce in the following act another way. They are not guided by the realisation of transfer functions in z which in turn are derived from the transfer functions of analog reference filters in p. These structures aim at directly imitating the structure of analog reference filters in the discrete-time domain.

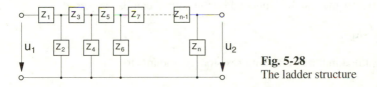

Fig. 5-28
The ladder structure

Ladder filters exhibit excellent properties with respect to coefficient sensitivity, which renders them very appropriate to act as a model for digital filter implementations. The good properties of this structure are based on the fact that ladder filters do not reflect the system differential equation as a whole but its decomposition into a system of lower-order differential equations. Ladder structures consist of a cascade of alternating series and shunt impedances as shown in Fig. 5-28. Each impedance Z_i may represent a single element (capacitor, inductor or resistor) or a combination of several elements.

Stability is automatically created because all elements of the circuit are passive and hence the whole network possesses the property of passivity. Further improvement of the filter sensitivity, especially in the passband, can be achieved with implementations using lossless ladder networks that are terminated by resistors on both sides. We will investigate this case in more detail in the next section.

A starting point for the preservation of the good properties of the analog reference filter in the discrete-time domain is to convert the electrical circuit diagram of the analog filter, whose behaviour is governed by Kirchhoff's current and voltage laws, into a signal flowgraph or a more implementation-oriented block diagram.

The signal quantities appearing in these flowgraphs or block diagrams may be voltages, currents or even wave quantities. In the next step, mathematical operators such as integrators, for instance, are replaced by corresponding discrete-time equivalents, and continuous-time properties of elements such as capacitors, inductors and resistors are converted into appropriate discrete-time representations.

5.7.1 Lossless Networks

Lossless filters inserted between resistive terminations (Fig. 5-29) have interesting properties with respect to the coefficient sensitivity in the passband of the filter. An example of this filter type is the widely used LC ladder filter whose design is supported by numerous design tables available in the literature [53, 47]. Let us first consider the power balance at the input and output of the two-port.

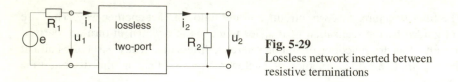

Fig. 5-29
Lossless network inserted between resistive terminations

The maximum power that can be launched by the source can be expressed as

$$P_{max} = |e|^2 / 4R_1 \quad .$$

The power consumed in the terminating resistor R_2 amounts to

$$P_2 = |u_2|^2 / R_2 \quad .$$

An important characteristic of a two-port network is its transmittance, denoted as S_{21}. The squared magnitude of S_{21} is the ratio of the actually consumed power in R_2 to the maximum possible power that the source can deliver to R_2.

$$|S_{21}|^2 = \frac{P_2}{P_{max}} = 4 \frac{R_1}{R_2} \frac{|u_2|^2}{|e|^2}$$

The transmittance itself describes the transfer behaviour of the network between the source voltage e and the voltage at the output of the network u_2 with respect to magnitude and phase.

$$S_{21} = 2 \sqrt{\frac{R_1}{R_2}} \frac{u_2}{e} \tag{5.34}$$

From the definition of the transmittance it follows automatically that the magnitude of S_{21} will never exceed unity. S_{21} reaches this maximum if
1. the network is lossless and thus the whole input power $u_1 i_1$ is passed to the resistor R_2 and
2. the network transforms the real resistance R_2 into the real internal resistance of the source R_1 (matching between source and load).
The network structure enforces an upper limit of the magnitude of the frequency response $H(j\omega)$ independently of the values of the network elements.

$$|S_{21}(j\omega)| = 2 \sqrt{R_1/R_2} \, |H(j\omega)| \leq 1$$

Networks with this property are called structure-induced bounded [57]. Let us now assume that a network has the named property, and for an arbitrary frequency ω_k we have

$$|S_{21}(j\omega_k)| = 1 \ .$$

If the ith element of the network changes its value W_i, by ageing or temperature changes for instance, the magnitude of the transmittance can only decrease. The solid line in Fig. 5-30 shows a corresponding characteristic which has a maximum at W_{i0} equivalent with a slope that is zero. Thus the first-order sensitivity of the transmittance with respect to tolerances of the network elements vanishes around the nominal value W_{i0}.

$$\left. \frac{\partial |S_{21}(j\omega_k)|}{\partial W_i} \right|_{W_i = W_{i0}} = 0 \qquad \forall i, \forall k \tag{5.35}$$

Each element of the network shows the described behaviour. A sensitivity characteristic, as indicated by the dotted line in Fig. 5-30, cannot occur in a structure-induced bounded network. Selective filters often have a number of maxima with $|S_{21}| = 1$ distributed over the passband. Between these maxima the gain only deviates marginally from unity. We can therefore expect that structure-induced bounded networks exhibit excellent sensitivity properties in the whole passband because the derivative (5.35) assumes low values or vanishes completely in this frequency range.

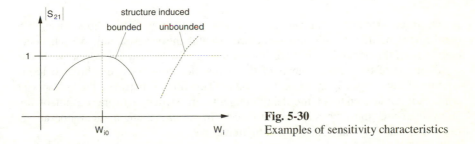

Fig. 5-30
Examples of sensitivity characteristics

5.7.2 Realisability

Mathematical algorithms that describe the behaviour of discrete-time systems may be represented in the form of flowgraphs. In order to be realisable, a flowgraph has to meet the following two conditions:

- The sequence of the mathematical operations that are to be performed must be clearly identifiable.
- It must be guaranteed that all operations can recur periodically with the frequency $f_s = 1/T$.

The first condition leads to the requirement that there must not be any delay-free directed loops in the flowgraph. A loop is called directed if the arrowheads in all branches have the same orientation with respect to a given orientation of the loop. Loop I in Fig. 5-31 is directed and does not contain any delay element. In this case it is not possible to clearly identify in which sequence the mathematical operations have to be performed. Thus the flowgraph is not realisable.

The condition of periodicity of all operations leads to the requirement that the total delay in any loop (directed or not) must be equal to an integer multiple (zero, positive or negative) of T. The total delay is calculated as sum of the delays in all branches of the loop. The delay in a branch that has the same orientation as the loop is assumed positive, otherwise it is assumed negative. Delays of fractions of T may occur in a flowgraph if, for instance, a number of mathematical operations has to be performed sequentially within one period T. The total delay in a loop, however, must be in any case an integer multiple of T. Examples are shown again in Fig. 5-31. Loop III has a total delay of $3/2\,T$, the delay in loop IV amounts to $-T/2$. Both loops are not realisable by definition. The only loop that fulfils the realisability condition is loop II with a total delay of $-T$.

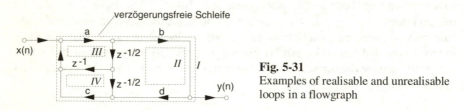

Fig. 5-31
Examples of realisable and unrealisable loops in a flowgraph

All filter structures that we have treated up to now fulfil the realisability condition. Common to all these structures is that the input signal to each delay element in the recursive part of the filter is always calculated as a linear combination of the output signals of the delay elements. Thus all directed loops contain delays (e.g. Fig. 5-17 or Fig. 5-24). The time T between the instants at which the input signals of the delay elements are switched to the output of the delay elements must be sufficient to perform all the mathematical operations needed to calculate these linear combinations.

Transformations that convert continuous-time systems into discrete-time systems have to meet certain conditions to come to a useful result: one is stability, which is assured by having poles within the unit circle in the z-plane. In this section we have become acquainted with a second important condition. The discrete-time system that we obtain by a transformation must be realisable in the sense that we have discussed above.

5.7.3 Kirchhoff Ladder Filters

By means of a concrete example, we want to illustrate step by step how we can, starting with the Kirchhoff current and voltage laws and a resulting continuous-time signal flowgraph, finally obtain a block diagram of a discrete-time realisation of the filter. Fig. 5-32 shows a third-order ladder low-pass filter. We identify a lossless two-port inserted between resistive terminations with the well-known good sensitivity properties.

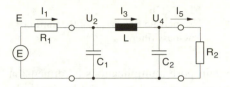

Fig. 5-32
Third-order ladder low-pass filter

The following current and voltage equations can be read from Fig. 5-32:

$$I_1 = \frac{E - U_2}{R_1} \qquad I_5 = \frac{U_4}{R_2} \qquad U_2 = \frac{I_1 - I_3}{j\omega C_1}$$

$$I_3 = \frac{U_2 - U_4}{j\omega L} \qquad U_4 = \frac{I_3 - I_5}{j\omega C_2} \; .$$

Figure 5-33a shows a translation of these equations into a flowgraph and a block diagram. Currents and voltages can now be interpreted as signal quantities which are related by additions and transfer functions in the connecting branches. This representation is independent of the actual physical implementation of the desired transfer behaviour. If we replace current and voltage by physical quantities like force and velocity, for instance, and elements like capacitors and inductors by masses and springs, this flowgraph could also be derived from a system that represents a third-order mechanical low-pass filter.

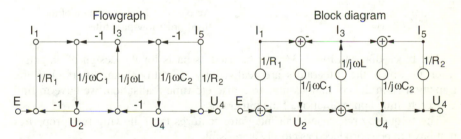

Fig. 5-33a Flowgraph and block diagram derived from a third-order ladder low-pass filter

If we normalise the impedances and admittances in Fig. 5-33a to an arbitrary resistance R, we obtain a flowgraph in which only voltages appear as signal quantities and all coefficients and transfer functions are dimensionless, which is advantageous for the further numerical treatment of the structure. This normalisation is achieved by multiplying all coefficients in the ascending branches, and dividing all coefficients in the descending branches, by R. This measure does not change the transfer behaviour at all.

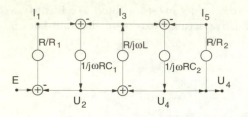

Fig. 5-33b
Normalised flowgraph of the third-order
low-pass filter

Both lateral branches in Fig. 5-33b contain pure multiplications by dimension-less constants. The three inner branches contain integrators, as can be concluded from the frequency response that is inverse proportional to the frequency ($\sim 1/j\omega$). Thus 4 adders, 3 integrators and 2 coefficient multipliers are required to realise the low-pass filter (Fig. 5-33c). The coefficients and time constants of the integrators in this realisation can be calculated as

$$k_1 = R / R_1$$
$$\tau_2 = R\,C_1$$
$$\tau_3 = L / R$$
$$\tau_4 = R\,C_2$$
$$k_5 = R / R_2 \ .$$

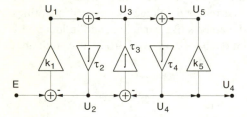
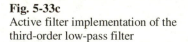

Fig. 5-33c
Active filter implementation of the
third-order low-pass filter

The block diagram Fig. 5-33c can be used as basis for the design of an active filter in which the integrators are realised by operational amplifiers. For the transition from the continuous-time to a discrete-time realisation, we have to find a discrete-time approximation of the integrator.

The frequency response of an integrator features the following two properties that have to be reproduced as closely as possible:

• The phase is constantly 90°.
• The magnitude is inversely proportional to the frequency.

Transfer function and frequency response of the continuous-time integrator can be expressed as

$$H(p) = \frac{1}{p\tau} \qquad \rightarrow \qquad H(j\omega) = \frac{1}{j\omega\tau} \ .$$

τ denotes the time constant of the integrator. The constant phase can be easily realised by a discrete-time algorithm, while the reproduction of a magnitude inversely proportional to the frequency causes problems, since the magnitude function must be periodical with the sampling frequency and representable by a trigonometric function.

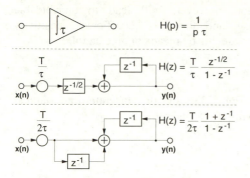

Fig. 5-34
Possible discrete-time realisations of an integrator

Fig. 5-34 shows two possible candidates for the discrete-time realisation of the integrator block. Transfer function and frequency response of these algorithms can be expressed as

$$H(z) = \frac{T}{\tau}\frac{z^{-1/2}}{1-z^{-1}} \qquad \rightarrow \qquad H\!\left(e^{j\omega T}\right) = \frac{T}{2\tau}\frac{1}{j\sin(\omega T/2)} \qquad (5.36a)$$

and

$$H(z) = \frac{T}{2\tau}\frac{1+z^{-1}}{1-z^{-1}} \qquad \rightarrow \qquad H\!\left(e^{j\omega T}\right) = \frac{T}{2\tau}\frac{1}{j\tan(\omega T/2)} \; . \qquad (5.36b)$$

Both realisations possess a constant phase of 90°. A good approximation of the magnitude of the analog integrator is only given at low frequencies where sine and tangent can be approximated by their respective arguments. As the frequency ω is replaced by the sine and tangent terms in (5.36), the frequency axis is distorted in both cases: (5.36a) means stretching, (5.36b) compression of the frequency axis. A way to circumvent this problem is to predistort the frequency response of the analog reference filter such that the discrete-time implementation has exactly the desired cutoff frequencies.

Starting from the circuit diagram of the analog implementation, we have mathematically found a way to derive a discrete-time block diagram of a filter which has a ladder filter as model. At first glance, the only drawback of the discrete-time solution seems to be that the frequency axis is distorted which can be defused by proper measures.

More serious, however, is the fact that replacing the integrators in Fig. 5-33c by either discrete-time approximation depicted in Fig. 5-34 does not result in realisable systems in the sense discussed in Sect. 5.7.2. In one case, there exist loops in the block diagram with a total delay that is not an integer multiple of T. In the other case we find delayless loops. This result can be generalised: flowgraphs and block diagrams which are based on voltage, current and the corresponding Kirchhoff laws lead to unrealisable discrete-time block diagrams. We will show in the next section that this problem can be solved by the transition to wave quantities. The only way out in the present approach is to add or remove delays to guarantee realisability. This operation, however, will lead to unavoidable deviations of the frequency response, and the good coefficient sensitivity behaviour will be lost. We will demonstrate this by means of an example.

Example

Replacing the analog integrator in Fig. 5-33c by the discrete-time version according to (5.36a) leads to the following block diagram.

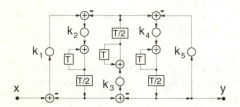

Fig. 5-35a
Block diagram of a third-order ladder low-pass filter

The filter coefficients in this representation are calculated as follows:

$$k_1 = R \,/\, R_1$$
$$k_2 = T \,/\, \tau_1 = T/RC_1$$
$$k_3 = T \,/\, \tau_2 = TR/L \qquad\qquad\qquad (5.37)$$
$$k_4 = T \,/\, \tau_3 = T/RC_2$$
$$k_5 = R \,/\, R_2 \ .$$

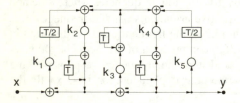

Fig. 5-35b
Block diagram of the discrete-time ladder filter after equivalent rearrangement

A number of equivalent rearrangements in Fig. 5-35a lead to the form of Fig. 5-35b. The lateral branches in this block diagram contain delays of $-T/2$. This leads to loops that violate the realisability conditions.

If the sampling frequency is high compared to the cutoff frequency of the low-pass filter, we can add delays of T/2 in both lateral branches (see Fig. 5-35c) without changing the frequency response too dramatically [4]. As a consequence, the left branch only contains the multiplication by the coefficient k_1, the right branch the multiplication by k_5. This modification leads to a realisable filter structure.

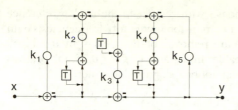

Fig. 5-35c
Block diagram after removal of the
delays in the lateral branches

We now want to compare the magnitude responses of the analog ladder filter according to Fig. 5-32 and the corresponding discrete-time implementation according to Fig. 5-35c. We choose a third-order low-pass filter with the following design specification:

-   characteristic: Chebyshev, 1.256 dB ripple
-   normalised cutoff frequency: $f_c T = 0.1$
-   normalisation impedance: $R = R_1 = R_2$

Filter design table [47] yields the following values for the capacitors and the inductor:

$$C_1 = C_2 = 2.211315/(2\pi f_c R)$$
$$L = 0.947550 R/2\pi f_c \quad .$$

Substitution in (5.37) leads to the following coefficients of the discrete-time filter:

$$k_1 = k_5 = 1$$
$$k_2 = k_4 = 2\pi\, 0.1\,/\, 2.211315 = 0.284138$$
$$k_3 = 2\pi\, 0.1\,/\, 0.947550 = 0.663098 \quad .$$

Figure 5-36 compares the magnitude characteristics of the analog and the discrete-time realisation. The deviations in the passband are caused by the manipulations of the block diagram which assure realisability. The property of boundedness of the magnitude gets lost since the curve visibly exceeds unity in the vicinity of the cutoff frequency. The frequency axis is stretched as a consequence of the frequency response of the used integrator approximation (5.36a).

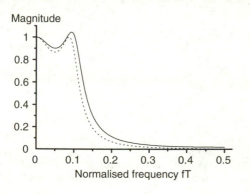

Fig. 5-36
Magnitude of a third-order ladder filter
(Chebyshev 1.256 dB ripple, cutoff
frequency $f_c T = 0.1$)
dashed line: analog filter
solid line: digital ladder filter

The deviation of the magnitude response with respect to the analog reference filter as depicted in Fig. 5-36 is relatively small for low cutoff frequencies. With increasing cutoff frequency, however, the removal of the delays in the block diagram becomes more and more noticeable. In the area of $f_c T = 0.25$, the original frequency response of the analog filter can not be recognised any more.

5.7.4 Wave Digital Filters

The considerations in the previous section showed that flowgraphs based on current and voltage as signal parameters (see Fig. 5-33a) result in nonrealisable discrete-time systems. By the modifications to remove this drawback, the resulting digital filter looses the good properties of the reference filter, and the frequency response more or less deviates from the specification. The transition to wave parameters can solve the problem. Fettweis [20, 22, 23] showed that this approach does not only result in realisable systems but also leads to filters with excellent stability properties. Under certain circumstances, passivity and thus stability can be guaranteed even if the unavoidable rounding operations are taken into account.

Wave parameters have their origin in transmission line theory, and form, in a somewhat modified form, the basis of wave digital filters. Because of their great importance for the understanding of wave digital filters, wave parameters are considered in more detail in the following section.

It must be noted that we cannot treat all aspects of wave digital filter theory in this textbook. The low-pass filter according to Fig. 5-32 is to serve as an example again in to order to finally arrive at a realisable discrete-time flowgraph. A detailed overview of theory and practice of wave digital filters can be found in [22].

5.7.4.1 Wave Parameters

Wave parameters have their origin in transmission line theory. The voltage u at an arbitrary position on the line can be assumed to result from the superposition of a voltage wave propagating in the positive x-direction u_+ and a voltage wave in opposite direction u_-. The same is true for current waves on the line (see Fig. 5-37).

$$u = u_+ + u_-$$
$$i = i_+ + i_-$$

(5.38)

$$u_+ = i_+ R_L$$
$$u_- = -i_- R_L$$

$$u = u_+ + u_-$$
$$i = i_+ + i_-$$

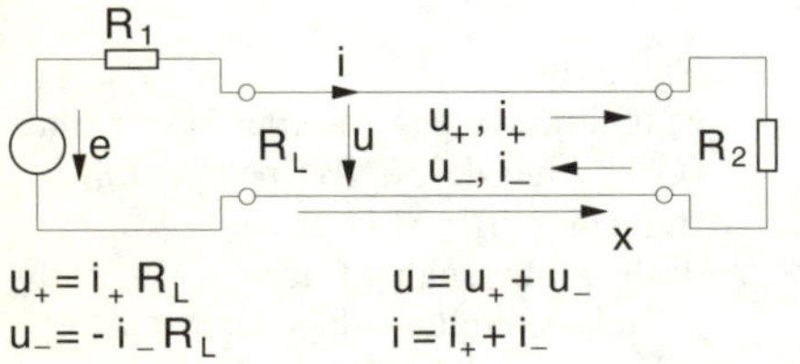

Fig. 5-37
Definition of wave parameters derived from the theory of transmission lines

Current and voltage waves are related by the characteristic impedance R_L of the line.

$$u_+/i_+ = R_L \qquad\qquad u_-/i_- = -R_L \qquad\qquad\qquad (5.39)$$

Using (5.38) and (5.39), we can calculate the voltage waves propagating in both directions at an arbitrary position on the line from the voltage and the current measured at this position.

$$u_+ = (u + R_L\, i)/2 \qquad\qquad u_- = (u - R_L\, i)/2 \qquad\qquad (5.40)$$

These relations are independent of the length of the line. In principle they are also valid for a line length equalling zero. In this case R_L is no more the characteristic impedance of a line but a parameter that can be freely chosen. u_+ and u_- have the meaning of quantities that describe the electrical state of a port in the same way as the pair u and i. For our purpose we define the wave parameters for simplicity without the factor ½ and denote them by the letters a and b. Instead of R_L we write simply R.

$$a = u + R\,i \qquad\qquad b = u - R\,i \qquad\qquad\qquad (5.41a)$$

$$u = (a+b)/2 \qquad\qquad i = (a-b)/2R \qquad\qquad\qquad (5.41b)$$

R denotes the so-called port resistance. The parameters a and b defined in this way are called voltage wave parameters since they have the dimension of a voltage. Similarly we can define current wave parameters

$$a = u/R + i \qquad\qquad b = u/R - i$$

or power wave parameters

$$a = u/\sqrt{R} + i\,\sqrt{R} \qquad\qquad b = u/\sqrt{R} - i\,\sqrt{R}$$

from which we can calculate the power of the incident and reflected waves by squaring.

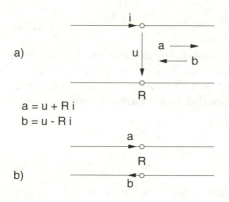

a)

$$a = u + R\,i$$
$$b = u - R\,i$$

b)

Fig. 5-38
Definition of wave parameters:
a) characterisation of a port by current and voltage
b) the corresponding flowgraph based on wave parameters

All three parameter definitions are suitable for the following considerations, but it has become custom to use voltage waves. Figure 5-38a depicts a port with the assignment of current, voltage, port resistance and the wave parameters a and b. Figure 5-38b shows the equivalent wave flowgraph. It is important to note that the indication of the port resistance in the flowgraph is absolutely mandatory, as otherwise the parameters a and b are not defined. Without the knowledge of the port resistance it is impossible to reckon back the electrical state of the port in terms of voltage u and current i from the wave parameters a and b.

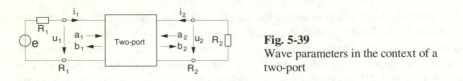

Fig. 5-39
Wave parameters in the context of a two-port

Figure 5-39 shows a two-port network inserted between two resistive terminations. In the following we calculate the wave parameters associated with the ports of this network. The waves propagating towards the network are denoted as a_i, the waves reflected by the network as b_i. The choice of the port impedances R_i is arbitrary in principle. In practice, however, one will select these particular parameter values of R_i such that they lead to the simplest overall expressions for our problem. This is the case if the port impedances are chosen equal to the terminating resistances. Based on this assumption, we calculate the four incident and reflected waves as depicted in Fig. 5-39 using (5.41a).

wave a_1:
$$a_1 = u_1 + R_1\, i_1$$
$$i_1 = \left(e - u_1\right)/R_1$$
$$a_1 = e \tag{5.42a}$$

The incident wave at port 1 equals the voltage e of the source.

wave b_1:
$$b_1 = u_1 - R_1\, i_1$$
$$i_1 = \left(e - u_1\right)/R_1$$
$$b_1 = 2u_1 - e \tag{5.42b}$$

The reflected wave at port 1 depends on the input impedance of the corresponding network port. If this impedance equals R_1, which is the matching case, then $u_1 = e/2$, and the reflected wave b_1 vanishes. Otherwise b_1 is terminated in the internal resistor of the source which acts as a sink in this case.

wave a_2:
$$a_2 = u_2 + R_2\, i_2$$
$$i_2 = -u_2/R_2$$
$$a_2 = 0 \tag{5.42c}$$

Because of the matching termination, there is no incident wave at port 2.

wave b_2:
$$b_2 = u_2 - R_2\, i_2$$
$$i_2 = -u_2/R_2$$
$$b_2 = 2u_2 \qquad\qquad\qquad (5.42\text{d})$$

The reflected wave at port 2 is the part of the source signal that is transmitted over the network. Its value is the double of the voltage at port 2. Figure 5-40 summarises the calculated wave flows and depicts a flowgraph which is equivalent to the circuit diagram Fig. 5-39.

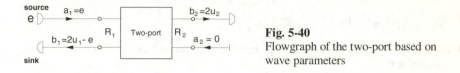

Fig. 5-40

Flowgraph of the two-port based on wave parameters

Fig. 5-40 introduces two new symbols, a wave source and a wave sink. The equivalent of a voltage source contains both elements, since it generates an outgoing wave and terminates an incoming wave in its internal resistor. The terminating resistance at the right side of the two-port is characterised by a sink only.

If the two-port is a selective filter and we consider the stopband of the frequency response, then the incident wave a_1 is, in essence, reflected at port 1. In the passband, the two-port network matches the internal resistor of the source R_1 more or less perfectly with the terminating resistor R_2. As a consequence, the wave a_1 is transmitted without, or with low, attenuation over the network and is terminated as wave b_2 in the resistor R_2, whereas b_1 vanishes. Thus the two-port shares the incident wave a_1 among the reflected waves b_1 and b_2.

Incident and reflected waves of a two-port are related by so-called scattering matrices:

$$\begin{pmatrix} b_1 \\ b_2 \end{pmatrix} = \begin{pmatrix} S_{11} & S_{12} \\ S_{21} & S_{22} \end{pmatrix}\begin{pmatrix} a_1 \\ a_2 \end{pmatrix}. \qquad\qquad (5.43)$$

The elements S_{ii} are reflection coefficients, which are also called reflectances. They define the share of the incident wave a_i at port i that is reflected at this port. The elements S_{ij} describe the transfer behaviour from port j to port i. They are called transmittances.

$$S_{11} = \left.\frac{b_1}{a_1}\right|_{a_2=0} \qquad\qquad S_{12} = \left.\frac{b_1}{a_2}\right|_{a_1=0}$$

$$\qquad\qquad\qquad\qquad\qquad\qquad\qquad\qquad (5.44)$$

$$S_{21} = \left.\frac{b_2}{a_1}\right|_{a_2=0} \qquad\qquad S_{22} = \left.\frac{b_2}{a_2}\right|_{a_1=0}$$

The parameter S_{21} has a special meaning. If the network is properly terminated ($a_2 = 0$), then (5.42a) and (5.42d) apply and we get

$$S_{21} = 2\frac{u_2}{e} \,.\tag{5.45}$$

Apart from the factor 2, S_{21} is the transfer function of the two-port network terminated by R_1 and R_2. Furthermore, apart from the factor $\sqrt{(R2/R1)}$, (5.45) is identical to the definition of the transmittance (5.34). The difference is due to the fact that (5.34) is based on the power balance and therefore on power waves, whereas (5.45) is derived from voltage waves.

If the two-port is lossless, the sum of the power of the incident waves must equal the sum of the power of the reflected waves.

$$|a_1|^2\big/R_1 + |a_2|^2\big/R_2 = |b_1|^2\big/R_1 + |b_2|^2\big/R_2$$

Using (5.43) yields the following relations between the coefficients of the scattering matrix:

$$|S_{11}|^2 + (R_1/R_2)|S_{21}|^2 = 1\tag{5.46a}$$

$$|S_{22}|^2 + (R_2/R_1)|S_{12}|^2 = 1\tag{5.46b}$$

$$S_{11}S_{12}^*\big/R_1 + S_{21}S_{22}^*\big/R_2 = 0\tag{5.46c}$$

Equation (5.46a) points to an interesting property of the reflectance of a lossless two-port. Transmittance S_{21} and reflectance S_{11} are complementary with respect to each other. Thus S_{11} can be considered a transfer function, too, which has its passband in that frequency range, where S_{21} has its stopband and vice versa. The output b_1 in the wave flowgraph Fig. 5-40 can therefore be used as the output of a filter with a complementary frequency response. The same applies to the pair S_{22} and S_{12}. From (5.46a) and (5.46b) we can derive further interesting properties of the scattering parameters for the lossless case.

$$\begin{aligned}|S_{11}| &\le 1 & |S_{21}| &\le \sqrt{R_2/R_1}\\ |S_{22}| &\le 1 & |S_{12}| &\le \sqrt{R_1/R_2}\end{aligned}\tag{5.47}$$

All frequency responses derived from the scattering parameters are bounded. It can therefore expected that filters based on scattering parameters of lossless networks (inserted between resistive terminations) possess the excellent properties concerning coefficient sensitivity discussed in Sect. 5.7.2. Additionally taking (5.46c) into account yields the following interesting relations:

$$|S_{11}| = |S_{22}|\tag{5.48a}$$

$$|S_{21}| = \frac{R_2}{R_1}|S_{12}| \,.\tag{5.48b}$$

Apart from a constant factor, the magnitudes of S_{12} and S_{21} are identical. Thus port 1 and port 2 can reverse roles.

Example

In the following we derive the flowgraph of a simple voltage divider circuit (Fig. 5-41) based on voltage waves. Since the two-port network directly connects source and terminating resistor, voltage and current at both ports are related as follows:

$$u_1 = u_2 \qquad\qquad i_1 = -i_2 \; .$$

Furthermore we have according to (5.41b)

$$u_1 = (a_1 + b_1)/2 \qquad\qquad i_1 = (a_1 - b_1)/2R_1$$
$$u_2 = (a_2 + b_2)/2 \qquad\qquad i_2 = (a_2 - b_2)/2R_2 \; .$$

From these relations we can determine the reflected waves b_1 and b_2 in terms of the incident waves a_1 and a_2.

$$b_1 = \frac{2R_1}{R_1 + R_2} a_2 - \frac{R_1 - R_2}{R_1 + R_2} a_1$$

$$b_2 = \frac{2R_2}{R_1 + R_2} a_1 - \frac{R_1 - R_2}{R_1 + R_2} a_2$$

$$(5.49)$$

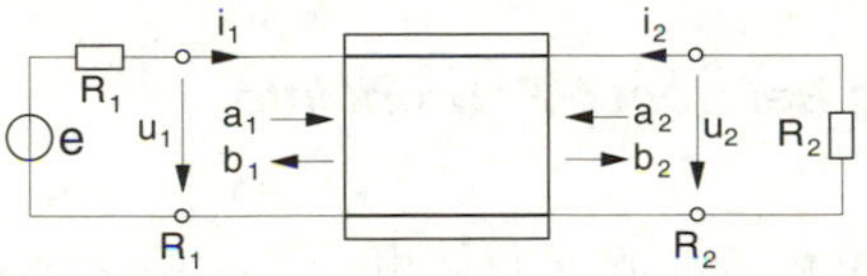

Fig. 5-41
Two-port representation of the voltage divider circuit

Using the abbreviation

$$r = \frac{R_1 - R_2}{R_1 + R_2} \; ,$$

(5.49) can be written as

$$b_1 = -r\, a_1 + (1 + r)\, a_2$$
$$b_2 = (1 - r)\, a_1 + r\, a_2$$

$$(5.50)$$

which leads to the following scattering matrix describing the wave-flow behaviour of the simple two-port in Fig. 5-41.

$$S = \begin{pmatrix} -r & 1+r \\ 1-r & r \end{pmatrix}$$

Figure 5-42 shows a possible realisation of (5.50) in the form of a block diagram. This structure is called a two-port adapter. It links ports with different port resistances R_i. If the internal resistance of the source R_1 and the terminating resistance R_2 are matched ($r = 0$), no reflections occur at the ports. Incident wave a_1 leaves the two-port as b_2, and a_2 passes the two-port accordingly and comes out as b_1.

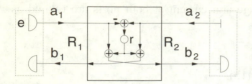

Fig. 5-42
Block diagram representation of the voltage divider circuit

The transfer behaviour of the voltage divider is described by the scattering parameter S_{21} as can be seen from (5.45).

$$S_{21} = 2\frac{u_2}{e} \quad \text{or} \quad \frac{u_2}{e} = \frac{1}{2}S_{21}$$

With $S_{21} = 1 - r$ from above, we can write

$$\frac{u_2}{e} = \frac{1}{2}(1-r) = \frac{1}{2}\left(1 - \frac{R_1 - R_2}{R_1 + R_2}\right) = \frac{R_2}{R_1 + R_2} \; ,$$

which is the expected result.

5.7.4.2 Decomposition of the Analog Reference Network into One-Ports and Multi-Ports

The starting point for the design of the Kirchhoff ladder filter has been the establishment of the corresponding node and branch equations of the filter network. The theory of wave digital filters uses a different approach. In the first step, all elements of the electrical reference network are described by their properties with respect to wave parameters. Each element is considered a one-port with an associated port resistance R_{Port} (see Fig. 5-43).

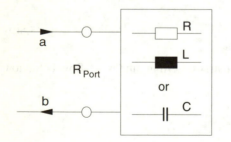

Fig. 5-43
Definition of the wave one-port

The circuit element is now characterised by its reflectance, which is the ratio of reflected to incident wave. In case of a resistor, we for simplicity choose $R_{\text{Port}} = R$.

The reflectance, and consequently the reflected wave, become zero (see the example of R_2 in Fig 5-41 and Fig. 5-42). The reflectance of capacitors and inductors is complex, frequency dependent and does not vanish for any frequency. For the choice of R_{Port} we will have to consider relations between the sampling period T of the target digital filter and the time constants $\tau = R_{Port}C$ or $\tau = L/R_{Port}$, as we will show in the next section.

In the next step, all network elements are linked together by adapters to form the wave flowgraph of the whole filter network. We have already introduced the two-port adapter (Fig. 5-42) which links together ports with different port resistances (see the example in the previous section where the adapter describes the matching behaviour of a voltage source with internal resistor R_1 and a terminating resistor R_2). For more complex networks, however, we need adapters with three or more ports. We attempt now to decompose the ladder filter according to Fig. 5-32 into one-ports that represent the elements of the circuit and adapters that describe how these elements are linked.

Fig. 5-44
Decomposition of a ladder filter into partial two-ports.

Figure 5-44 depicts the ladder filter modified into the form of a cascade of two-ports. Each two-port contains exactly one element, which is arranged either as a series or as a shunt element. For each of these two-ports, we can give a scattering matrix. In [59] these scattering matrices are used as the basis for the design of RC-active filters with low component sensitivity. For our purpose, however, we want to go one step further. We modify the flowgraph in such a way that the network elements appear in the form of one-ports. This can be accomplished by rearrangement as shown in Fig. 5-45. Note that we still have the same ladder filter that we started with.

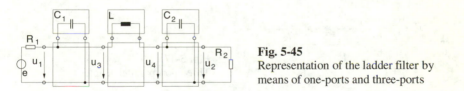

Fig. 5-45
Representation of the ladder filter by means of one-ports and three-ports

The circuit diagram now consists of a series connection of three-ports. The remaining unattached ports are terminated by one-ports which represent the voltage source with internal resistance R_1, the terminating resistance R_2, and the components of the lossless filter network C_1, C_2 and L. If we convert Fig. 5-45 into a wave flowgraph, we have to distinguish between two types of three-port adapters: C_1 and C_2 are linked to the connecting network via parallel adapters, the inductor L via a serial adapter. Figure 5-46 introduces the two new symbols that represent these adapter types and shows the result of the conversion.

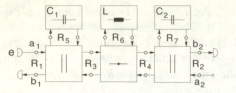

Fig. 5-46
Wave flowgraph of the ladder filter with three-ports as basic building blocks

Each of the adapters in Fig. 5-46 has to link three ports with different port resistances. After the interconnection of all blocks, we identify seven independent port resistances that have to be specified:

- R_1 and R_2 are determined by the internal resistance of the source and the terminating resistance of the network.
- R_5, R_6 and R_7 will be determined in such a way that the discrete-time realisation of the frequency dependent reflectance of the capacitors and of the inductor becomes as simple as possible.
- The choice of R_3 and R_4 seems to be arbitrary at first glance. We will see, however, that there are constraints with respect to the realisability of the discrete-time filter structure that have to be observed.

The general relation between incident and reflected waves of a three-port can be expressed as:

$$b_1 = S_{11}\,a_1 + S_{12}\,a_2 + S_{13}\,a_3$$
$$b_2 = S_{21}\,a_1 + S_{22}\,a_2 + S_{23}\,a_3 \tag{5.51}$$
$$b_3 = S_{31}\,a_1 + S_{32}\,a_2 + S_{33}\,a_3 \ .$$

In the following, we determine the scattering coefficients of both three-port adapter types and give possible realisations of these adapters in the form of block diagrams. We proceed in the same way as we did for the derivation of the two-port adapter in the previous section. We start with the parallel adapter.

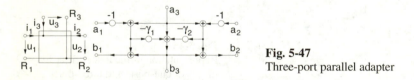

Fig. 5-47
Three-port parallel adapter

As an example, we show the calculation of the scattering coefficient S_{23} in more detail. This coefficient describes the situation that a voltage source is connected to port 3, and port 1 and port 2 are terminated with R_1 and R_2 respectively. So there are no incident waves at port 1 and port 2. From (5.51) it follows:

$$S_{23} = \frac{b_2}{a_3}\bigg|_{a_1 = 0, a_2 = 0} \ . \tag{5.52a}$$

From the definition of the wave parameters (5.41) we have:

$$b_2 = u_2 - R_2\, i_2$$
$$a_3 = u_3 + R_3\, i_3 \ .$$

(5.52b)

According to Fig. 5-47, we can write down the following node and mash equations for the parallel adapter.

$$u_1 = u_2 = u_3$$
$$i_1 + i_2 + i_3 = 0$$

(5.52c)

The terminations with R_1 and R_2 yield the following additional equations:

$$u_1 = -R_1\, i_1 \qquad \text{and} \qquad u_2 = -R_2\, i_2 \ .$$

(5.52d)

The set of equations (5.52a,b,c,d) finally results in:

$$S_{23} = \frac{2G_3}{G_1 + G_2 + G_3} \ .$$

G_n are the reciprocals of the port resistances R_n. In the same way we can calculate the remaining scattering coefficient.

$$b_1 = \frac{G_1 - G_2 - G_3}{G_1 + G_2 + G_3}\, a_1 + \frac{2G_2}{G_1 + G_2 + G_3}\, a_2 + \frac{2G_3}{G_1 + G_2 + G_3}\, a_3$$

$$b_2 = \frac{2G_1}{G_1 + G_2 + G_3}\, a_1 + \frac{G_2 - G_1 - G_3}{G_1 + G_2 + G_3}\, a_2 + \frac{2G_3}{G_1 + G_2 + G_3}\, a_3$$

$$b_3 = \frac{2G_1}{G_1 + G_2 + G_3}\, a_1 + \frac{2G_2}{G_1 + G_2 + G_3}\, a_2 + \frac{G_3 - G_1 - G_2}{G_1 + G_2 + G_3}\, a_3$$

Using the abbreviations

$$\gamma_1 = \frac{2G_1}{G_1 + G_2 + G_3} \qquad\qquad \gamma_2 = \frac{2G_2}{G_1 + G_2 + G_3}$$

we can simplify these relations. The transfer behaviour of the three-port parallel adapter is fully specified by the two coefficients γ_1 and γ_2.

$$b_1 = (\gamma_1 - 1)a_1 + \gamma_2 a_2 + (2 - \gamma_1 - \gamma_2)a_3$$

$$b_2 = \gamma_1 a_1 + (\gamma_2 - 1)a_2 + (2 - \gamma_1 - \gamma_2)a_3$$

(5.53a)

$$b_3 = \gamma_1 a_1 + \gamma_2 a_2 + (1 - \gamma_1 - \gamma_2)a_3$$

The block diagram in Fig. 5-47 shows a possible realisation of (5.53). Two multipliers and six adders are required. From (5.53) we can derive the following scattering matrix:

$$S = \begin{pmatrix} \gamma_1 - 1 & \gamma_2 & 2 - \gamma_1 - \gamma_2 \\ \gamma_1 & \gamma_2 - 1 & 2 - \gamma_1 - \gamma_2 \\ \gamma_1 & \gamma_2 & 1 - \gamma_1 - \gamma_2 \end{pmatrix} . \tag{5.53b}$$

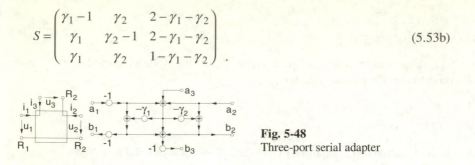

Fig. 5-48
Three-port serial adapter

The three-port serial adapter can be treated in the same way. According to Fig. 5-48 we can write down the following node and mash equations:

$$i_1 = -i_2 = -i_3$$

$$u_1 - u_2 - u_3 = 0 .$$

Again using the definition of the wave parameters (5.41), the equations describing the respective port terminations and the abbreviations

$$\gamma_1 = \frac{2R_1}{R_1 + R_2 + R_3} \qquad \qquad \gamma_2 = \frac{2R_2}{R_1 + R_2 + R_3}$$

finally result in the following relations between the incident and reflected waves:

$$b_1 = (1 - \gamma_1)a_1 + \gamma_1 a_2 + \gamma_1 a_3$$

$$b_2 = \gamma_2 a_1 + (1 - \gamma_2)a_2 - \gamma_2 a_3 \tag{5.54a}$$

$$b_3 = (2 - \gamma_1 - \gamma_2)a_1 - (2 - \gamma_1 - \gamma_2)a_2 - (1 - \gamma_1 - \gamma_2)a_3 .$$

The transfer behaviour of the adapter is fully determined again by the two coefficients γ_1 and γ_2. The block diagram in Fig. 5-48 depicts a possible implementation of (5.54). Two multipliers and six adders are required. The following scattering matrix can be derived from (5.54):

$$S = \begin{pmatrix} 1 - \gamma_1 & \gamma_1 & \gamma_1 \\ \gamma_2 & 1 - \gamma_2 & -\gamma_2 \\ 2 - \gamma_1 - \gamma_2 & -2 + \gamma_1 + \gamma_2 & -1 + \gamma_1 + \gamma_2 \end{pmatrix} . \tag{5.54b}$$

5.7.4.3 The Reflectance of the Capacitor and the Inductor

What is missing in the wave flowgraph of Fig. 5-46 to complete the picture are the one-ports that describe the reflection behaviour of the capacitors and the inductor. The current and voltage of these elements are related by the respective impedances $j\omega L$ and $1/j\omega C$.

$$u_L = j\omega L\, i_L \qquad\qquad i_C = j\omega C\, u_C$$

Using (5.41a) and assuming an arbitrary port resistance R_{Port} yields the following ratios of reflected to incident waves:

$$\frac{b_L}{a_L} = \frac{u_L - R_{Port}\, i_L}{u_L + R_{Port}\, i_L} = \frac{j\omega - R_{Port}/L}{j\omega + R_{Port}/L} \tag{5.55a}$$

$$\frac{b_C}{a_C} = \frac{u_C - R_{Port}\, i_C}{u_C + R_{Port}\, i_C} = \frac{1/(R_{Port}C) - j\omega}{1/(R_{Port}C) + j\omega} \, . \tag{5.55b}$$

The transfer functions (5.55) describe the behaviour of first-order all-pass filters. The magnitude of these transfer functions is unity. The phase can be expressed as:

$$b_L(\omega) = 2\arctan(\omega L\, /\, R_{Port}) + \pi \tag{5.56a}$$

$$b_C(\omega) = 2\arctan(\omega R_{Port}\, C) \, . \tag{5.56b}$$

(Note: In the literature, reflected waves and the phase response of a system are both denoted by the letter b!)

The group delay of the all-pass filter is the derivative of the phase with respect to the frequency ω.

$$\tau_L(\omega) = \frac{db_L(\omega)}{d\omega} = \frac{2L\, /\, R_{Port}}{1 + (\omega L\, /\, R_{Port})^2} \tag{5.57a}$$

$$\tau_C(\omega) = \frac{db_C(\omega)}{d\omega} = \frac{2R_{Port}C}{1 + (\omega R_{Port}C)^2} \tag{5.57b}$$

The ladder filter of Fig. 5-32 can therefore be realised using three first-order all-pass filters that are linked by three-port parallel and serial adapters (Fig. 5-49). In order to convert this continuous-time flowgraph into a discrete-time realisation of the filter, we have to find a discrete-time approximation of the all-pass filter.

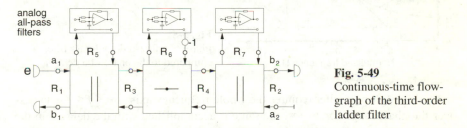

Fig. 5-49
Continuous-time flow-graph of the third-order ladder filter

The course of the group delay according to (5.57) is relatively constant up to the cutoff frequency $\omega_c = 1/(R_{Port}C)$ or $\omega_c = R_{Port}/L$. The group delay at $\omega = 0$

amounts to $\tau_g = 2R_{Port}C$ or $\tau_g = 2L/R_{Port}$. In the discrete-time domain it is very easy to realise a constant delay corresponding to the sample period T. The port resistances R_{Port} are chosen now in such a way that the all-pass filters have a group delay of T at $\omega = 0$.

$$R_{PortL} = 2L / T \tag{5.58a}$$

$$R_{PortC} = T / 2C \tag{5.58b}$$

If we replace the continuous-time all-pass filter by a delay of T, we obtain a discrete-time realisation of the ladder filter (see Fig. 5-50). A capacitor is thus represented by a delay in the flowgraph, an inductor by a delay plus an inverter.

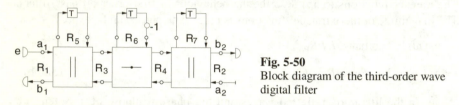

Fig. 5-50
Block diagram of the third-order wave digital filter

Which deviation with respect to the frequency response do we expect if the all-pass filter is replaced by a simple delay? The magnitude is constant and equals unity in both cases, but there is a difference with respect to the phase.

$$b_{delay}(\omega) = \omega T = 2\pi fT \tag{5.59a}$$

$$b_{allpass}(\omega) = 2\arctan(\omega T / 2) = 2\arctan(\pi fT) \tag{5.59b}$$

Figure 5-51 compares both phase responses.

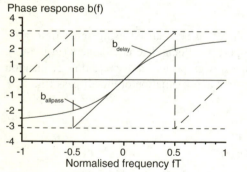

Phase response b(f)

Normalised frequency fT

Fig. 5-51
Phase of a first-order all-pass filter compared to the phase of a delay T

While the all-pass filter requires the whole frequency axis $-\infty \leq f \leq +\infty$ to pass over the phase range $-\pi$ to $+\pi$, the same is accomplished with the delay T in the frequency range $-f_s/2 \leq f \leq +f_s/2$. It can therefore be expected in the discrete-time realisation according to Fig. 5-50 that the frequency response of the analog reference filter is compressed to the range $-f_s/2 \leq f \leq +f_s/2$ and periodically continued. By comparison of (5.59a) and (5.59b) it is easy to understand that we obtain the frequency response of the discrete-time filter by substitution of the frequency variable ω by the term

$$\omega' = \frac{2}{T}\tan(\omega T / 2) \qquad (5.60)$$

Because of this distortion of the frequency axis, it is important to design the analog reference filter in such a way that characteristic frequencies, as the passband or stopband edges for instance, are shifted towards higher frequencies using (5.60). After compression of the frequency response in the discrete-time realisation, these frequencies will assume exactly the desired values. Fig. 5-52 compares the magnitude of the analog reference filter and of the corresponding wave digital filter implementation. The passband ripple remains unchanged. Because of the compression of the frequency axis, the cutoff frequency is shifted to a lower frequency and the slope of the magnitude plot in the transition band becomes steeper.

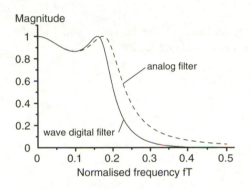

Fig. 5-52
Magnitude of a third-order ladder filter (analog and wave digital filter) Chebyshev 1.256 dB ripple, cutoff frequency $0.2\,f_s$

5.7.4.4 The Role of the Bilinear Transform

Application of the so-called bilinear transform (5.61), which transforms a transfer function in the continuous-time frequency variable p into a corresponding transfer function in z, leads exactly to the behaviour described in the previous section: capacitors and inductors are converted into delays, and the frequency axis is distorted according to (5.60).

$$p = \frac{2}{T}\frac{z-1}{z+1} \qquad (5.61)$$

This transform also plays an important role in the design of IIR filters, as we will show in Chap. 6.

Let us first apply this transform to an inductance L. Complex current and voltage are related as

$$U = pLI \ .$$

Substitution of (5.61) leads to the following relation in the z-domain:

$$U = \frac{2L}{T} I \frac{z-1}{z+1} \tag{5.62a}$$

$$U = RI\Psi . \tag{5.62b}$$

The term $2L/T$ has the dimension of a resistance and is determined by the value of the inductance L. The term $(z-1)/(z+1)$ is denoted by the letter ψ, which represents a new dimensionless frequency variable. Formally, we can now define the impedance of the inductor in the discrete-time domain as

$$Z_L = \frac{U}{I} = R\Psi = R\frac{z-1}{z+1} . \tag{5.63a}$$

With $R = T/2C$, we obtain for the capacitor

$$Z_C = \frac{U}{I} = R/\Psi = R\frac{z+1}{z-1} . \tag{5.63b}$$

In the following, we calculate the reflectance of the capacitor and inductor using the impedance definition (5.63). From (5.41a) we have

$$\frac{b}{a} = \frac{U - R_{\text{Port}}I}{U + R_{\text{Port}}I} = \frac{U/I - R_{\text{Port}}}{U/I + R_{\text{Port}}} = \frac{Z - R_{\text{Port}}}{Z + R_{\text{Port}}} .$$

Substitution of (5.63) yields

$$\frac{b_L}{a_L} = \frac{R\dfrac{z-1}{z+1} - R_{\text{Port}}}{R\dfrac{z-1}{z+1} + R_{\text{Port}}} = \frac{z\dfrac{R - R_{\text{Port}}}{R + R_{\text{Port}}} - 1}{z - \dfrac{R - R_{\text{Port}}}{R + R_{\text{Port}}}}$$

$$\frac{b_C}{a_C} = \frac{R\dfrac{z+1}{z-1} - R_{\text{Port}}}{R\dfrac{z+1}{z-1} + R_{\text{Port}}} = \frac{z\dfrac{R - R_{\text{Port}}}{R + R_{\text{Port}}} + 1}{z + \dfrac{R - R_{\text{Port}}}{R + R_{\text{Port}}}} .$$

If we choose R_{Port} equal to R in the impedance definition (5.63), we obtain the following results:

$$\frac{b_L}{a_L} = -z^{-1} \qquad\qquad \frac{b_C}{a_C} = z^{-1} .$$

Thus by appropriate choice of the port impedance, the bilinear transform turns capacitors and inductors into simple delays in the flowgraph. We consider now the frequency response of the impedances Z_L and Z_C. We replace z by $e^{j\omega T}$ in (5.63) and obtain

$$Z_L(\omega) = R \frac{e^{j\omega T} - 1}{e^{j\omega T} + 1} = jR\tan(\omega T / 2)$$

$$Z_C(\omega) = R \frac{e^{j\omega T} + 1}{e^{j\omega T} - 1} = -jR / \tan(\omega T / 2) \ .$$

The impedances of the capacitor and inductor are no longer directly or inversely proportional to the frequency ω, but are functions of $\tan(\omega T/2)$. The frequency response of these impedances and thus of the whole discrete-time filter is distorted according to (5.60). To sum up, we can conclude that the approximation of the analog first-order all-pass filter by a simple delay T has the same effect as applying the bilinear transform for the transition from the continuous-time to a discrete-time representation of the frequency-dependent impedances of the capacitor and inductor.

5.7.4.5 Observance of the Realisability Condition

Five of the seven port impedances in Fig. 5-50 are determined by the internal resistance of the source R_1, the terminating resistance of the network R_2 and the three elements of the ladder network C_1, C_2 and L. The choice of R_3 and R_4 seems to be arbitrary at first glance. It turns out, however, that they have to be determined in such a way that the filter becomes realisable. A look at the block diagrams of the three-port adapters (Fig. 5-47 and Fig. 5-48) shows that for each port there exists a direct path between input a_i and output b_i. If we directly connect two adapters, as is the case twice in our example, we find delayless loops in the overall filter structure. The problem can be solved by making at least one of the involved ports reflection-free. This means that no share of the incident wave a_i at port i reaches the output b_i. b_i must therefore be composed of shares of the incident waves a_j $(i \neq j)$ at the remaining ports only. A reflection-free port i manifests itself by a vanishing reflectance S_{ii} in the scattering matrix. With respect to the scattering matrix of the three-port parallel adapter (5.52), for instance, this means that for a reflection-free port 2 the parameter γ_2 becomes unity. The resulting scattering matrix (5.64) possesses only one free parameter γ_1.

$$S = \begin{pmatrix} \gamma_1 - 1 & 1 & 1 - \gamma_1 \\ \gamma_1 & 0 & 1 - \gamma_1 \\ \gamma_1 & 1 & -\gamma_1 \end{pmatrix} \qquad (5.64)$$

The port impedance R_2 of the reflection-free port is no longer independent. It results from the condition $\gamma_2 = 1$ and is fixed by the other ports by the relation

$$R_2 = R_1 \ / \ / R_3 = \frac{R_1 \ R_3}{R_1 + R_3} \ . \qquad (5.65)$$

The adapter coefficient γ_1 is related to the independent coefficients R_1 and R_3 in the following way:

$$\gamma_1 = R_3/(R_1 + R_3) \ . \tag{5.66}$$

Fig. 5-53 shows symbol and block diagram of the corresponding three-port parallel adapter according to (5.64), with a reflection-free port 2. It is fully defined by the coefficient γ_1. The implementation requires one multiplier and four adders. If we track the signal paths in the block diagram, we do not detect a direct path between the input and output of port 2 any longer.

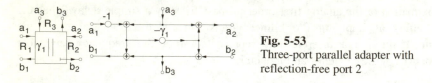

Fig. 5-53
Three-port parallel adapter with
reflection-free port 2

Similar considerations can be made for the three-port serial adapter. If the parameter γ_2 of the scattering matrix (5.54) is unity, port 2 becomes reflection-free. The resulting scattering matrix (5.67) possesses only one free parameter γ_1.

$$S = \begin{pmatrix} 1-\gamma_1 & \gamma_1 & \gamma_1 \\ 1 & 0 & -1 \\ 1-\gamma_1 & \gamma_1-1 & \gamma_1 \end{pmatrix} \tag{5.67}$$

The port impedance R_2 of the reflection-free port is no longer independent. Because of the condition $\gamma_2 = 1$, it is fixed by the impedance of the other ports by the relation

$$R_2 = R_1 + R_3 \ . \tag{5.68}$$

The adapter coefficient γ_1 is related to the independent coefficients R_1 and R_3 in the following way:

$$\gamma_1 = R_1/(R_1 + R_3) \ . \tag{5.69}$$

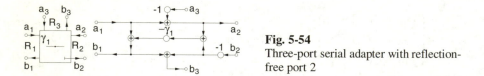

Fig. 5-54
Three-port serial adapter with reflection-
free port 2

Fig. 5-54 shows symbol and block diagram of the corresponding three-port serial adapter according to (5.67) with a reflection-free port 2. It is fully defined by the coefficient γ_1. The implementation requires one multiplier and four adders. Again there is no direct path between the input and output of port 2 any longer.

Example

We calculate the adapter coefficients of the example shown in Fig. 5-52. For a third-order Chebyshev filter with 1.256 dB ripple, we take the following values for the capacitors and the inductor from the filter design table [47]:

$$C_1 = C_2 = 2,211315 / (2\pi f_c R)$$

$$L = 0,947550 R / 2\pi f_c$$

Assuming a cutoff frequency of $0.2 f_s$ ($0.2/T$), we obtain the following port resistances:

$$R_1 = R_2 = R$$

$$R_5 = R_7 = T/2C_{1/2} = 0.284138R$$

$$R_6 = 2L/T = 1.508073R$$

$$R_3 = R_4 = R_1//R_5 = R_2//R_7 = 0.221267R \; .$$

In order to guarantee realisability, we choose parallel adapters with a reflection-free port. The serial adapter in the middle is independently configurable at all ports (Fig. 5-55).

$$\gamma_1 = \gamma_2 = R_5 / (R_1 + R_5) = R_7 / (R_2 + R_7) = 0.221267$$

$$\gamma_3 = \gamma_4 = 2R_3 / (R_3 + R_4 + R_6) = 2R_4 / (R_3 + R_4 + R_6) = 0.226870$$

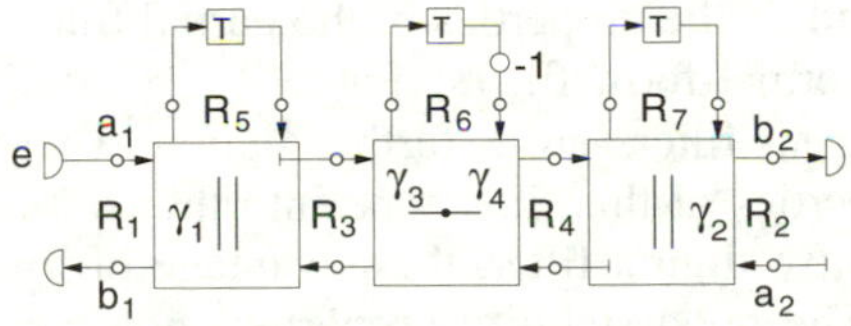

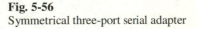

Fig. 5-55
Block diagram of a third-order wave digital filter, taking into account realisability

The structure can be further simplified. The serial adapter in the middle is symmetrical since $R_3 = R_4$ and hence $\gamma_3 = \gamma_4$. In this case, one coefficient γ is sufficient to completely describe its behaviour. Figure 5-56 depicts the structure of the symmetrical serial adapter. One multiplier and six adders are needed. Assuming $\gamma = \gamma_1 = \gamma_2$, we can derive the following scattering matrix from (5.54):

$$S = \begin{pmatrix} 1-\gamma & \gamma & \gamma \\ \gamma & 1-\gamma & -\gamma \\ 2-2\gamma & -2+2\gamma & -1+2\gamma \end{pmatrix} . \qquad (5.70)$$

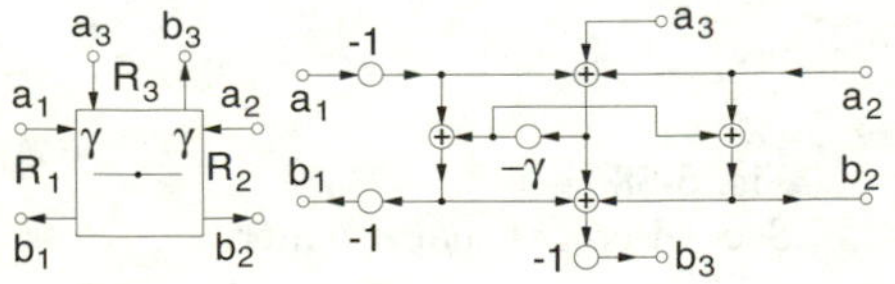

Fig. 5-56
Symmetrical three-port serial adapter

Figure 5-57 shows the block diagram of the corresponding symmetrical parallel adapter. One multiplier and six adders are required. Assuming again $\gamma = \gamma_1 = \gamma_2$, it follows from (5.52):

$$S = \begin{pmatrix} \gamma - 1 & \gamma & 2 - 2\gamma \\ \gamma & \gamma - 1 & 2 - 2\gamma \\ \gamma & \gamma & 1 - 2\gamma \end{pmatrix}.$$

(5.71)

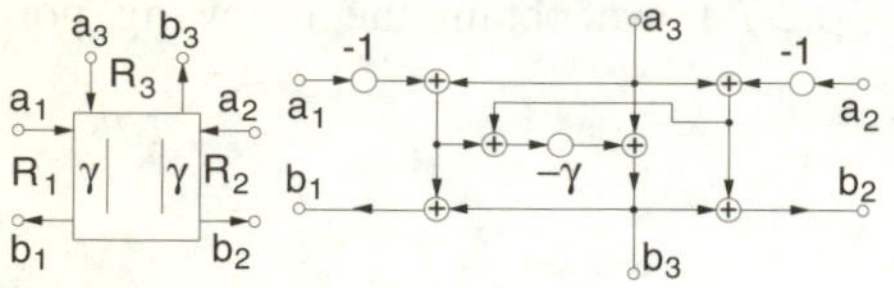

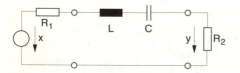

Fig. 5-57
Symmetrical three-port parallel adapter

In summary, the flowgraph of the whole filter is composed of two three-port parallel adapters with a reflection-free port, one symmetrical three-port serial adapter and three delays. The implementation requires three multipliers, fourteen adders and three delays. It is noticeable that wave digital filters in general require more adders than the corresponding direct form.

5.7.4.6 Second-Order Wave Digital Filter Block

The characteristics of filters that are realised as a parallel or series arrangement of second-order filter blocks essentially depend on the properties of the partial filters. Apart from direct-form (Fig. 5-20) and normal-form filters (Fig. 5-27), second-order filter blocks derived from wave digital filters are a further option. These introduce, at least partially, the good properties of this filter type into the overall filter. Compared to a full realisation as wave digital filter, the advantage of low parameter sensitivity gets partially lost. The good stability properties, however, are fully kept.

In order to derive a second-order filter block based on wave digital filters, we choose a simple resonant circuit, as depicted in Fig. 5-58, as an analog reference filter. This circuit acts as a bandpass filter with the following transfer function:

$$H(p) = \frac{R_2 C p}{1 + (R_1 + R_2)Cp + LCp^2} .$$

By means of two three-port serial adapters, this filter can be easily converted into a wave digital filter structure (Fig. 5-59). Using the scattering matrix (5.67)

Fig. 5-58
Second-order bandpass filter

we can derive the following relationships between the state variables z_1 and z_2, the signals v_1 and v_2 in the connecting branches and the input and output signals of the filter x and y.

$$z_1(n+1) = (1-\gamma_1)x(n) + (\gamma_1 - 1)v_2(n) - \gamma_1 z_1(n)$$

$$z_2(n+1) = (\gamma_2 - 1)v_1(n) + \gamma_2 z_2(n)$$

$$v_1(n) = x(n) + z_1(n) \qquad\qquad v_2(n) = -z_2(n)$$

$$y(n) = \gamma_2 v_1(n) + \gamma_2 z_2(n)$$

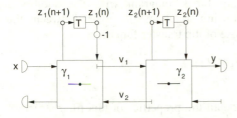

Fig. 5-59
Wave digital filter flowgraph of a second-order resonant circuit

By elimination of v_1 and v_2 we obtain a state-space representation of the filter that can be easily represented by a block diagram (Fig. 5-60).

$$z_1(n+1) = z_2(n) - \gamma_1\big(z_1(n) + z_2(n) + x(n)\big) + x(n)$$

$$z_2(n+1) = -z_1(n) + \gamma_2\big(z_1(n) + z_2(n) + x(n)\big) - x(n) \qquad (5.72)$$

$$y(n) = \gamma_2\big(z_1(n) + z_2(n) + x(n)\big)$$

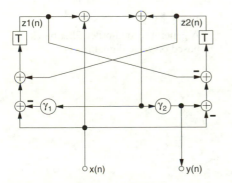

Fig. 5-60
Block diagram of a second-order band-pass filter based on wave digital filters

The following state-space coefficients can be derived from (5.72):

$$A = \begin{pmatrix} -\gamma_1 & 1-\gamma_1 \\ \gamma_2 - 1 & \gamma_2 \end{pmatrix} \qquad b = \begin{pmatrix} 1-\gamma_1 \\ \gamma_2 - 1 \end{pmatrix} \qquad c^T = (\gamma_2 \quad \gamma_2) \qquad d = \gamma_2 .$$

Using (5.30) we can calculate the transfer function of this filter.

$$H(z) = \gamma_2 \frac{z^2 - 1}{z^2 + (\gamma_1 - \gamma_2)z + 1 - \gamma_1 - \gamma_2}$$

Figure 5-60 can be easily extended to cover the more general case of a second-order filter block that is characterised by the coefficients b_0, b_1 and b_2 in the numerator polynomial of the transfer function.

$$H(z) = \frac{b_0 z^2 + b_1 z + b_2}{z^2 + a_1 z + a_2}$$

Figure 5-61 shows the corresponding block diagram with the additional coefficients b_0, b_1 and b_2. The following relationships exist between the coefficients of the block diagram and those of the transfer function:

$$b_1 = \beta_1 - \beta_2 \qquad\qquad b_2 = \beta_0 - \beta_1 - \beta_2 \qquad\qquad b_0 = \beta_0$$
$$a_1 = \gamma_1 - \gamma_2 \qquad\qquad a_2 = 1 - \gamma_1 - \gamma_2 \;.$$

In Chap. 8 we will investigate in more detail the properties of this filter structure [45].

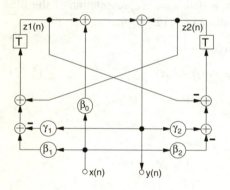

Fig. 5-61
A general second-order filter block based on wave digital filters

6 Design of IIR Filters

6.1 Introduction

The different mathematical expressions for the transfer functions of FIR and IIR filters also result in very different design methods for both filter types. While a given frequency response is approximated in case of an IIR filter by a rational fractional transfer function of z, with poles and zeros (5.18), the FIR filter is simply characterised by zeros and a polynomial of z (5.2).

In case of IIR filters we can take advantage of their relationship to continuous-time filters made up of networks of discrete elements. Both have impulse (or unit-sample) responses of infinite length and are described by rational fractional transfer functions in the frequency domain. We can start with the actual filter design procedure completely in the continuous-time domain by making use of the whole range of classical approximation methods, such as Chebyshev, Butterworth or Bessel for instance, and of numerous design tables available in the literature. The transformation of the obtained coefficients of the analog reference filters into those of the target discrete-time filters can be easily performed using a pocket calculator. The kernel of commercial digital filter design software is in general an analog filter design program with an appendage to calculate the digital filter coefficients.

It seems confusing at first glance that various methods are available for the transition from the continuous-time to the discrete-time domain, as a look at the headings of this chapter shows. Why does not exist one universal method to transfer all characteristics of the analog filter in the time and frequency domain to corresponding discrete-time realisations? Our first clue to this problem was the fact that the frequency response of the continuous-time filter is a rational fractional function of $j\omega$ while the frequency response of the discrete-time system is periodic and a function of $e^{j\omega T}$. Thus we cannot expect to obtain identical frequency responses within the frequency range $-\pi/T \leq \omega \leq +\pi/T$ for both cases. Also in the time domain, discrete-time and continuous-time systems show marked differences concerning the mathematical description of the system behaviour. The relations between the input and output signals of the system are expressed in the one case by an integral, the convolution integral (1.1), in the other case by a sum, the convolution sum (3.3). It can be generally stated that the characteristics of continuous-time and discrete-time systems converge more the smaller the chosen sampling period T. This statement is true for both the time and the frequency domain. In the limiting case when T approaches zero, the exponential $e^{j\omega T}$ can be replaced by its argument and the sum converges to an integral.

As a rule, it is desirable to choose the sampling frequencies of digital systems to be not much higher than is required for processing the respective signals in accordance with the sampling theorem. Sampling rate, and thus processing speed, have an important influence on hardware costs. Examples for the increase of costs are the need for higher signal processing speed, faster digital logic, higher A/D and D/A conversion rate and lower memory access times. Another effect can be observed in the context of non-real-time applications where data have to be stored. As the amount of sampled data increases with the sampling rate, larger storage capacities, in terms of RAM, hard disk or CD for instance, are required, which is a further cost factor.

In the following section we will show that, under certain circumstances, convolution sum and convolution integral yield comparable system behaviour. The limiting conditions needed, however, to reach this result are hard to satisfy in practice.

6.2 Preservation Theorem of the Convolution

The behaviour of continuous-time LTI systems is described by the convolution integral according to (1.1).

$$y_a(t) = \int_{-\infty}^{+\infty} x_a(\tau)\, h_a(t-\tau)\, d\tau \tag{6.1}$$

Assuming that the input signal $x_a(t)$ is band-limited and fulfils the requirements of the sampling theorem, we can derive a relation that permits the numerical calculation of samples of the output signal $y_a(t)$ at discrete instants nT. The starting point of our considerations is the following relation between the analog input and output signals in the frequency domain:

$$Y_a(j\omega) = X_a(j\omega)\, H_a(j\omega) \ .$$

Since $X_a(j\omega)$ is assumed to be band-limited, $Y_a(j\omega)$ is also band-limited. For the calculation of the samples $y_a(n)$ of the output signal we can therefore use (4.4a).

$$y_a(n) = \frac{1}{2\pi} \int_{-\pi/T}^{+\pi/T} X_a(j\omega)\, H_a(j\omega)\, e^{j\omega Tn}\, d\omega$$

According to (4.4b) we can express the band-limited spectrum $X_a(j\omega)$ by the samples $x_a(n)$ of the input signal.

$$y_a(n) = \frac{1}{2\pi} \int_{-\pi/T}^{+\pi/T} T \sum_{k=-\infty}^{+\infty} x_a(k)\, e^{-j\omega Tk}\, H_a(j\omega)\, e^{j\omega Tn}\, d\omega$$

Interchanging summation and integration yields

$$y_a(n) = T \sum_{k=-\infty}^{+\infty} x_a(k) \ \frac{1}{2\pi} \int_{-\pi/T}^{+\pi/T} H_a(j\omega) \, e^{j\omega T(n-k)} \, d\omega \ .$$

The integral term represents, according to (4.1), the samples $h_a(nT)$ of the impulse response of the analog filter. Since the integral covers only the frequency range $|\omega| < \pi/T$, the calculation of the samples only makes use of the spectral components up to half the sampling frequency. The integral thus yields the samples of a low-pass limited version of the impulse response of the analog filter which we denote in the following by the index LP.

$$y_a(n) = T \sum_{k=-\infty}^{+\infty} x_a(k) \, h_{aLP}(n-k) = \sum_{k=-\infty}^{+\infty} x_a(k) \, T \, h_{aLP}(n-k) \tag{6.2}$$

Under the following two conditions, the convolution sum (6.2) thus yields the same sample values $y_a(n)$ of the output signal as the convolution integral (6.1):
1. The spectrum of the input signal must be limited to the Nyquist frequency.
2. The spectrum of the impulse response of the analog filter must also be limited to the Nyquist frequency before it is sampled to obtain the unit-sample response of the discrete-time filter to be realised.

A comparison with the discrete convolution that we introduced in Sect. 3.1 shows that the following relation exists between the discrete-time unit-sample response $h(n)$ and the impulse response $h_{aLP}(t)$ of the band-limited impulse response of the corresponding analog filter:

$$h(n) = T \, h_{aLP}(nT) \ .$$

The second requirement avoids a possible aliasing and thus distortion of the frequency response in the baseband. Such aliasing occurs, as shown in Chap. 4, by the periodic continuation of the spectrum if the impulse response is sampled and the sampling theorem is not fulfilled.

The preservation theorem of the convolution provides, in principle, the desired method to numerically imitate the convolution integral in the discrete-time domain. At the same time, we obtain a transfer function in the frequency domain that is identical to the one of the corresponding analog filter in the frequency range $|\omega| < \pi/T$. We merely have to design a digital filter that has a unit-sample response corresponding to the samples of the low-pass limited impulse response of the analog reference filter. In most practical applications, however, this second requirement of the preservation theorem can only hardly be met or cannot be fulfilled at all.

The fewest problems can be expected if the analog reference filter is a low-pass filter whose frequency response has reached such a high attenuation at $f_s/2$ that aliasing does not cause noticeable distortion of the frequency response in the baseband. This condition is approximately fulfilled by higher-order low-pass and bandpass filters whose cutoff frequency lies well below the Nyquist frequency.

In all other cases, an additional sharp low-pass filter is required to limit the frequency response of the analog filter strictly to $f_s/2$ before the impulse response is sampled. Unfortunately, this low-pass filter, whose characteristic is contained in the sampled impulse response, has to be implemented in the discrete-time domain, too, which may increase the complexity of the digital filter considerably. The filter order of the discrete-time system would exceed the order of the original analog reference filter by far. Even if this extra complexity is tolerated, the described procedure is not an ideal solution since filters with sharp transition bands cause considerable phase distortions which would destroy the exact reproduction of the system behaviour in the time domain. A further drawback is the fact that the additional low-pass filter has a transition band with finite width which cannot be used any more. This is especially annoying for the design of high-pass and bandstop filters which, in the ideal case, have no attenuation at the Nyquist frequency.

The discussion so far makes clear that it is difficult to exactly reproduce the frequency response of an analog filter by a discrete-time realisation. Experience with practical applications shows, however, that such an exact reproduction is seldom needed. In the following, we will show how rational fractional transfer functions of z can be derived for all filter types that preserve the order of the analog reference filter, and approximate the characteristics of this filter with an accuracy sufficient for most practical cases.

6.3 Approximation in the Time Domain

We start with a group of approximation methods that violate the sampling theorem in both requirements of the preservation theorem of the convolution, but maintain the filter order of the analog reference filter. The reaction of the analog filter with respect to special input signals is nevertheless exactly reproduced in the time domain. These methods are therefore summarised under the term "approximation in the time domain". The named violation of the sampling theorem has the consequence that these methods are not universally valid, but each of them yields exact results for a special class of input signals.

The approximation methods introduced in this section allow the dimensioning of discrete-time filters that exactly reproduce the reaction of analog filters to three special test functions as summarised in Table 6-1 and depicted in Fig. 6-1: impulse, unit step and ramp.

	Analog filter	Digital filter
Impulse	$x(t) = T\,\delta(t)$	$x(n) = \delta(n)$
Unit step	$x(t) = u(t)$	$x(n) = u(n)$
Ramp	$x(t) = (t\,/\,T)\,u(t)$	$x(n) = n\,u(n)$

Table 6-1
Continuous-time test functions and their discrete-time equivalents

Exact reproduction in this context means: the analog filter is stimulated by one of the three test functions. The corresponding discrete-time filter is stimulated by a related test function as shown in Table 6-1. The reactions of the analog filter $y_a(t)$ and of the discrete-time filter $y(n)$ to these test functions are related as

$$y(n) = y_a(nT) \ . \tag{6.3}$$

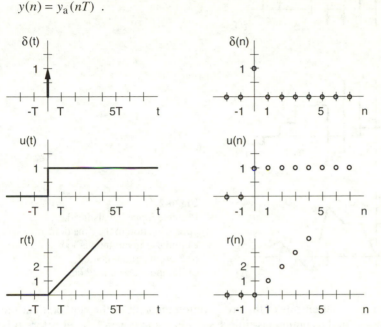

Fig. 6-1 Test signals for the filter approximation in the time domain

The methods introduced in this section have special importance in the field of system simulation. They allow, for instance, to calculate and graphically represent the impulse and step response of analog filters. Since the filters in discussion are linear and shift-invariant, the presented approximations are not limited to the aforementioned test functions. They are also valid for signals that are combinations of these elementary functions. Figure 6-2 shows, for each of the three classes, possible input signals that lead, by application of the appropriate design rules, to an exact discrete-time reconstruction of the output of the corresponding analog filter. A prerequisite that this works is that the impulses (Fig. 6-2a), steps (Fig. 6-2b) and creases (Fig. 6-2c) exactly fit into the time grid corresponding to the sample period of the simulating discrete-time system.

6.3.1 Impulse-Invariant Design

According to (6.3), the design rule of this method can be expressed as follows:

$$h(n) = T \, h_a(nT) \ . \tag{6.4}$$

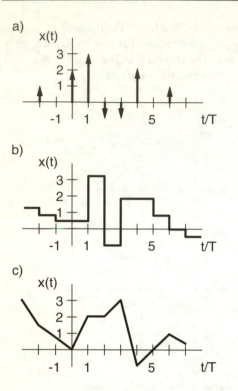

a) x(t)

b) x(t)

c) x(t)

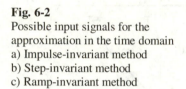

Fig. 6-2
Possible input signals for the
approximation in the time domain
a) Impulse-invariant method
b) Step-invariant method
c) Ramp-invariant method

In contrast to the requirements of the preservation theorem of the convolution, the impulse response of the continuous-time system is not low-pass limited before it is sampled. Thus we tacitly tolerate the occurring aliasing. The frequency response of the resulting discrete-time filter is obtained by Fourier transform of the sampled impulse response.

$$H\left(e^{j\omega T}\right) = \sum_{n=-\infty}^{+\infty} h(n)\, e^{-j\omega Tn}$$

$$H\left(e^{j\omega T}\right) = T \sum_{n=-\infty}^{+\infty} h_a(nT)\, e^{-j\omega Tn}$$

Using (4.2b) yields

$$H\left(e^{j\omega T}\right) = \sum_{k=-\infty}^{+\infty} H_a(j\omega + jk2\pi / T) \ . \tag{6.5}$$

The frequency response of the discrete-time filter is obtained through periodical continuation of the frequency response of the analog filter with the period $2\pi/T$ (6.5). We can distinguish between three cases:

1. **$h(t)$ is low-pass limited to the frequency range $|\omega| < \pi/T$.**
   In this case, the frequency responses of analog and discrete-time filters are completely identical since no aliasing occurs because of the periodical continuation. Such a filter, however, is not realisable with finite cost, so case 2 is more likely.
2. **$h(t)$ represents an analog low-pass or bandpass filter which exhibits finite loss beyond the Nyquist frequency.**
   In this case, the frequency response is more or less distorted depending on the degree of aliasing. Bad attenuation in the stopband and significant deviations of the DC or overall gain are the consequence.
3. **$h(t)$ represents a high-pass, bandstop or allpass filter.**
   In this case, the periodical continuation of the frequency response is not possible since the sum (6.5) does not converge. This can be readily verified if one tries to perform the periodical continuation graphically by hand for these filter types. High-pass, bandstop and allpass filters are therefore excluded from this design method. This statement is closely related to the fact that the impulse response of these filters contains a delta function which cannot be represented by a finite sample value.

Before we illustrate the characteristics of the impulse invariant design in the frequency domain by means of examples, we have to derive the algorithm to convert the transfer function $H(p)$ of the analog reference filter into the transfer function $H(z)$ of the impulse-invariant discrete-time filter. The starting point of the following considerations is the partial-fraction representation of the transfer function $H(p)$ of the analog filter.

$$H(p) = \frac{\sum_{r=0}^{M} b_r p^r}{\sum_{i=0}^{N} a_i p^i} = K + \sum_{k=1}^{N} \frac{C_k}{p - p_{\infty k}} \tag{6.6}$$

N is the filter order and $p_{\infty k}$ the kth pole of the filter. C_k is the kth partial-fraction coefficient. By inverse Laplace transform of $H(p)$ we obtain the impulse response $h_a(t)$ of the analog filter.

$$h_a(t) = K\,\delta(t) + \sum_{k=1}^{N} C_k\, e^{p_{\infty k} t}\, u(t) \tag{6.7}$$

The terms in the summation formula in (6.7) represent the exponentially decaying characteristic oscillations of the system. The unit-step function takes into account that $h_a(t) = 0$ for $t < 0$. In the case of high-pass, allpass and bandstop filters, for which the degrees of numerator and denominator polynomial are equal, an additional delta function appears in the impulse response. Since the delta function is not a function in the true sense of the word, as pointed out in Sect. 1.1, there exists no meaningful sample sequence for this part of the impulse response. That is why the impulse invariant design method can only be applied to low-pass and bandpass filters. In the case that one is only interested in the simulation of the decaying characteristic oscillations of the system, the present method may also be applied to high-pass, allpass and bandstop filters by simply ignoring the delta function. Using (6.4), we obtain the unit-sample response of the discrete-time filter to be designed.

$$h(n) = T \sum_{k=1}^{N} C_k \, e^{P_{\infty k} nT} \, u(n)$$

With the help of the Table 3-2, we can determine the transfer function $H(z)$ from the unit-step response $h(n)$ above, as

$$H(z) = T \sum_{k=1}^{N} C_k \, \frac{z}{z - e^{P_{\infty k} T}} \; . \tag{6.8}$$

Equation (6.8) is a partial-fraction representation of the transfer function which can be readily converted to the standard rational fractional form. The impulse-invariant design method guarantees stability since poles of the analog filter in the left half of the p-plane (Re $p_{\infty k} < 0$) result in poles of the discrete-time filter with magnitude less than unity, as can be verified by inspection of (6.8).

Example

A resonant circuit with the transfer function

$$H(P) = \frac{P}{P^2 + 0.1P + 1} \qquad f_r = 0.2 / T \qquad (\omega_r = 0.4\pi / T)$$

is to be realised as a discrete-time system. The Q-factor of the circuit amounts to ten, the resonant frequency is $0.2 f_s$. The poles of the filter are the zeros of the denominator polynomial.

$$P^2 + 0.1P + 1 = 0$$
$$P_{\infty 1} \approx -0.05 + j \qquad P_{\infty 2} \approx -0.05 - j$$

The transfer function can therefore be expressed as

$$H(P) = \frac{P}{(P + 0.05 - j)(P + 0.05 + j)} \; . \tag{6.9}$$

For the transition to the partial fraction representation, we have to determine the partial fraction coefficients A_1 and A_2:

$$H(P) = \frac{A_1}{P + 0.05 - j} + \frac{A_2}{P + 0.05 + j} \; .$$

Multiplying out and comparison of the coefficients with (6.9) yields

$$A_1 = 0.5 + j0.025 \qquad A_2 = 0.5 - j0.025$$

$$H(P) = \frac{0.5 + j0.025}{P + 0.05 - j} + \frac{0.5 - j0.025}{P + 0.05 + j} \quad .$$

The relation

$$P = \frac{p}{\omega_r} = \frac{p}{0.4\pi / T}$$

leads to the unnormalised form

$$H(p) = \frac{(0.628 + j0.031) / T}{p - (-0.063 + j1.26) / T} + \frac{(0.628 - j0.031) / T}{p - (-0.063 - j1.26) / T} \quad .$$

For the transition to the impulse-invariant discrete-time system, the following coefficients can be taken from this transfer function:

$$C_1 = (0.628 + j0.031) / T \qquad\qquad C_2 = (0.628 - j0.031) / T$$

$$p_{\infty 1} = (-0.063 + j1.26) / T \qquad\qquad p_{\infty 2} = (-0.063 - j1.26) / T \quad .$$

Substitution of these coefficients in (6.6) leads to the desired transfer function

$$H(z) = \frac{(0.628 + j0.031)z}{z - e^{-0.063 + j1.26}} + \frac{(0.628 - j0.031)z}{z - e^{-0.063 - j1.26}}$$

or, in rational fractional representation,

$$H(z) = \frac{1.26z^2 - 0.42z}{z^2 - 0.58z + 0.88} = \frac{1.26 - 0.42z^{-1}}{1 - 0.58z^{-1} + 0.88z^{-2}} \quad .$$

Figure 6-3 depicts the frequency response of this filter in comparison to the frequency response of the analog reference filter. The aliasing especially affects

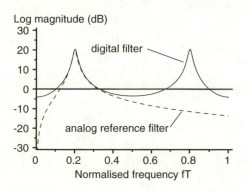

Fig. 6-3
Frequency response of the discrete-time realisation of an impulse invariant resonant circuit

the low-frequency range where the high attenuation of the analog filter cannot be achieved.

Figure 6-4 shows a further example where we approximate a second-order low-pass filter. The gain matches well at the resonant frequency, but is too low at low frequencies and too high in the range of higher frequencies. The example of a simple first-order low-pass filter is shown in Fig. 6-5. The overall gain is too high compared to the reference filter.

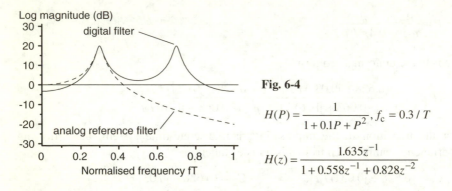

Fig. 6-4

$$H(P) = \frac{1}{1 + 0.1P + P^2}, f_c = 0.3 / T$$

$$H(z) = \frac{1.635z^{-1}}{1 + 0.558z^{-1} + 0.828z^{-2}}$$

The discussion above shows that the impulse invariant design method is only of limited use for the design of discrete-time filters with prescribed frequency response. The main application area of this method is the simulation of analog systems in the time domain. It is well suited to simulate the response of analog filters to short pulses (impulse response) or to periodical or nonperiodical streams of pulses (Fig. 6-2a). This can be done on a computer, for instance, which also allows the comfortable representation of the results in graphical form. For impulses as input signals, the response of the discrete-time system exactly matches the behaviour of the analog counterpart. For other types of signals we can expect deviations. This is due to the fact, as explained earlier, that both the considered input signal (delta function) and the impulse response of the analog filter violate the sampling theorem. Another approximation method that is optimised for a special type of input signal will be introduced in the next section.

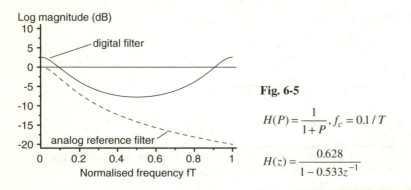

Fig. 6-5

$$H(P) = \frac{1}{1 + P}, f_c = 0.1 / T$$

$$H(z) = \frac{0.628}{1 - 0.533z^{-1}}$$

6.3.2 Step-Invariant Design

By this method, a discrete-time filter is designed in such a way that its unit-step response exactly matches the step response of the analog reference filter at the sampling instants.

$$a(n) = a_a(nT) \tag{6.10}$$

Unit-step and unit-sample response of the discrete-time system are related as

$$h(n) = a(n) - a(n-1) \ . \tag{6.11}$$

The corresponding relation for continuous-time systems reads

$$a_a(t) = \int_0^t h_a(\tau) \, d\tau \ . \tag{6.12}$$

Combining (6.10), (6.11) and (6.12) yields the following relation between the unit-sample response of the discrete-time filter and the impulse response of the analog reference filter:

$$h(n) = T \frac{1}{T} \int_{(n-1)T}^{nT} h_a(\tau) \, d\tau \ . \tag{6.13}$$

Comparison with (6.4) shows that the unit-sample response does not reflect the course of the analog impulse response any more. The samples are rather calculated as the average of the impulse response taken over the respective past sampling period T. This moving averaging smoothes the curve and has the effect of weighting the impulse response by a low-pass characteristic of the form

$$H_W(j\omega) = \frac{1 - e^{-j\omega T}}{j\omega T} = e^{-j\omega T/2} \frac{\sin \omega T/2}{\omega T/2} \ .$$

Figure 6-6 depicts the frequency response of this low-pass characteristic.

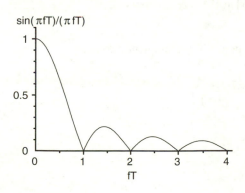

$\sin(\pi fT)/(\pi fT)$

Fig. 6-6
Low-pass characteristic resulting from the moving averaging of the impulse response

Taking into account this low-pass weighting, we obtain by periodical continuation the following relation between the frequency responses of the analog and discrete-time filters:

$$H\left(e^{j\omega T}\right) = \sum_{k=-\infty}^{+\infty} H_a(j\omega + jk2\pi/T)\, e^{-j(\omega T/2 + k\pi)}\, \frac{\sin(\omega T/2 + k\pi)}{\omega T/2 + k\pi} \; . \qquad (6.14)$$

The low-pass weighting of the transfer function has the advantageous side-effect that aliasing is strongly reduced when the frequency response is periodically continued. As a consequence, high-pass, allpass and bandstop filters can now also be handled with this method since (6.14) is guaranteed to converge in any case as long as $H_a(j\omega)$ is finite.

The algorithm to convert the transfer function $H(p)$ of the analog filter into the transfer function $H(z)$ of the discrete-time filter is similar to the one that we derived for the impulse-invariant design. The starting point of the following derivation is the general form of the impulse response of an Nth-order system (6.7).

$$h_a(t) = K\delta(t) + \sum_{k=1}^{N} C_k\, e^{P_{\infty k} t}\, u(t)$$

Since high-pass, allpass and bandstop filters are allowed now, the delta function term $K\,\delta(t)$ in the impulse response must be taken into account in the following mathematical treatment. According to (1.4b), we obtain the step response of the filter by integration of the impulse response.

$$a_a(t) = K\, u(t) + \sum_{k=1}^{N} \frac{C_k}{P_{\infty k}}\left(e^{P_{\infty k} t} - 1\right) u(t)$$

According to (6.10), the unit-step response of the corresponding discrete-time filter is obtained as

$$a_a(n) = K\, u(n) + \sum_{k=1}^{N} \frac{C_k}{P_{\infty k}}\left(e^{P_{\infty k} Tn} - 1\right) u(n) \; .$$

z-transformation yields the unit-step response in the frequency domain:

$$A(z) = K\frac{z}{z-1} + \sum_{k=1}^{N} \frac{C_k}{P_{\infty k}}\left(\frac{z-1}{z - e^{P_{\infty k} T}} - \frac{z}{z-1}\right) \; .$$

Since we have the relation

$$A(z) = H(z)\, U(z) = H(z)\frac{z}{z-1}$$

between unit-sample response $H(z)$ and unit-step response $A(z)$ in the frequency domain, with $U(z)$ being the z-transform of the unit step $u(n)$, we can express the transfer function of the step-invariant discrete-time filter as

$$H(z) = K + \sum_{k=1}^{N} \frac{C_k (e^{P_{\infty k}T} - 1)}{P_{\infty k}} \cdot \frac{1}{z - e^{P_{\infty k}T}} \cdot \tag{6.15}$$

Stability is guaranteed since poles in the left half of the p-plane are mapped again inside the unit circle in the z-plane.

Example

We consider again the resonant circuit that we have already used as an example in the previous section, and we can thus make use of the poles and partial fraction coefficients $p_{\infty 1}$, $p_{\infty 2}$, C_1 and C_2 calculated there. Substitution of these coefficients in (6.15) leads to the following transfer function in z of the step-invariant discrete-time filter in the partial-fraction form

$$H(z) = \frac{0.445 + j0.356}{z - e^{-0.063 + j1.26}} + \frac{0.445 - j0.356}{z - e^{-0.063 - j1.26}} ,$$

or, by multiplying out in the standard rational fractional form,

$$H(z) = \frac{0.89z - 0.89}{z^2 - 0.58z + 0.88} = \frac{0.89z^{-1} - 0.89z^{-2}}{1 - 0.58z^{-1} + 0.88z^{-2}} \cdot$$

Figure 6-7 shows the good correspondence between the frequency response of the analog reference filter and the step-invariant discrete-time filter, especially in the range of lower frequencies. The still-present aliasing causes the deviations that can be observed near the Nyquist frequency.

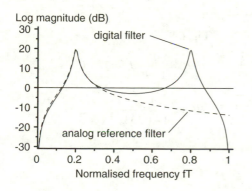

Fig. 6-7
Frequency response of the discrete-time realisation of a step-invariant resonant circuit

The example of a second-order low-pass filter as depicted in Fig. 6-8 shows similar behaviour. The approximation in the range of low frequencies is excellent. In the area of the Nyquist frequency, alias spectral components partly compensate themselves which results in a sharper drop in the gain than can be found in the frequency response of the analog reference filter. Figure 6-9 shows the example of a first-order high-pass filter. In this case, the aliasing obviously causes a too-high overall gain.

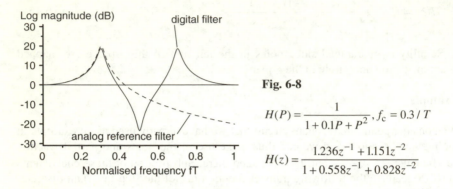

Fig. 6-8

$$H(P) = \frac{1}{1 + 0.1P + P^2}, f_c = 0.3 / T$$

$$H(z) = \frac{1.236z^{-1} + 1.151z^{-2}}{1 + 0.558z^{-1} + 0.828z^{-2}}$$

The step-invariant design method is well suited to simulate the response of analog systems to stepwise changes of the input signal in the time domain. The stimulating signal may be a single step, which allows the observation of the step response of the filter, or a series of steps (Fig. 6-2b), as in the case of digital clock or data signals, which would allow, for example, the simulation of digital transmission systems on the computer. The approximation of the frequency response in the lower frequency range is much better than in the case of the impulse-invariant design. The DC gains of the analog reference filter $H(p = 0)$ and of the step-invariant discrete-time filter $H(z = 1)$ are even identical.

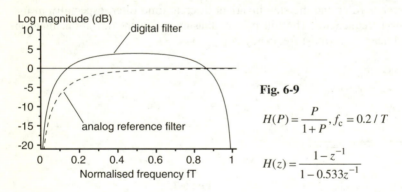

Fig. 6-9

$$H(P) = \frac{P}{1 + P}, f_c = 0.2 / T$$

$$H(z) = \frac{1 - z^{-1}}{1 - 0.533z^{-1}}$$

6.3.3 Ramp-Invariant Design

By this method, a discrete-time filter is designed in such a way that its response to the ramp function

$$r(n) = n \, u(n)$$

exactly matches the response of the analog reference filter to the function

$$r(t) = \frac{t}{T}u(t)$$

at the sampling instants. The starting point for the derivation of the design rule is again the transfer function of the analog filter in partial-fraction representation (6.6).

$$H(p) = K + \sum_{k=1}^{N} \frac{C_k}{p - p_{\infty k}}$$

We omit the mathematical details and give the result directly in the form of the corresponding transfer function of the ramp-invariant discrete-time filter.

$$H(z) = K + \sum_{k=1}^{N} \frac{C_k}{p_{\infty k}}\left(\frac{e^{p_{\infty k}T} - 1}{p_{\infty k}T}\frac{z-1}{z - e^{p_{\infty k}T}} - 1 \right) \qquad (6.16)$$

The inputs to (6.16) are again the coefficients of the partial-fraction representation of the analog reference filter C_k and $p_{\infty k}$. This approximation method is suitable if the input signal of the analog system to be simulated is of the type shown in Fig. 6-2c. These signals could be, for instance, approximations of continuous-time signals by chordal segments where the endpoints of the chords must lie on the sampling grid.

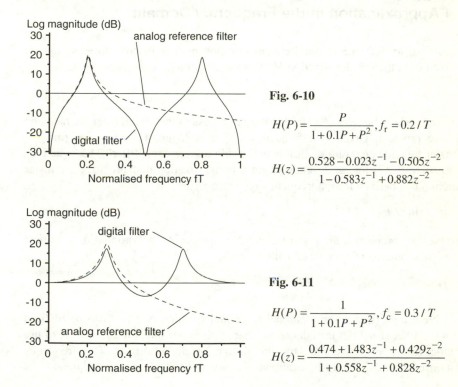

Fig. 6-10

$$H(P) = \frac{P}{1+0.1P + P^2}, f_r = 0.2 / T$$

$$H(z) = \frac{0.528 - 0.023z^{-1} - 0.505z^{-2}}{1 - 0.583z^{-1} + 0.882z^{-2}}$$

Fig. 6-11

$$H(P) = \frac{1}{1+0.1P + P^2}, f_c = 0.3 / T$$

$$H(z) = \frac{0.474 + 1.483z^{-1} + 0.429z^{-2}}{1 + 0.558z^{-1} + 0.828z^{-2}}$$

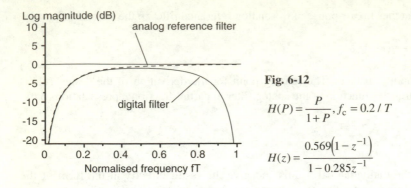

Log magnitude (dB)

Fig. 6-12

$$H(P) = \frac{P}{1+P}, f_c = 0.2 / T$$

$$H(z) = \frac{0.569\left(1 - z^{-1}\right)}{1 - 0.285z^{-1}}$$

Figure 6-10 and Fig. 6-11 show the frequency responses of second-order bandpass and low-pass filters which have been designed using (6.16). It turns out that the approximation of the analog reference filter cannot be further improved with respect to the results obtained by the step-invariant design. In the case of the first-order high-pass filter, according to Fig. 6-12, it can be observed that the gain of the filter is now in the right order of magnitude, compared with the step-invariant design as depicted in Fig. 6-9.

6.4 Approximation in the Frequency Domain

According to Sect. 3.3.1, the frequency response of a discrete-time system is derived from the transfer function $H(z)$ by substitution of (6.17).

$$z = e^{j\omega T} \tag{6.17}$$

z is replaced by the term $e^{j\omega T}$. In the design process of the digital filter we have the inverse problem. Given is a frequency response $H(j\omega)$. How can we find the transfer function $H(z)$ of a realisable and stable discrete-time filter? The basic idea is to solve (6.17) for $j\omega$ in order to get a transformation that provides a transfer function of z derived from a frequency response $H(j\omega)$.

$$j\omega = \ln z / T \tag{6.18a}$$

We have no problem if the given frequency response is a polynomial or a rational fractional function of $e^{j\omega T}$, since with

$$H(e^{j\omega T}) \qquad \rightarrow \qquad H\left(e^{\ln z / T \times T}\right) = H(z)$$

we obtain a polynomial or a rational fractional function of z that can be realised with the filter structures introduced in Chap. 5. It is our intention, however, to use analog filters as a reference for the design of digital filters. The frequency response of analog filters is a rational fractional function of $j\omega$. Application of

(6.18a) would yield rational fractional transfer functions of $\ln z / T$ which, unfortunately, are not realisable with any of the known structures.

$$H(j\omega) \qquad \rightarrow \qquad H(\ln z / T)$$

We can generalise (6.18a) to the transformation

$$p = \ln z / T \ , \tag{6.18b}$$

which would allow us to start from the transfer function $H(p)$ of the analog reference filter.

$$H(p) \qquad \rightarrow \qquad H(\ln z / T)$$

How can (6.18b), which is a transformation from the p-plane of the continuous-time domain into the z-plane of the discrete-time domain, be modified in such a way that
1. the desired frequency response is maintained and
2. the resulting transfer function in z is stable and realisable?

Series expansion of the logarithm in (6.18b) and an appropriate truncation of the series is a possibility to achieve rational fractional functions in z. We investigate in the following two candidates of series expansions which result in transformations with different properties.

6.4.1 Difference Method

A possible series expansion of the logarithm has the following form:

$$\ln z = \frac{(z-1)}{z} + \frac{(z-1)^2}{2z^2} + \frac{(z-1)^3}{3z^3} + \frac{(z-1)^4}{4z^4} + \dots$$

$$\text{with } \mathrm{Re}\, z > 1/2 \ .$$

Truncating this series after the first term yields the following transformation from the p-plane into the z-plane:

$$p = \frac{z-1}{z} \Big/ T = \frac{1-z^{-1}}{T} \ . \tag{6.19}$$

This transformation corresponds to the method which is often applied for the numerical solution of differential equations, where derivatives are replaced by differences. Figure 6-13 illustrates the situation. The derivative (tangent) at $t = nT$ is approximated by the secant through the points $x(nT)$ and $x((n-1)T)$. The error of this approximation depends heavily on the course of the function in the considered interval.

$$x'(t)\Big|_{t=nT} \approx \frac{x(t) - x(t-T)}{T}\Bigg|_{t=nT} = \frac{x(n) - x(n-1)}{T}$$

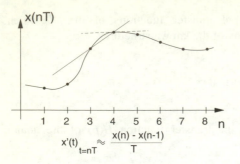

Fig. 6-13
Approximation of the first derivative by
a difference

Laplace transform of this relation yields

$$p\,X(p) \approx \frac{X(p) - X(p)\,\mathrm{e}^{-pT}}{T}$$

$$p \approx \frac{1 - \mathrm{e}^{-pT}}{T} \ .$$

Since $\mathrm{e}^{pT} = z$, the frequency variable p can be approximated by the rational term in z (6.19):

$$p \approx \frac{1 - z^{-1}}{T} = \frac{1}{T}\frac{z-1}{z} \ .$$

According to the rules of the z-transform, this relation means calculation of the difference between the current and the past sample (backward difference) and division by the sampling period T, which is in fact the slope of the secant and thus an approximation of the derivative. With decreasing T, the approximation improves, since the secant more closely approaches the tangent.

We consider now the stability of discrete-time systems whose transfer functions are obtained from analog reference filters using (6.19). As the poles p_∞ of stable analog filters lie in the left half of the p-plane, the transformation (6.19) results in the following inequality which must be fulfilled by poles z_∞ in the z-domain:

$$\operatorname{Re} p_\infty = \operatorname{Re} \frac{1 - z_\infty^{-1}}{T} < 0 \ .$$

From this inequality we can derive a condition for the location of stable poles in the z-plane:

$$(\operatorname{Re} z_\infty - 0.5)^2 + (\operatorname{Im} z_\infty)^2 < 0.25 \ .$$

The poles of the analog filter are mapped into a circle with the centre $z = 0.5$ and a radius of 0.5 (Fig. 6-14). It becomes clear that stability of the discrete-time filter is guaranteed since all stable poles in the left half of the p-plane are transformed inside the unit circle of the z-plane. Unfortunately with transformation (6.19) we can reach only a very small area of the unit circle. We cannot map any pole at all, wherever its stable reference pole may lie in the p-plane, into the left half of the unit circle, which represents the frequency range $f_s/4 \ldots f_s/2$. In the right half of the unit circle, the possibility of locating poles close to the unit circle is very limited. That means that poles with a high Q-factor, as required for sharp filters, can only be realised far below the sampling frequency in the vicinity of $\omega = 0$ or $z = 1$. If we want to approximate the frequency response of an analog filter with sufficient precision, we must choose a sampling frequency much higher (by a factor of ten or more) than the highest pole frequency of the filter. In practice, this means in most cases that the sampling frequency is also much higher than required to satisfy the sampling theorem for the signals to be processed.

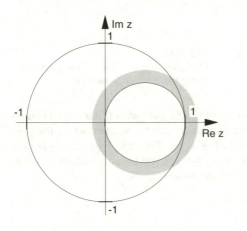

Fig. 6-14
Mapping of the left half of the p-plane into the z-plane in the case of the difference method

Two examples will illustrate the described behaviour. We consider again the second-order low-pass filter that we used earlier as an example:

$$H(P) = \frac{1}{P^2 + 0.1P + 1}, f_c = 0.3/T \qquad (\omega_c = 0.6\pi/T) \qquad (6.20)$$

$$P = \frac{p}{\omega_c} = \frac{p}{2\pi f_c} = \frac{p}{0.6\pi/T} \quad .$$

Using design rule (6.19) results in the following expression for the normalised frequency P:

$$P = \frac{(1 - z^{-1})/T}{0.6\pi/T} = \frac{1 - z^{-1}}{0.6\pi} \quad .$$

Substitution in the transfer function (6.20) yields the transfer function in z of the discrete-time filter:

$$H(z) = \frac{1}{\left(\dfrac{1 - z^{-1}}{0.6\pi}\right)^2 + 0.1\left(\dfrac{1 - z^{-1}}{0.6\pi}\right) + 1} \quad .$$

Multiplying out the brackets leads to the standard rational fractional form

$$H(z) = \frac{0.75z^2}{z^2 - 0.46z + 0.21} = \frac{0.75}{1 - 0.46z^{-1} + 0.21z^{-2}} \ .$$

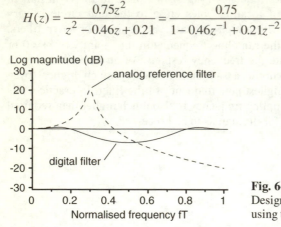

Fig. 6-15

Design of a second-order low-pass filter using the difference method

Figure 6-15 illustrates that the behaviour of a pole pair with Q-factor 10 cannot be recognised any more. A far better approximation can be reached when the pole is shifted to lower frequencies, as shown in Fig. 6-16. The cutoff frequency is now $f_c = 0.005/T$. However, the Q-factor is still too low, as the transformation (6.19) is not able to locate the pole pair close enough to the unit circle.

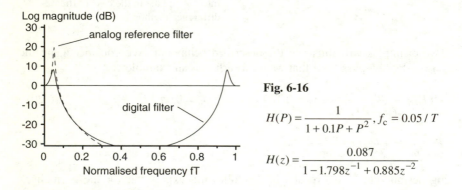

Fig. 6-16

$$H(P) = \frac{1}{1 + 0.1P + P^2}, f_c = 0.05 / T$$

$$H(z) = \frac{0.087}{1 - 1.798z^{-1} + 0.885z^{-2}}$$

6.4.2 Bilinear Transform

A further series expansion of the logarithm is of the following form:

$$\ln z = 2\left(\frac{z-1}{z+1} + \frac{1}{3}\left(\frac{z-1}{z+1}\right)^3 + \frac{1}{5}\left(\frac{z-1}{z+1}\right)^5 \cdots \right)$$

with $\mathrm{Re}\, z > 0$.

Truncation of the series after the first term leads to the following transformation from the p-plane into the z-plane:

$$p = \frac{2}{T}\frac{z-1}{z+1} \qquad\qquad (6.21)$$

This transformation, commonly referred to as bilinear transform, is also based on the approximation of the derivative by a secant through two points. The assumption made here is that the secant through two points is a good approximation of the average of the slopes of the curve at these two points (Fig. 6-17).

$$\left.\frac{x'(t)+x'(t-T)}{2}\right|_{t=nT} \approx \left.\frac{x(t)-x(t-T)}{T}\right|_{t=nT} = \frac{x(n)-x(n-1)}{T} \qquad (6.22)$$

Laplace transform of relation (6.22) yields

$$\frac{pX(p)+pX(p)\,\mathrm{e}^{-pT}}{2} \approx \frac{X(p)-X(p)\,\mathrm{e}^{-pT}}{T}$$

$$\frac{p\,(1+\mathrm{e}^{-pT})}{2} \approx \frac{1-\mathrm{e}^{-pT}}{T} \; .$$

Since $z = \mathrm{e}^{pT}$, we have

$$\frac{p\,(1+z^{-1})}{2} \approx \frac{1-z^{-1}}{T} \; ,$$

and, after solving for p,

$$p \approx \frac{2}{T}\frac{z-1}{z+1} \; .$$

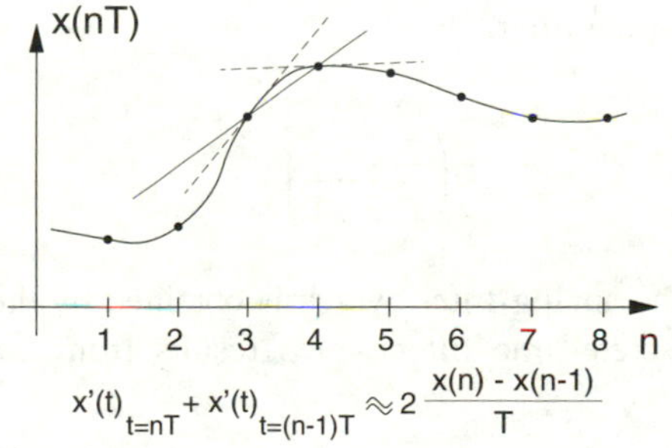

$$\left. x'(t)\right|_{t=nT} + \left. x'(t)\right|_{t=(n-1)T} \approx 2\,\frac{x(n) - x(n-1)}{T}$$

Fig. 6-17
Approximation of the first derivative in the case of the bilinear transform

The transformation is only usable if it leads to stable discrete-time filters. As the poles p_∞ of stable analog filters lie in the left half of the p-plane, the bilinear transform (6.21) results in the following inequality which must be fulfilled by the poles z_∞ in the z-domain.

$$\mathrm{Re}\, p_\infty = \mathrm{Re}\,\frac{2}{T}\frac{z_\infty -1}{z_\infty +1} < 0$$

From this inequality, we can derive a condition for the mapping of stable analog poles into the z plane.

$$(\operatorname{Re} z_\infty)^2 + (\operatorname{Im} z_\infty)^2 < 1$$

The left half of the *p*-plane is mapped into the interior of the unit circle in the *z*-plane, as depicted in Fig. 6-18. This means that a stable continuous-time filter always results in a stable discrete-time filter.

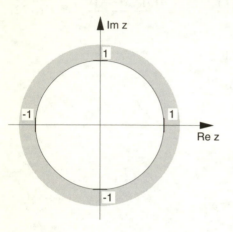

Fig. 6-18
Mapping of the left half of the *p*-plane into the unit circle of the *z*-plane using the bilinear transform

The application of the difference method that we introduced in the last section results in a poor approximation of the frequency response of the analog reference filter. In the following, we investigate the corresponding behaviour of the bilinear transform which converts the transfer function $H(p)$ of the analog filter into the transfer function in z of a related discrete-time filter.

$$H(p) \qquad \xrightarrow{\text{bilinear transform}} \qquad H\!\left(\frac{2}{T}\frac{z-1}{z+1}\right)$$

We compare the frequency response of the analog filter, which is obtained by the substitution $p = j\omega$, with that of the discrete-time filter which results from the substitution $z = e^{j\omega T}$.

$$H(j\omega) \qquad \xrightarrow{\text{bilinear transform}} \qquad H\!\left(\frac{2}{T}\frac{e^{j\omega T}-1}{e^{j\omega T}+1}\right)$$

With

$$\frac{2}{T}\frac{e^{j\omega T}-1}{e^{j\omega T}+1} = \frac{2}{T}\frac{e^{j\omega T/2}-e^{-j\omega T/2}}{e^{j\omega T/2}+e^{-j\omega T/2}} = \frac{2}{T}\tanh(j\omega T/2) = j\frac{2}{T}\tan(\omega T/2)$$

we obtain the following comparison between the frequency responses of analog and discrete-time system:

analog system   bilinear transform   discrete - time system

$$H(j\omega) \qquad\qquad \rightarrow \qquad\qquad H\left(j\frac{2}{T}\tan(\omega T / 2)\right) \quad.$$

It is obvious that the frequency response of the discrete-time system can be derived from the frequency response of the analog reference filter by a simple distortion of the frequency axis. ω is substituted by ω' which can be expressed as

$$\omega' = \frac{2}{T}\tan(\omega T / 2) \quad. \tag{6.23}$$

If ω progresses in the range from 0 up to half the sampling frequency π/T, ω' progresses from 0 to ∞. The bilinear transform thus compresses the whole course of the frequency response of the analog filter in the range $-\infty \le \omega \le +\infty$ to the frequency range $-\pi/T \le \omega \le +\pi/T$. Due to the tangent function, this compressed range is then periodically continued with the period $2\pi/T$. The described behaviour has the following consequences to be considered for the design of digital filters:

Aliasing
Since the frequency response of the analog filter is compressed prior to the periodical continuation, overlapping of spectra (aliasing), which caused problems in the case of the impulse- and step-invariant designs, is ruled out when the bilinear transform is applied. As a result, passband ripple and stopband attenuation of the analog reference filter are exactly maintained.

Filter Order
Due to the compression of the frequency axis of the analog reference filter, the sharpness of the transition band(s) increases compared to the original frequency response. This means that, if the bilinear transform is applied, a given tolerance scheme can be satisfied with a lower filter order than we needed for the corresponding analog filter. The relations to calculate the required filter order, which we introduced in Chap. 2, are nevertheless still valid. Instead of the original stopband and passband edge frequencies, we have to use the corresponding frequencies obtained by the frequency mapping (6.23). The closer the transition band is located to the Nyquist frequency, the higher the saving in terms of filter order. In the next section we will calculate an example of a filter design which will demonstrate the reduction of the filter order by using the bilinear transform.

Choice of the Cutoff Frequencies
The most important characteristic frequencies of filters are the edge frequencies of the tolerance scheme or the 3-dB cutoff frequency. These should be maintained in any case if the transfer function of the analog reference filter is converted into the transfer function of a corresponding discrete-time system. Due to the compression of the frequency axis caused by the bilinear transform, all characteristic frequencies

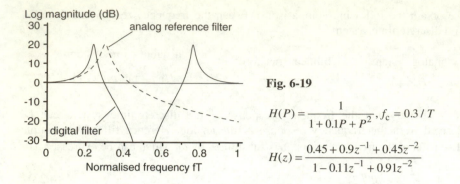

Fig. 6-19

$$H(P) = \frac{1}{1 + 0.1P + P^2}, f_c = 0.3/T$$

$$H(z) = \frac{0.45 + 0.9z^{-1} + 0.45z^{-2}}{1 - 0.11z^{-1} + 0.91z^{-2}}$$

are shifted towards lower frequencies, which is demonstrated in Fig. 6-19 by means of the cutoff frequency of a second-order low-pass filter.

This behaviour can be compensated for by shifting all characteristic frequencies to higher frequencies using (6.23). The analog filter design is then performed using these corrected values. Application of the bilinear transform to obtain a discrete-time equivalent of the analog reference filter shifts all characteristic frequencies back to the desired position. Figure 6-20 illustrates this approach.

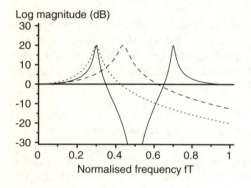

Fig. 6-20
Design of a second-order low-pass filter using the bilinear transform
dotted line: continuous-time reference filter
dashed line: shift of the cutoff frequency using (6.23)
solid line: resulting discrete-time system after bilinear transform

Phase Response

The imitation of the flat group-delay responses of Bessel filters is not possible if the bilinear transform is applied. Bessel filters exhibit an almost linearly increasing phase in the passband. Due to the nonlinear compression of the frequency axis, the slope of the phase curve increases more and more when approaching the Nyquist frequency. This behaviour is equivalent to an ever-increasing group delay. We come back to the solution of this problem in Sect. 6.4.4.

Example

We consider again the second-order low-pass filter that we have already used as an example:

$$H(P) = \frac{1}{P^2 + 0.1P + 1}, f_c = 0.3 / T \qquad (\omega_c = 0.6\pi / T) \qquad (6.24)$$

$$P = \frac{p}{\omega_c} = \frac{p}{2\pi f_c} = \frac{p}{0.6\pi / T} \quad .$$

Using design rule (6.21) results in the following expression for the normalised frequency P:

$$P = \frac{2/T}{0.6\pi / T} \frac{z-1}{z+1} = \frac{1}{0.3\pi} \frac{z-1}{z+1} \quad .$$

Substitution in the transfer function (6.24) yields the transfer function in z of the discrete-time filter:

$$H(z) = \frac{1}{\left(\dfrac{1}{0.3\pi} \dfrac{z-1}{z+1}\right)^2 + 0.1\left(\dfrac{1}{0.3\pi} \dfrac{z-1}{z+1}\right) + 1} \quad .$$

Multiplying out the brackets leads to the standard rational fractional form

$$H(z) = \frac{0.45z^2 + 0.9z + 0.45}{z^2 - 0.11z + 0.91} = \frac{0.45 + 0.9z^{-1} + 0.45z^{-2}}{1 - 0.11z^{-1} + 0.91z^{-2}} \quad .$$

Figure 6-19 depicts the corresponding magnitude response. The effect of the compression of the frequency axis is clearly visible. The gain at $\omega = 0$ and in the maximum, however, is maintained exactly.

6.4.3 Practical Filter Design Using the Bilinear Transform

For reasons that we will consider in more detail in Chap. 8, the cascade arrangement of second-order filter blocks as shown in Fig. 5-20 is of special importance. It is therefore a frequent task in practice to convert analog second-order filter blocks to corresponding discrete-time realisations using the bilinear transform. All first- and second-order filter blocks that can occur in the context of Butterworth, Chebyshev, Cauer and Bessel filters, also taking into account the possible filter transformations leading to high-pass, bandpass and bandstop filter variants, have normalised transfer functions of the following general form:

$$H(P) = \frac{B_0 + B_1 P + B_2 P^2}{1 + A_1 P + A_2 P^2} \quad . \qquad (6.25)$$

In order to obtain the unnormalised representation, we have to substitute $P = p/\omega_c$. Application of the bilinear transform results in

$$P = \frac{p}{\omega_c} = \frac{2}{\omega_c T} \frac{z-1}{z+1} \quad . \qquad (6.26)$$

Substitution of (6.26) in (6.25) yields the desired transfer function in z.

$$H(z) = \frac{\dfrac{4B_2}{(\omega_c T)^2}\left(\dfrac{z-1}{z+1}\right)^2 + \dfrac{2B_1}{\omega_c T}\left(\dfrac{z-1}{z+1}\right) + B_0}{\dfrac{4A_2}{(\omega_c T)^2}\left(\dfrac{z-1}{z+1}\right)^2 + \dfrac{2A_1}{\omega_c T}\left(\dfrac{z-1}{z+1}\right) + 1} \tag{6.27}$$

(6.27) can be readily rearranged to the normal form (6.28).

$$H(z) = V\frac{1+b_1 z^{-1} + b_2 z^{-2}}{1+a_1 z^{-1} + a_2 z^{-2}} \tag{6.28}$$

With the introduction of the notations

$$\begin{aligned}
C_1 &= 2A_1 / (\omega_c T) & D_1 &= 2B_1 / (\omega_c T) \\
C_2 &= 4A_2 / (\omega_c T)^2 & D_2 &= 4B_2 / (\omega_c T)^2
\end{aligned} \tag{6.29}$$

the coefficients of (6.28) can be calculated using the following relations.

Second-Order Filter Blocks

$$\begin{aligned}
V &= (B_0 + D_1 + D_2) / (1 + C_1 + C_2) \\
b_1 &= 2(B_0 - D_2) / (B_0 + D_1 + D_2) \\
b_2 &= (B_0 - D_1 + D_2) / (B_0 + D_1 + D_2) \\
a_1 &= 2(1 - C_2) / (1 + C_1 + C_2) \\
a_2 &= (1 - C_1 + C_2) / (1 + C_1 + C_2)
\end{aligned} \tag{6.30}$$

First-order Filter Blocks

$$\begin{aligned}
V &= (B_0 + D_1) / (1 + C_1) \\
b_1 &= (B_0 - D_1) / (B_0 + D_1) \\
b_2 &= 0 \\
a_1 &= (1 - C_1) / (1 + C_1) \\
a_2 &= 0
\end{aligned} \tag{6.31}$$

In the case of low-pass filters, the coefficients A_1, A_2 and B_2 can be directly taken from the filter tables in the Appendix. For other filter types like high-pass, bandpass or bandstop, these coefficients are calculated according to the relations for the respective filter transformation as derived in Sect. 2.7.4.

Example

We apply the above relations to determine the filter coefficients of an example which is specified by the following low-pass tolerance scheme:

passband ripple   :   1 dB $\rightarrow$ $v_p = 0.891$ $\rightarrow$ $v_{pnorm} = 1.96$
stopband attenuation   :   30 dB $\rightarrow$ $v_s = 0.032$ $\rightarrow$ $v_{snorm} = 0.032$
passband edge frequency : $f_p = 0.35/T$
stopband edge frequency : $f_s = 0.40/T$.

The filter characteristic is Butterworth. First we determine the design parameters N and ω_c of the analog filter without pre-mapping of the frequency axis so that we will be able to compare these results later with the parameters of the analog reference filter for the design of the discrete-time filter. The necessary order of the filter is determined using (2.14a).

$$N = \frac{\log \dfrac{1.96}{0.032}}{\log \dfrac{0.40}{0.35}} = 30.8 \qquad \text{or rounded up to} \quad N = 31$$

The 3-dB cutoff frequency is obtained by using relation (2.14b).

$$\omega_c = \omega_p \times 1.96^{1/31}$$
$$\omega_c = 2\pi \times 0.35 / T \times 1.022$$
$$\omega_c = 0.715\pi / T = 2.25 / T \qquad\qquad f_c = 0.3577 / T$$

An analog filter satisfying the above tolerance scheme is thus determined by the two design parameters $N = 31$ and $\omega_c = 2.25/T$.

For the design of the analog reference filter as the basis for a digital filter design, the edge frequencies of the tolerance scheme must be first transformed using relation (6.23).

$$\omega'_p = 2 / T \tan(2\pi \times 0.35 / T \times T / 2) = 3.93 / T$$
$$\omega'_s = 2 / T \tan(2\pi \times 0.40 / T \times T / 2) = 6.16 / T$$

The filter order of the discrete-time filter is calculated as

$$N = \frac{\log \dfrac{1.96}{0.032}}{\log \dfrac{6.16}{3.93}} = 9.16 \qquad \text{or rounded up to} \quad N = 10 \ .$$

Equation (2.14b) is used again to calculate the 3-dB cutoff frequency.

$$\omega_c = \omega_p \times 1.96^{1/10} = 3.93 / T \times 1.070$$
$$\omega_c = 4.205 / T \qquad\qquad f_c = 0.669 / T$$

As already predicted, the order of the reference filter for the digital filter design is considerably lower than that of an analog filter satisfying the same tolerance scheme. The 3-dB cutoff frequency is shifted to higher values, which finally results in the correct position when the bilinear transform is applied.

The starting point for the design of the discrete-time filter are the normalised coefficients of the analog reference filter. For a tenth-order Butterworth filter, which consists of five second-order filter blocks, we find the following coefficients in the corresponding table in the Appendix.

$$A_{11} = 0.312869 \qquad A_{21} = 1.0$$
$$A_{12} = 0.907981 \qquad A_{22} = 1.0$$
$$A_{13} = 1.414214 \qquad A_{23} = 1.0$$
$$A_{14} = 1.782013 \qquad A_{24} = 1.0$$
$$A_{15} = 1.975377 \qquad A_{25} = 1.0$$
$$B_{0v} = 1 \qquad B_{1v} = B_{2v} = 0$$
$$V_{tot} = 1$$

Relation (6.29) and (6.30) are now used to calculate the coefficients of the five partial discrete-time filters, which, arranged in cascade, have the desired frequency response.

	Filter 1	Filter 2	Filter 3	Filter 4	Filter 5
V	0.462	0.482	0.527	0.603	0.727
a_1	0.715	0.746	0.815	0.934	1.126
a_2	0.132	0.182	0.291	0.479	0.783
b_1	2	2	2	2	2
b_2	1	1	1	1	1

Figure 6-21 shows the magnitude responses of the five partial filters, which all have a common asymptote with a slope of 12 dB/octave. While filter 4 and filter 5 show resonant peaks in the range of the cutoff frequency, the remaining second-order blocks exhibit a monotonous rise of the attenuation.

The cascade arrangement of the five partial filters yields the magnitude response shown in Fig. 6-22a and Fig. 6-22b. As a result we obtain the expected flat response of a Butterworth filter which is fully compliant with the prescribed tolerance limits.

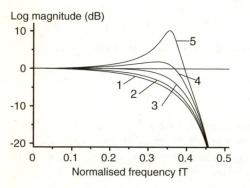

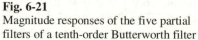

Fig. 6-21
Magnitude responses of the five partial filters of a tenth-order Butterworth filter

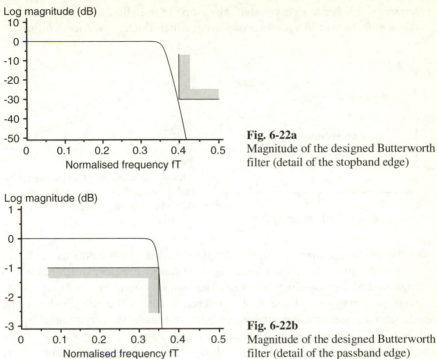

Fig. 6-22a
Magnitude of the designed Butterworth filter (detail of the stopband edge)

Fig. 6-22b
Magnitude of the designed Butterworth filter (detail of the passband edge)

6.4.4 IIR Filter with Bessel Characteristics

Bessel filters exhibit an approximately constant group delay or a linear phase in a wide range of frequencies. If we apply the bilinear transform to the transfer function, this property gets lost, unfortunately. We start by considering the influence of the bilinear transform on the ideal phase response of a linear phase filter, which can be expressed as

$$b(\omega) = \omega \, t_0 \ .$$

We obtain the corresponding phase response after bilinear transform by substitution of (6.23).

$$b(\omega) = (2t_0 \, / \, T) \tan(\omega T \, / \, 2) \tag{6.32}$$

Differentiation of (6.32) with respect to ω yields the corresponding group-delay characteristic.

$$\tau(\omega) = t_0 \left(1 + \tan^2(\omega T \, / \, 2) \right)$$

The group delay is approximately constant for very low frequencies only. The closer the frequency approaches $f_s / 2$, the more the impact of the compression of the frequency axis becomes noticeable. The slope of the phase increases, which is equivalent with an ever-increasing group delay which finally approaches infinity.

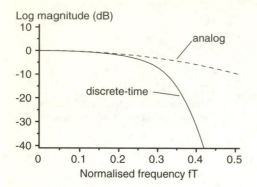

Fig. 6-23
Magnitude response of an analog fifth-order Bessel filter and of a related discrete-time filter obtained by bilinear transform

We choose the example of a fifth-order Bessel filter (3-dB cutoff frequency at $0.3 f_s$) to demonstrate the consequences of the bilinear transform with respect to the magnitude and group-delay response. The dashed line in Fig. 6-23 represents the magnitude response of the analog reference filter. The magnitude of the discrete-time transfer function obtained by bilinear transform is represented by the solid line. The compression of the frequency axis causes a sharper slope of the curve in the stopband. Figure 6-24 shows the corresponding group-delay responses. The analog filter exhibits the flat response that is typical for Bessel approximations. The curve for the related discrete-time filter starts with the same value at $\omega = 0$ but rises rapidly towards the Nyquist frequency.

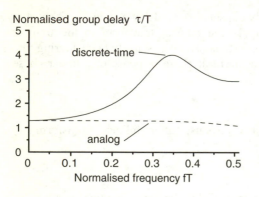

Fig. 6-24
Group delay response of an analog fifth-order Bessel filter and of a related discrete-time filter obtained by bilinear transform

A way out of this problem is to predistort the phase response of the analog reference filter in such a way that the bilinear transform yields a linear phase again. An appropriate target phase response for the analog reference filter is the inverse tangent:

$$b(\omega) = (2t_0 / T) \arctan(\omega T / 2) \ .$$

If the bilinear transform is applied to a transfer function with the above phase, we obtain a strictly linear phase, as can be verified by substitution of (6.23). In the following, we normalise the frequency ω to the delay t_0 ($\Omega = \omega t_0$):

$$b(\Omega) = (2t_0 / T)\, \arctan(\Omega T / 2t_0) \ .$$

With the notation $\mu = 2t_0/T$ we finally arrive at

$$b(\Omega) = \mu \arctan(\Omega / \mu) \ . \tag{6.33}$$

According to (2.50), the phase of a transfer function of the form

$$H(P) = \frac{1}{1 + P + A_2 P^2 + A_3 P^3 + A_4 P^4 + \ \dots}$$

can be expressed as

$$b(\Omega) = \arctan \frac{\Omega - A_3 \Omega^3 + A_5 \Omega^5 - \dots}{1 - A_2 \Omega^2 + A_4 \Omega^4 - \dots} \ . \tag{6.34}$$

In case of an analog filter design, (6.34) has to approximate a linear phase of the form $b(\Omega) = \Omega$ in the best possible way. Our analog reference filter for the digital filter design, however, has to approximate the predistorted phase response (6.33)

$$\mu \arctan(\Omega / \mu) = \arctan \frac{\Omega - A_3 \Omega^3 + A_5 \Omega^5 - \dots}{1 - A_2 \Omega^2 + A_4 \Omega^4 - \dots}$$

or

$$f(\Omega) = \tan\big(\mu \arctan(\Omega / \mu)\big) = \frac{\Omega - A_3 \Omega^3 + A_5 \Omega^5 - \dots}{1 - A_2 \Omega^2 + A_4 \Omega^4 - \dots} \ . \tag{6.35}$$

According to [21] $f(\Omega)$ can be expanded into a continued fraction that, if truncated after the Nth term, leads to a rational fractional function of degree N. (6.36) shows the systematics of this continued-fraction representation, which is truncated after the fifth partial quotient in this example ($N = 5$):

$$f(\Omega) = \tan\big(\mu \arctan(\Omega / \mu)\big) \approx \cfrac{1}{\cfrac{1}{\Omega} - \cfrac{1 - 1/\mu^2}{\cfrac{3}{\Omega} - \cfrac{1 - 4/\mu^2}{\cfrac{5}{\Omega} - \cfrac{1 - 9/\mu^2}{\cfrac{7}{\Omega} - \cfrac{1 - 16/\mu^2}{\cfrac{9}{\Omega} - \ \dots\dots}}}}} \ . \tag{6.36}$$

For $N = 1 \dots 3$ we obtain the following rational fractional approximations of $f(\Omega)$:

$N = 1$ $\qquad$ $f(\Omega) \approx \Omega$

$N = 2$ $\qquad$ $f(\Omega) \approx \dfrac{3\Omega}{3 - \left(1 - 1/\mu^2\right)\Omega^2}$

$N = 3$ $\qquad$ $f(\Omega) \approx \dfrac{15\Omega - \left(1 - 4/\mu^2\right)\Omega^3}{15 - \left(6 - 9/\mu^2\right)\Omega^2}$.

These rational fractional approximations of $f(\Omega)$ directly provide the desired filter coefficients A_i, as comparison with (6.35) shows.

$N = 1$ $\qquad$ $H(P) = \dfrac{1}{1 + P}$

$N = 2$ $\qquad$ $H(P) = \dfrac{3}{3 + 3P + \left(1 - 1/\mu^2\right)P^2}$ $\qquad$ (6.37)

$N = 3$ $\qquad$ $H(P) = \dfrac{15}{15 + 15P + \left(6 - 9/\mu^2\right)P^2 + \left(1 - 4/\mu^2\right)P^3}$

The reader will find in the Appendix of this book a Pascal routine which allows the calculation of the coefficients A_i for any filter order N. The parameter μ has to satisfy a certain condition in order to guarantee stability of the filter. From the theory of Hurwitz polynomials (polynomials with poles exclusively in the left half of the P-plane) it is known that all coefficients of the continued-fraction expansion must be positive. This leads to the condition

$1 - (N - 1)^2 / \mu^2 > 0$

or

$\mu = 2t_0 / T > N - 1$. $\qquad$ (6.38)

The delay t_0 of the discrete-time Bessel filter must therefore exceed a certain minimum value in order to obtain stable filters with all poles located in the left half of the P-plane.

6.4.4.1 Design with Prescribed Delay t_0

If the delay t_0 is given as design parameter, the following steps have to be performed for the practical filter design:
1. Design parameters are the filter order N, the sampling period T of the target discrete-time filter and a delay $t_0 > (N-1)T/2$. The design parameter μ is also fixed by these input data.
2. The coefficients A_i of the analog reference filter are determined using (6.36) or by means of the Pascal routine in the Appendix.
3. If a cascade realisation is aimed at, the poles of the transfer function have to be determined numerically by finding the roots of the denominator polynomial of

the transfer function. This allows one to establish the partial first- and second-order transfer function of the cascade arrangement.

4. The bilinear transform is applied to obtain the transfer function in z of the desired discrete-time filter. The normalised frequency P is replaced by the following term:

$$P = p\,t_0 = \frac{2}{T}\frac{z-1}{z+1}t_0 = \mu\frac{z-1}{z+1} \qquad \text{with } \mu = 2t_0\,/\,T\ .$$

6.4.4.2 Design with Prescribed 3-dB Cutoff Frequency

If the 3-dB cutoff frequency is to serve as filter design parameter, a relationship has to be derived between the parameter μ and the desired cutoff frequency ω_{3dB}. In case of the analog Bessel filter, the product of cutoff frequency ω_{3dB} and delay time t_0 is a constant for a given filter order N (2.58).

$$\omega_{3dB}\,t_0 = \Omega_{3dB}(N)$$

The corresponding values $\Omega_{3dB}(N)$ can be found in Table 2-2. The transfer functions of the analog reference filters for the digital filter design (6.37) have the additional parameter μ so that the product of delay and cutoff frequency will also be a function of this parameter.

$$\omega'_{3dB}\,t_0 = \Omega'_{3dB}(N,\mu)$$

ω'_{3dB} is the 3-dB cutoff frequency of the analog reference filter, which has to be shifted to higher values according to (6.23) in order to obtain the correct value after application of the bilinear transform.

$$2\,/\,T \times \tan\!\left(\omega_{3dB}T\,/\,2\right)t_0 = \Omega'_{3dB}(N,\mu)$$

With $\mu = 2\,t_0/T$ we obtain

$$\mu\tan\!\left(\omega_{3dB}\,T\,/\,2\right) = \Omega'_{3dB}(N,\mu)$$

$$\omega_{3dB}\,T\,/\,2 = \arctan\!\left(\Omega'_{3dB}(N,\mu)\,/\,\mu\right) \qquad\qquad (6.39)$$

$$\omega_{3dB}\,t_0 = \mu\arctan\!\left(\Omega'_{3dB}(N,\mu)\,/\,\mu\right)$$

$$\Omega_{3dB} = \mu\arctan\!\left(\Omega'_{3dB}(N,\mu)\,/\,\mu\right)\ . \qquad\qquad (6.40)$$

Ω_{3dB}, which is the product of delay and cutoff frequency of the discrete-time filter, is not a constant as in the case of analog filters (refer to section 2.6), but depends on the parameter μ and thus on the delay t_0. Ω_{3dB} approaches the values in Table 2-2 for large values of μ but is smaller in practical situations, as illustrated in Fig. 6-25. A relation between Ω_{3dB} and the given cutoff frequency $f_{3dB}\,T$, which would be helpful for the practical filter design, can only be determined numerically. For the derivation of a graphical representation of this desired relation we use (6.39) in the slightly modified form

$$f_{3\text{dB}}T = \arctan\big(\Omega'_{3\text{dB}}(N,\mu)\,/\,\mu\big)\big/\pi\ ,\tag{6.39a}$$

and (6.40)

$$\Omega_{3\text{dB}} = \mu\arctan\big(\Omega'_{3\text{dB}}(N,\mu)\,/\,\mu\big)\ .$$

We let the parameter μ progress in a fine grid in the range $N{-}1 < \mu < 1000$. The lower limit $N{-}1$ takes condition (6.38) into account. For each value of μ, we determine numerically the normalised cutoff frequenc $\Omega'_{3\text{dB}}$ of the analog reference filter and calculate with this result, using (6.39a) and (6.40), a pair $(\Omega_{3\text{dB}}|f_{3\text{dB}}\,T)$ which forms a point of the desired curve. Figure 6-25 depicts the relation between $\Omega_{3\text{dB}}$ and $f_{3\text{dB}}\,T$ for filter orders up to $N = 12$ in form of a nomogram which may be used as a design tool.

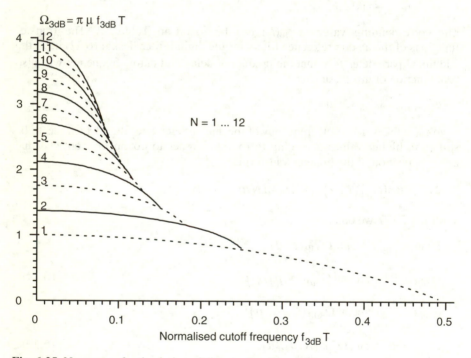

Fig. 6-25 Nomogram for the design of discrete-time Bessel filters

Figure 6-25 illustrates that the range of allowed cutoff frequencies is more or less restricted depending on the order of the filter. While $f_{3\text{dB}}\,T = 0.25$ is the upper limit for a second-order filter, the allowed range reduces to about 0.08 for a twelfth-order filter. This restriction is a consequence of condition (6.38), which guarantees the stability of the filter.

Input for the practical filter design are the three parameters
- filter order N,
- 3-dB cutoff frequency f_{3dB}, and
- sampling period T of the target digital filter.

The following design steps have to be performed. Figure 6-25 is used to determine the parameter μ. With f_{3dB} T and N given, we read the corresponding value for Ω_{3dB} from the appropriate curve. Since

$$\Omega_{3dB} = \omega_{3dB}\, t_0 = 2\pi f_{3dB}\, t_0 = \pi f_{3dB}\, T\, 2t_0\, / \, T = \pi f_{3dB} T \mu$$

and hence

$$\mu = \frac{\Omega_{3dB}}{\pi\, f_{3dB}\, T}\ ,$$

it is easy to calculate the design parameter μ. As μ is known at this point, we now use design steps 2, 3 and 4 as described in Sect. 6.4.4.1.

Example

We have the task of designing a digital Bessel filter with a cutoff frequency at f_{3dB} $T = 0.14$. The filter order should be as high as possible to obtain the best possible selectivity. According to Fig. 6-25, this design target can be achieved with a filter order not exceeding $N = 4$. For Ω_{3dB} we read the value 1.55 from the corresponding curve. The design parameter μ can now be calculated as:

$$\mu = \frac{1.55}{0.14\pi} = 3.52\ .$$

μ is greater than $N-1$ so that the stability criterion (6.38) is satisfied. With the knowledge of μ and the filter order N, we can determine the coefficients of the analog reference filter using Subroutine Bessel in the Appendix:

$A_0 = 1$
$A_1 = 1$
$A_2 = 0,3594$
$A_3 = 0,0530$
$A_4 = 0,0024$.

This yields the transfer function

$$H(P) = \frac{1}{1 + P + 0.3594 P^2 + 0.0530 P^3 + 0.0024 P^4}\ .$$

The roots of the denominator polynomial must be determined numerically.

$$P_{\infty 1} = \quad -12.651$$
$$P_{\infty 2} = \quad -2.913$$
$$P_{\infty 3/4} = \quad -3.271 \pm j\,0.790$$

This result allows us to establish the transfer function in the cascade form

$$H(P) = \frac{12.651}{P+12.651} \frac{2.913}{P+2.913} \frac{11.324}{P^2 + 6.542P + 11.324} \,.$$

The transition from the transfer function $H(P)$ to the corresponding transfer function $H(z)$ of the discrete-time filter happens by means of the transformation

$$P = 3.52 \frac{z-1}{z+1} \,.$$

$$H(z) = 0.086 \frac{1+z^{-1}}{1+0.565z^{-1}} \frac{1+z^{-1}}{1-0.0944z^{-1}} \frac{1+2z^{-1}+z^{-2}}{1-0.0458z^{-1}+0.0145z^{-2}}$$

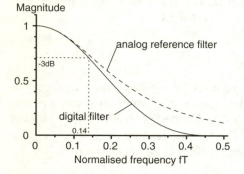

Fig. 6-26
Magnitude response of a fourth-order digital Bessel filter with a 3-dB cutoff frequency at $0.14\,f_s$.

Figure 6-26 compares the magnitude response of the analog reference filter with the one of the resulting discrete-time filter. The prescribed 3-dB cutoff frequency is met exactly. The group-delay curve of the analog reference filter is monotonously falling (Fig. 6-27). This behaviour is able to compensate for the rise of the group delay, caused by the bilinear transform, in a wide frequency range which results in an almost flat response up to $0.3\,f_s$ in our example.

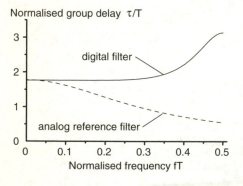

Fig. 6-27
Group delay response of a fourth-order digital Bessel filter with a 3-dB cutoff frequency at $0.14\,f_s$.

7 Design of FIR Filters

7.1 Introduction

For FIR filters, no design techniques exist in the continuous time domain that could be used as a basis for the design of related digital filters as in the case of IIR filters. There are no design tables available with normalised coefficients of prototype filters that could be easily adapted to special design specifications. So each filter design has to pass, in principle, a complete mathematical design procedure. According to the chosen approximation method, a number of more or less complex algorithms are available, iterative or analytic, that in most cases require computer support.

It is the aim of all design methods to approximate a given frequency response as "close" as possible by a finite number of FIR filter coefficients. The starting point of all these methods is the assumption of idealised frequency responses or tolerance specifications in the passband and stopband. The finite length of the unit-sample response and hence the finite number of filter coefficients limits the degree of approximation to the given specification. Low variation of the magnitude (ripple) in the passband, high attenuation in the stopband and sharp cutoff are competing design parameters in this context.

There are a number of possible criteria for an optimum choice of coefficients that lead to "close" approximations of the design target. Four such criteria are considered in more detail in this chapter. Two of them aim at minimising an error measure defined with respect to the deviation of the target magnitude response from the actually realised response, whereas the two remaining criteria aim at matching the characteristics of the desired and realised magnitude responses at a limited number of frequencies:

- The filter coefficients are to be determined in such a way that the mean square deviation between desired and actually realised magnitude responses is minimised.
- The filter coefficients are to be determined in such a way that the magnitude of the maximum deviation between desired and actually realised amplitude responses is minimised.
- The filter coefficients have to be determined in such a way that desired and actually realised amplitude response coincide in few points with respect to the gain and a number of derivatives (in case of a low-pass design, for instance, at two points: $\omega = 0$ and $\omega = \pi/T$).
- n filter coefficients are to be determined in such a way that desired and actually realised magnitude responses coincide at n points. A system of n equations has to be solved to determine the coefficients in this case.

7.2 Linear-Phase Filters

The main advantage of the FIR filter structure is the realisability of exactly linear-phase frequency responses. That is why almost all design methods described in the literature deal with filters with this property. Since the phase response of linear-phase filters is known, the design procedures are reduced to real-valued approximation problems, where the coefficients have to be optimised with respect to the magnitude response only. In the following, we derive some useful relations in the context of linear-phase filters in the time and frequency domain.

An Nth-order FIR filter is determined by $N + 1$ coefficients and has, according to (5.3), the following frequency response:

$$H\left(e^{j\omega T}\right) = \sum_{r=0}^{N} b_r\, e^{-j\omega rT} \quad . \tag{7.1}$$

Extracting an exponential term out of the sum that corresponds to a delay of $NT/2$ (half the length of the shift register of the FIR filter) yields

$$H\left(e^{j\omega T}\right) = e^{-j\omega NT/2} \sum_{r=0}^{N} b_r\, e^{-j\omega(r-N/2)T} \quad . \tag{7.2}$$

For the coefficients b_r there exist two possible symmetry conditions that lead to the situation that, apart from the extracted linear-phase delay term, there are no further frequency-dependent phase variations contributing to the overall frequency response. These are:

$$b_r = b_{N-r} \tag{7.3a}$$

$$b_r = -b_{N-r} \quad . \tag{7.3b}$$

Condition (7.3a) leads to symmetrical unit-sample responses as shown in Fig. 7-1a and Fig. 7-1b. Condition (7.3b) yields antisymmetrical unit-sample responses as

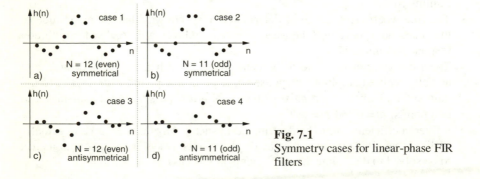

Fig. 7-1
Symmetry cases for linear-phase FIR filters

depicted in Fig. 7-1c and Fig. 7-1d. Taking the symmetry conditions (7.3) into account, the complex-valued sum in (7.2) can by expressed by real sine or cosine terms respectively.

$$H\left(e^{j\omega T}\right) = e^{-j\omega NT/2} \sum_{r=0}^{N} b_r \cos(r - N/2)\,\omega T$$

$$\text{for } b_r = b_{N-r}$$

(7.4a)

and

$$H\left(e^{j\omega T}\right) = j e^{-j\omega NT/2} \sum_{r=0}^{N} -b_r \sin(r - N/2)\,\omega T$$

$$\text{for } b_r = -b_{N-r}$$

(7.4b)

The summation terms in (7.4a) and (7.4b) are purely real and thus represent a zero-phase filter with the frequency response $H_0(\omega)$. The zero-phase frequency response must not be confused with the amplitude (magnitude) response since $H_0(\omega)$ may also assume negative values. The following relationships exist between frequency response and zero-phase frequency response.

$$H\left(e^{j\omega T}\right) = e^{-j\omega NT/2} \, H_0(\omega) \qquad \text{for } b_r = b_{N-r}$$

(7.5a)

and

$$H\left(e^{j\omega T}\right) = j e^{-j\omega NT/2} \, H_0(\omega) \qquad \text{for } b_r = -b_{N-r}$$

(7.5b)

Relations (7.4b) and (7.5b) contain, beside the linear-phase delay term, the additional factor j that causes a frequency-independent phase shift of 90°. Filters with antisymmetrical unit-sample responses are therefore well suited for FIR filter realisations of differentiators, integrators, Hilbert transformers and quadrature filters.

The relations (7.4) can be further simplified since half of the terms in the respective sums are identical in pairs because of the symmetry condition. The cases of even and odd filter orders must be treated separately so that we can distinguish between four cases in total (refer to Fig. 7-1) which feature certain symmetry properties in the frequency domain.

Case 1   Symmetrical unit-sample response, *N* even (Fig. 7-1a)

$$H_0(\omega) = \sum_{n=0}^{Nt} B(n)\cos n\omega T$$

with $Nt = N/2$, $B(0) = b(Nt)$

and $B(n) = 2b(Nt - n) = 2b(Nt + n)$ for $n = 1 \dots Nt$

symmetry condition: $H_0(\omega) = H_0(2\pi/T - \omega)$

(7.6a)

Case 2   Symmetrical unit-sample response, N odd (Fig. 7-1b)

$$H_0(\omega) = \sum_{n=0}^{Nt} B(n)\cos(n-1/2)\,\omega T$$

with $Nt = (N+1)/2$

and $B(n) = 2b(Nt-n) = 2b(Nt+n-1)$ for $n = 1 \ldots Nt$ \hfill (7.6b)

symmetry conditions: $H_0(\omega) = -H_0(2\pi/T - \omega)$

$$H_0(\pi/T) = 0$$

Case 3   Antisymmetrical unit-sample response, N even (Fig. 7-1c)

$$H_0(\omega) = \sum_{n=0}^{Nt} B(n)\sin n\omega T$$

with $Nt = N/2$

and $B(n) = 2b(Nt-n) = -2b(Nt+n)$ for $n = 1 \ldots Nt$ \hfill (7.6c)

symmetry conditions: $H_0(\omega) = -H_0(2\pi/T - \omega)$

$$H_0(0) = 0 \ \text{ and } \ H_0(\pi/T) = 0$$

Case 4   Antisymmetrical unit-sample response, N odd (Fig. 7-1d)

$$H_0(\omega) = \sum_{n=0}^{Nt} B(n)\sin(n-1/2)\,\omega T$$

with $Nt = (N+1)/2$

and $B(n) = 2b(Nt-n) = -2b(Nt+n-1)$ for $n = 1 \ldots Nt$ \hfill (7.6d)

symmetry conditions: $H_0(\omega) = H_0(2\pi/T - \omega)$

$$H_0(0) = 0$$

The frequency response of a zero-phase filter can be therefore expressed as the sum of Nt sine or cosine terms. The coefficients $B(n)$ thus have the meaning of coefficients of a Fourier's decomposition of the periodical frequency response of the FIR filter. It has to be finally noted that the four considered symmetry cases exhibit different properties with respect to the realisable filter characteristics:

Case 1:   no restriction,

Case 2:   low-pass and bandpass filters only   since $H_0(\pi/T) = 0$,

Case 3:   bandpass filter only   since $H_0(0) = 0$ and $H_0(\pi/T) = 0$,

Case 4:   high-pass and bandpass filters only   since $H_0(0) = 0$.

7.3 Frequency Sampling

In case of the frequency sampling method, the $N + 1$ filter coefficients of a Nth-order FIR filter are determined in such a way that the frequency response coincides at $N + 1$ distinct frequencies ω_i with the values of the desired frequency response $H_D(\omega_i)$. Using (7.1) we can establish a set of $N + 1$ equations that allows the determination of the $N + 1$ filter coefficients b_n.

$$H_D(\omega_i) = \sum_{n=0}^{N} b_n \, e^{-j\omega_i nT}$$

$$i = 0 \dots N$$

(7.7)

In order to obtain the filter coefficients, the set of equations (7.7) has to be solved with respect to the coefficients b_n. A particularly effective method of solving this set of equations can be derived if the frequencies ω_i are equally spaced over the range $\omega = 0 \dots 2\pi/T$.

$$\omega_i = i \frac{2\pi}{(N+1)T}$$

$$i = 0 \dots N$$

(7.8)

Substitution of (7.8) in (7.7) yields a mathematical relationship between the filter coefficients and the samples of the desired frequency response that is commonly referred to as "Discrete Fourier Transform" (DFT).

$$H_D(\omega_i) = \sum_{n=0}^{N} b_n \, e^{-j2\pi in/(N+1)}$$

$$i = 0 \dots N, \quad \omega_i = i2\pi / [(N+1)T]$$

(7.9)

Relationship (7.9) represents a $(N + 1)$ point DFT, which allows us to calculate $N + 1$ samples of the frequency response from $N + 1$ filter coefficients which, in case of the FIR filter, are identical with the unit-sample response. For our problem we have to derive an inverse DFT (IDFT) in order to calculate conversely the coefficients b_n from the samples of the frequency response $H_D(\omega_i)$. We start with multiplying both sides of (7.9) by the term $e^{j2\pi im/(N+1)}$ and summing both sides with respect to the index i. After exchanging the sums on the right side we obtain

$$\sum_{i=0}^{N} H_D(\omega_i) e^{j2\pi im/(N+1)} = \sum_{n=0}^{N} b_n \sum_{i=0}^{N} e^{-j2\pi i(n-m)/(N+1)}$$

$$\omega_i = i2\pi / [(N+1)T] \ .$$

(7.10)

Because of the periodicity and the symmetry properties of the complex exponential function, it can be readily shown that the right sum over i vanishes, with the exception of the case $m = n$ where the exponential term becomes unity. This simplifies (7.10) to the form

$$\sum_{i=0}^{N} H_{\mathrm{D}}(\omega_i) e^{j2\pi i m/(N+1)} = b_m (N+1)$$

$$m = 0 \ldots N ,$$

or

$$b_m = \frac{1}{N+1} \sum_{i=0}^{N} H_{\mathrm{D}}(\omega_i) e^{j2\pi i m/(N+1)} \tag{7.11}$$

$$\omega_i = i2\pi / [(N+1)T], \quad m = 0 \ldots N .$$

Relation (7.11) enables the calculation of the filter coefficients b_m from $N+1$ equally spaced samples of the desired frequency response $H_{\mathrm{D}}(\omega)$. For the practical filter design it would be advantageous if we could prescribe the real-valued zero-phase instead of the complex-valued linear-phase frequency response. By substitution of (7.5), relation (7.11) can be modified in this direction.

$$b_m = \frac{1}{N+1} \sum_{i=0}^{N} H_{0\mathrm{D}}(\omega_i) e^{j2\pi i(m-N/2)/(N+1)} \qquad \text{for } b_m = b_{N-m}$$

$$b_m = \frac{1}{N+1} \sum_{i=0}^{N} H_{0\mathrm{D}}(\omega_i) j\, e^{j2\pi i(m-N/2)/(N+1)} \qquad \text{for } b_m = -b_{N-m}$$

$$\omega_i = i2\pi / [(N+1)T], \quad m = 0 \ldots N$$

Because of the properties of the zero-phase frequency response for the various symmetry cases of the filter coefficients (7.6a–d), these relations can be further simplified, resulting in purely real expressions.

Case 1

$$b_m = b_{N-m} = \frac{1}{N+1} \left[H_{0\mathrm{D}}(0) + 2 \sum_{i=1}^{N/2} H_{0\mathrm{D}}(\omega_i) \cos[2\pi i(m - N/2)/(N+1)] \right] \tag{7.12a}$$

$$m = 0 \ldots Nt, \quad \omega_i = i2\pi / [(N+1)T]$$

Case 2

$$b_m = b_{N-m} = \frac{1}{N+1} \left[H_{0\mathrm{D}}(0) + 2 \sum_{i=1}^{(N-1)/2} H_{0\mathrm{D}}(\omega_i) \cos[2\pi i(m - N/2)/(N+1)] \right]$$

$$m = 0 \ldots Nt, \quad \omega_i = i2\pi / [(N+1)T] \tag{7.12b}$$

Case 3

$$b_m = -b_{N-m} = \frac{1}{N+1}\left[2\sum_{i=1}^{N/2}H_{0D}(\omega_i)\sin[2\pi i(m-N/2)/(N+1)]\right] \qquad (7.12c)$$

$$m = 0 \dots Nt, \quad \omega_i = i2\pi /[(N+1)T]$$

Case 4

$$b_m = -b_{N-m} = \frac{1}{N+1}\left[(-1)^{m-(N+1)/2}H_{0D}(\pi/T)\right.$$

$$\left. +2\sum_{i=1}^{(N-1)/2}H_{0D}(\omega_i)\sin[2\pi i(m-N/2)/(N+1)]\right] \qquad (7.12d)$$

$$m = 0 \dots Nt, \quad \omega_i = i2\pi /[(N+1)T]$$

Relation (7.12) enables the calculation of the filter coefficients for the various possible symmetry cases from equally spaced samples of the zero-phase frequency response. By means of three examples we will demonstrate the power and the drawbacks of the frequency sampling method.

Example

A 32nd-order low-pass filter with a cutoff frequency at $f_cT = 0.25$ is to be designed by frequency sampling. The corresponding samples of the zero-phase frequency response are:

$H_{0D}(i) = 1$ for $i = 0 \dots 8$,

$H_{0D}(i) = 0$ for $i = 9 \dots 16$.

(7.12a) yields the following filter coefficients:

$b(0)$	=	$b(32)$	=	0.0209		
$b(1)$	=	$b(31)$	=	−0.0231		
$b(2)$	=	$b(30)$	=	−0.0193		
$b(3)$	=	$b(29)$	=	0.0261		
$b(4)$	=	$b(28)$	=	0.0180		
$b(5)$	=	$b(27)$	=	−0.0303		
$b(6)$	=	$b(26)$	=	−0.0170		
$b(7)$	=	$b(25)$	=	0.0365		
$b(8)$	=	$b(24)$	=	0.0163		

$b(9) = b(23) = -0.0463$

$b(10) = b(22) = -0.0158$

$b(11) = b(21) = 0.0643$

$b(12) = b(20) = 0.0154$

$b(13) = b(19) = -0.1065$

$b(14) = b(18) = -0.0152$

$b(15) = b(17) = 0.3184$

$b(16) = \qquad\qquad 0.5152$.

Unit-sample response h(n)

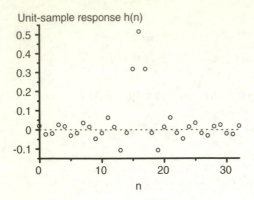

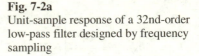

Fig. 7-2a
Unit-sample response of a 32nd-order
low-pass filter designed by frequency
sampling

Fig. 7-2a shows a graphical representation of the 33 filter coefficients and hence of the unit-sample response of the filter. Using (7.6a), we calculate the corresponding magnitude response, which is depicted in Fig. 7-2b.

It attracts attention that the achieved attenuation in the stopband of about 16 dB is rather poor, considering the relatively high filter order of $N = 32$. It must not be forgotten, however, that the specification of the filter leads to an extremely sharp transition band. The transition from the passband to the stopband happens between two neighbouring samples of the desired frequency response. We already pointed out earlier that high attenuation in the stopband and sharp cutoff are competing design targets.

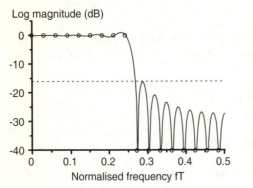

Log magnitude (dB)

Normalised frequency fT

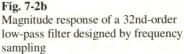

Fig. 7-2b
Magnitude response of a 32nd-order
low-pass filter designed by frequency
sampling

In order to achieve a better stopband attenuation, we replace the abrupt transition between passband and stopband by a certain transition area in which the gain drops linearly. If we choose one intermediate point within the transition band, we obtain a sample sequence of the frequency response of the form ..., 1, 1, 0.5, 0, 0, ..., and in case of two intermediate points we have the sequence ..., 1, 1, 0.67, 0.33, 0, 0, ... accordingly. Figure 7-3 shows the resulting magnitude responses, again for a filter order of $N = 32$ and a cutoff frequency at $f_c T = 0.25$.

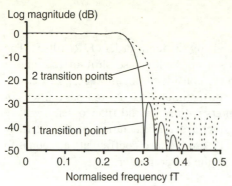

Log magnitude (dB)

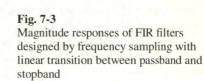

Fig. 7-3
Magnitude responses of FIR filters designed by frequency sampling with linear transition between passband and stopband

The stopband attenuation is noticeably improved. An optimum solution, however, is obviously not yet found, particularly since two interpolating values yield a worse result than only one. Lowering of the first lobe in the stopband would considerably improve the result. This can be achieved by varying the intermediate values until the optimum curves are achieved. The analytic method of applying the IDFT to the samples of the desired frequency response is thus supplemented by an iterative procedure which is applied to a limited number of values in the transition band. The optimum values in our example are 0.3908 for one and 0.5871 / 0.1048 for two intermediate values. Figure 7-4 depicts the corresponding magnitude responses. Two intermediate points in the transition band now yield a stopband attenuation more than 20 dB higher than only one.

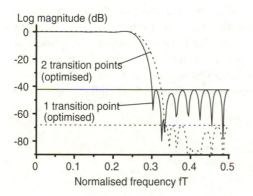

Log magnitude (dB)

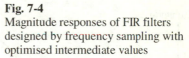

Fig. 7-4
Magnitude responses of FIR filters designed by frequency sampling with optimised intermediate values

The frequency sampling method is apparently not well suited for the design of selective filters with sharp transition bands. The calculation of the coefficients is easy using relationship (7.12), but additional iteration is necessary to arrive at acceptable results. Furthermore, there are restrictions concerning the choice of the edge frequencies since these have to coincide with the sampling grid of the frequency response. In the following two examples, we discuss zero-phase frequency responses that exhibit smooth curves.

Example

A differentiator is to be approximated by a tenth-order FIR filter. The corresponding transfer function features a constant phase shift of $\pi/2$ and a linearly rising magnitude. According to the considerations in Sect. 7.2, a case 4 filter is the right choice because the differentiator is a high-pass filter, in principle, with $H(0) = 0$. Since case 4 filters require an odd filter order, we have to choose $N = 11$. Fig. 7-5a depicts a graphical representation of the 12 filter coefficients as calculated using relation (7.12d). These coefficients are equal to the samples of the unit-sample response of the differentiator.

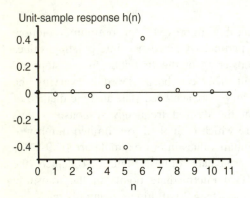

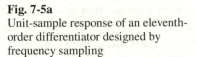

Fig. 7-5a
Unit-sample response of an eleventh-order differentiator designed by frequency sampling

Fig. 7-5b shows the sampling points, which lie on a straight line, and the plot of the approximated magnitude response of the FIR filter. In this case, the frequency sampling method yields an excellent result. The largest deviation with respect to the ideal curve can be observed in the vicinity of the Nyquist frequency. Since the magnitude curve is symmetrical about the frequency $\omega = \pi/T$, the slope becomes zero in this point. Because of the antisymmetry of the unit-sample response, the phase shift is exactly $\pi/2$.

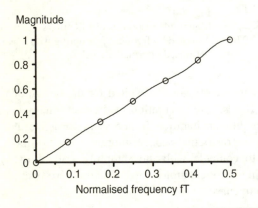

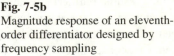

Fig. 7-5b
Magnitude response of an eleventh-order differentiator designed by frequency sampling

Example

The sin$x$/x-distortion of a D/A converter (Fig. 3-8) is to be compensated in the discrete-time domain by means of a digital filter. Since this distortion is linear-phase according to Sect. 3.4, a linear-phase FIR filter is the appropriate choice for our problem. The deviation of the approximation from the ideal curve must not exceed 0.25 dB in the frequency range $f = 0 \dots 0.45/T$. This specification can be met by a sixth-order filter. Figure 7-6 shows the resulting magnitude response compared to the target curve and the sample values that have been used for the design. Because of the symmetry condition for $H_0(\omega)$, the slope has to be zero at $\omega = \pi/T$, which causes the deviation to be mainly concentrated in this frequency range.

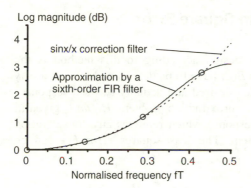
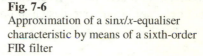

Fig. 7-6
Approximation of a sin$x$/x-equaliser characteristic by means of a sixth-order FIR filter

The previous two examples show that frequency sampling is a simple analytic method to approximate more or less smooth frequency responses by means of FIR filters. Selective filters designed with this method exhibit a bad attenuation behaviour in the stopband, which can only be improved by iterative variation of sample values in a defined transition band. The choice of the edge frequencies is restricted by the sampling grid of the zero-phase frequency response, which leads to severe restrictions, especially in the case of low filter orders. A filter specification in form of a tolerance scheme is not possible with this method.

We have seen, in the first example of a low-pass filter, that idealised frequency responses lead to bad results when frequency sampling is applied. The more realistic assumption of a finite slope in the transition band improves the results significantly. The approach in [40] goes one step further. On the basis of Chebyshev polynomials, it is possible to construct even more realistic frequency responses for which passband edge, stopband edge, passband ripple and stopband attenuation can be explicitly specified. Sampling of such frequency responses often leads to almost optimum results with regular ripple in the passband and stopband. The length of the filter and the degree of the involved Chebyshev polynomials can only be determined approximately from the filter specification. Some iteration steps are therefore required to find an optimum combination of these parameters.

We have included in the Appendix two Pascal subroutines that the reader may easily integrate in his own filter design programs. The Procedure CoeffIDFT is an implementation of (7.12), which allows the calculation of the filter coefficients from the equally spaced samples of the given zero-phase frequency response. A special form of the inverse discrete Fourier transform (IDFT) [46] is used that takes into account the special symmetry conditions in the time and frequency domain and is optimised to handle real sequences in both domains. The Procedure FirFreqResp enables the calculation of the zero-phase frequency response of FIR filters from the filter coefficients b_n with arbitrary resolution.

7.4 Minimisation of the Mean-Square Error

The starting point for the FIR filter design according to this method is the requirement that the filter coefficients $B(n)$ have to be determined in such a way that the mean square deviation between desired and actually realised responses is minimised, which is a least-squares approximation problem. $H_D(\omega)$ is given in the form of a zero-phase frequency response which removes any consideration of the phase from our design problem. The least-squares error E_{LS} can be expressed as

$$E_{LS} = \frac{T}{\pi} \int_0^{\pi/T} \left(H_0(\omega) - H_{0D}(\omega) \right)^2 \, d\omega \; . \tag{7.13}$$

Depending on the symmetry of the unit-sample response and on the filter order (even or odd), we substitute one of the equations (7.6a–d) in (7.13). In order to determine the minimum of E_{LS} we equate the derivatives of (7.13) with respect to the filter coefficients $B(n)$ to zero. Solving the resulting equations for $B(n)$ yields the desired set of coefficients.

We consider case 1 filters in more detail and substitute (7.6a) in (7.13).

$$E_{LS}\left[B(0), \ldots, B(Nt)\right] = \frac{T}{\pi} \int_0^{\pi/T} \left(\sum_{n=0}^{Nt} B(n) \cos n\omega T - H_{0D}(\omega) \right)^2 \, d\omega$$

for N even

We differentiate this equation with respect to the νth filter coefficient $B(\nu)$ and set the resulting term to zero.

$$\frac{dE_{LS}}{dB(\nu)} = \int_0^{\pi/T} 2 \left(\sum_{n=0}^{Nt} B(n) \cos n\omega T - H_{0D}(\omega) \right) \cos \nu\omega T \, d\omega = 0$$

$$\sum_{n=0}^{Nt} B(n) \int_0^{\pi/T} \cos n\omega T \cos v\omega T \, d\omega = \int_0^{\pi/T} H_{0D}(\omega)\cos v\omega T \, d\omega \qquad (7.14)$$

On the left side of (7.14), only the case $n = v$ gives a nonzero contribution to the summation. All other terms in the sum vanish since we integrate over integer periods of cosine functions.

$$B(v) \int_0^{\pi/T} \cos^2 v\omega T \, d\omega = \int_0^{\pi/T} H_{0D}(\omega)\cos v\omega T \, d\omega$$

The coefficients $B(n)$ of a case 1 filter are thus calculated as

$$B(0) = \frac{T}{\pi} \int_0^{\pi/T} H_{0D}(\omega) \, d\omega$$

$$B(v) = \frac{2T}{\pi} \int_0^{\pi/T} H_{0D}(\omega) \cos v\omega T \, d\omega \qquad (7.15a)$$

$$v = 1 \ldots Nt \ .$$

Applying the same procedure to case 2 filters yields the following expression for the calculation of the coefficients $B(v)$.

$$B(v) = \frac{2T}{\pi} \int_0^{\pi/T} H_{0D}(\omega) \cos(v - 1/2)\omega T \, d\omega \qquad (7.15b)$$

$$v = 1 \ldots Nt$$

Using the relations between the filter coefficients $b(n)$ and the Fourier coefficients $B(n)$ given in (7.6a) and (7.6b), a single equation can be derived for even and odd filter orders to determine the filter coefficients $b(n)$.

$$b(n) = h(n) = \frac{T}{\pi} \int_0^{\pi/T} H_{0D}(\omega) \cos(v - N/2)\omega T \, d\omega \qquad (7.16)$$

$$v = 0 \ldots N$$

We consider a simple low-pass filter as an example, which is specified as follows:

$$H_{0D}(\omega) = \begin{cases} 1 & \text{for } 0 \le \omega < \omega_c \\ 0 & \text{for } \omega_c < \omega \le \pi/T \end{cases} \qquad (7.17)$$

Using (7.16) we obtain the following filter coefficients.

$$b(v) = h(v) = \frac{T}{\pi} \int\limits_{0}^{\omega_c} \cos(v - N/2)\,\omega T\,\mathrm{d}\omega$$

$$b(v) = h(v) = \frac{\sin \omega_c T(v - N/2)}{\pi(v - N/2)} = \frac{\omega_c T}{\pi} \operatorname{sinc}\left(\omega_c T(v - N/2)\right) \qquad (7.18)$$

$$v = 0 \dots N$$

Depending on the filter order N, the sinc function (7.18) is a more-or-less good approximation of the infinite unit-sample response of an ideal low-pass filter as specified by (7.17). The truncation of the ideal unit-sample response to the $N + 1$ samples of the realised filter is called windowing.

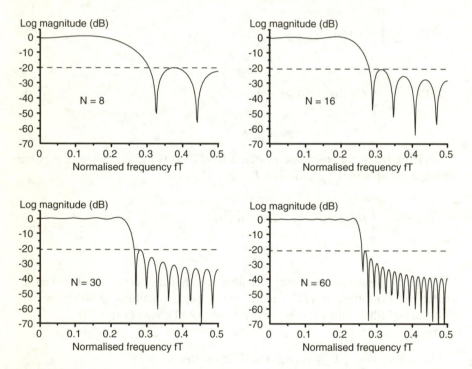

Fig. 7-7 Magnitude response of low-pass filters designed according to the least-squares error criterion $(f_c T = 0.25)$

Figure 7-7 depicts the magnitude responses of low-pass filters for various filter orders N, which have been designed using relation (7.18). It appears that the sharpness of the transition band goes up as the filter order increases. The stopband attenuation, however, characterised by the highest maximum in the stopband, is absolutely independent of the filter order. The attenuation of the first lobe in the stopband amounts to about 20 dB. This poor behaviour in the stopband is characteristic for the kind of windowing that we performed to truncate the length of the unit-sample response to $N + 1$ samples resulting in a FIR filter of order N. We have applied a so-called rectangular window that sharply cuts away the rest of the ideal unit-sample response and equally weights the samples within the window.

As in the case of the frequency sampling method, the magnitude response of the obtained filter oscillates around the ideal values in the passband and stopband. The largest deviation occurs in the area of the abrupt transition between passband and stopband. In the following, we discuss two methods that lead to better results with respect to passband ripple and stopband attenuation.

Direct solution

The first method is based on a further development of equation (7.14) and takes into account that, for a realisable filter, the graph of the magnitude must have a finite slope in the transition area between passband and stopband. The method assumes a certain width of the transition band, but ignores this frequency interval in the mean-square error expression (7.13). As a consequence of this approach, the relatively simple computation of the coefficients $B(n)$ using (7.18) is no longer possible. Rather, we have to solve a system of equations of the form

$$C\,b = h$$

in order to obtain the coefficient vector b.

The window method

The infinite unit-sample response of the ideal prototype filter is multiplied by special window functions which result in better filter characteristics than we have obtained by the simple rectangular windowing as described above. These windows have the property that they taper smoothly to zero at each end, with the desired effect that the height of the side lobes in the stopband is diminished.

7.4.1 Direct Solution

The average of the mean-square deviation taken over the whole frequency interval $\omega = 0 \dots \pi/T$, as expressed in (7.13), is to be minimised by an appropriate choice of the coefficients $B(n)$. Since the ideal sharp transition between passband and stopband as specified in (7.17) lies in this interval, the filter resources are largely used to approximate this abrupt drop in the gain. These resources are accordingly not available to optimise the filter characteristics with respect to passband ripple and stopband attenuation. An attenuation of the first side lobe in the stopband higher than about 21 dB is not attainable (refer to Fig. 7-7). A more realistic assumption about the magnitude curve in the transition band would certainly help. We have shown, in the context of the frequency sampling method, however, that it is difficult to predict an optimum transition curve between passband and stopband. It is therefore an obvious approach to leave out the transition band out of consideration completely when we write down the least-squares error expression (7.13) which is to be minimised.

$$E_{LS} = \frac{T}{\pi} \int_{\mathcal{A}} \left(H_0(\omega) - H_{0D}(\omega) \right)^2 d\omega$$

$$\mathcal{A} = \text{interval } 0 \dots \pi/T \text{ excluding the transition band}$$

So we optimise the characteristics in the passband and stopband according to the least-squares error criterion, but we do not make any effort to enforce a certain curve in the transition band, which remains unspecified. In consequence we do not integrate over complete periods of the cosine, as in (7.14), any more, so that the relatively simple equations to calculate the coefficients $B(n)$ do not survive. Relation (7.14) rather takes on the more complex form

$$\sum_{n=0}^{Nt} B(n) \left(\int_0^{\omega_p} \cos n\omega T \cos v\omega T \, d\omega + \int_{\omega_s}^{\pi/T} \cos n\omega T \cos v\omega T \, d\omega \right) =$$

$$\int_0^{\omega_p} H_{0D}(\omega) \cos v\omega T \, d\omega + \int_{\omega_s}^{\pi/T} H_{0D}(\omega) \cos v\omega T \, d\omega \qquad (7.19)$$

$$n = 0 \dots Nt \ .$$

Relation (7.19) can be easily written in the form of a matrix equation.

$$\boldsymbol{C}\,\boldsymbol{b} = \boldsymbol{h} \qquad (7.20)$$

The elements of the matrix $\boldsymbol{C}$ are calculated as

$$C_{vn} = \int_0^{\omega_p} \cos n\omega T \cos v\omega T \, d\omega + \int_{\omega_s}^{\pi/T} \cos n\omega T \cos v\omega T \, d\omega \qquad (7.21a)$$

$$v = 0 \dots Nt, \ n = 0 \dots Nt \ .$$

The matrix $\boldsymbol{C}$ is independent of the transfer function to be approximated. It only contains the information about the frequency ranges that are taken into account for the minimisation of the mean-square deviation between desired and realised magnitude responses. The vector $\boldsymbol{h}$ on the right side of (7.20) is determined as follows:

$$h_v = \int_0^{\omega_p} H_{0D}(\omega) \cos v\omega T \, d\omega + \int_{\omega_s}^{\pi/T} H_{0D}(\omega) \cos v\omega T \, d\omega \qquad (7.21b)$$

$$v = 0 \dots Nt \ .$$

In the case of low-pass filters, only the left integral contributes to the result since $H_{0D}(\omega)$ vanishes in the stopband $\omega = \omega_s \ldots \pi/T$.

$$h_v = \int_0^{\omega_p} \cos v\omega T \, d\omega \qquad\qquad v = 0 \ldots Nt \qquad\qquad (7.21c)$$

For high-pass filters we obtain, accordingly,

$$h_v = \int_{\omega_p}^{\pi/T} \cos v\omega T \, d\omega \qquad\qquad v = 0 \ldots Nt \ . \qquad\qquad (7.21d)$$

Solving (7.20) with respect to b yields the desired filter coefficients in vector form

$$b = \left(B_0, B_1, B_2, \ldots, B_{Nt}\right)^{\mathrm{T}} \ .$$

The evaluation of the integrals (7.21a–d) is elementary. Care has to be taken of the special cases $v = n$ and $v = n = 0$, which have to be treated separately.

$$C_{vn} = \begin{cases} 0.5\left(\dfrac{\sin(n-v)\omega_p T - \sin(n-v)\omega_s T}{(n-v)T} + \dfrac{\sin(n+v)\omega_p T - \sin(n+v)\omega_s T}{(n+v)T} \right) & \begin{array}{l} n \neq v \end{array} \\[4ex] 0.5\left(\omega_p + \pi/T - \omega_s + \dfrac{\sin(n+v)\omega_p T - \sin(n+v)\omega_s T}{(n+v)T} \right) & \begin{array}{l} n = v \\ n \neq 0, v \neq 0 \end{array} \\[4ex] \omega_p + \pi/T - \omega_s \qquad\qquad n = 0, v = 0 \end{cases}$$

$$\qquad\qquad\qquad (7.22a)$$
$$v = 0 \ldots Nt, \ \ n = 0 \ldots Nt$$

$$h_v = \begin{cases} \dfrac{\sin v\omega_p T}{vT} & \text{for } v \neq 0 \\[3ex] \omega_p & \text{for } v = 0 \end{cases} \qquad \text{for low - pass filters} \qquad\qquad (7.22b)$$

$$v = 0 \ldots Nt$$

$$h_v = \begin{cases} -\dfrac{\sin v\omega_p T}{vT} & \text{for } v \neq 0 \\[3ex] \pi/T - \omega_p & \text{for } v = 0 \end{cases} \qquad \text{for high - pass filters} \qquad\qquad (7.22c)$$

$$v = 0 \ldots Nt$$

The starting point for the treatment of case 2 filters, which cover odd filter orders, is the substitution of (7.6b) in (7.13). The set of equations to determine the filter coefficients $B(n)$ assumes the form

$$
\sum_{n=1}^{Nt} B(n) \left(\int_0^{\omega_p} \cos(n-1/2)\omega T \cos(v-1/2)\omega T \, d\omega \right.
$$

$$
\left. + \int_{\omega_s}^{\pi/T} \cos(n-1/2)\omega T \cos(v-1/2)\omega T \, d\omega \right) = \tag{7.23}
$$

$$
\int_0^{\omega_p} H_{0D}(\omega) \cos(v-1/2)\omega T \, d\omega + \int_{\omega_s}^{\pi/T} H_{0D}(\omega) \cos(v-1/2)\omega T \, d\omega
$$

$$
v = 0 \dots Nt \ .
$$

Evaluation of the integrals yields the following relations for the calculation of the matrix C and the vector h:

$$
C_{vn} = \begin{cases} 0.5 \left(\dfrac{\sin(n-v)\omega_p T - \sin(n-v)\omega_s T}{(n-v)T} \right. \\[2ex] \left. + \dfrac{\sin(n+v-1)\omega_p T - \sin(n+v-1)\omega_s T}{n+v-1)T} \right) & n \neq v \\[2ex] 0.5 \left(\omega_p + \pi/T - \omega_s + \dfrac{\sin(n+v-1)\omega_p T - \sin(n+v-1)\omega_s T}{(n+v-1)T} \right) & n = v \end{cases} \tag{7.24a}
$$

$$
v = 0 \dots Nt, \ n = 0 \dots Nt
$$

$$
h_v = \frac{\sin\left((v-1/2)\,\omega_p T\right)}{(v-1/2)\,T} \qquad \text{for low - pass filters} \tag{7.24b}
$$

$$
v = 0 \dots Nt
$$

$$
h_v = \frac{(-1)^{v+1} - \sin\left((v-1/2)\,\omega_p T\right)}{(v-1/2)\,T} \qquad \text{for high - pass filters} \tag{7.24c}
$$

$$
v = 0 \dots Nt \ .
$$

Example

The following example is intended to illustrate the consequences of disregarding the transition band in the approximation process. We consider a 30th-order FIR filter with a cutoff frequency at $f_c = 0.25/T$. We want to show the magnitude responses for normalised transition bandwidths of $\varDelta = 0$, $\varDelta = 0.4$ and $\varDelta = 0.8$. The input parameters for the calculation of the elements of the matrix C and the vector h using (7.22) are as follows:

①	$\varDelta = 0$	$f_p T = 0.25$	$f_s T = 0.25$	$Nt = 15$
②	$\varDelta = 0.4$	$f_p T = 0.23$	$f_s T = 0.27$	$Nt = 15$
③	$\varDelta = 0.8$	$f_p T = 0.21$	$f_s T = 0.29$	$Nt = 15$.

The determination of the filter coefficients is practicable only with computer support. In the first place we calculate the elements of C and h using (7.22). Note that C consists of 256 matrix elements. Afterwards we solve the linear set of equations (7.20) using one of the well known methods such as the Gauss elimination algorithm. Figure 7-8 shows the resulting magnitude responses.

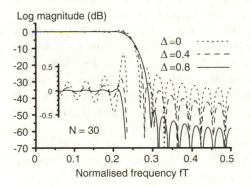

Fig. 7-8
Magnitude responses of FIR filters for various widths of the transition band $\varDelta$

The case $\varDelta = 0$ corresponds to the magnitude response for $N = 30$ shown in Fig. 7-7, which only has a stopband attenuation of about 20 dB. With increasing width of the unspecified transition band we observe an increase of the attainable stopband attenuation and a reduction of the ripple in the passband. The price to pay for these improvements is a reduced selectivity of the filter.

Example

A high-pass filter with the following specification is to be designed:

Filter order:	$N = 30$
Cutoff frequency	$f_c T = 0.25$
Normalised width of the transition band:	$\varDelta = 0.4$.

These data yield the following input parameters for the determination of the matrix C (7.22a) and the vector h (7.22c):

$$f_s T = 0.23 \qquad f_p T = 0.27 \qquad Nt = 15 \ .$$

Figure 7-9 depicts the magnitude response of the resulting high-pass filter.

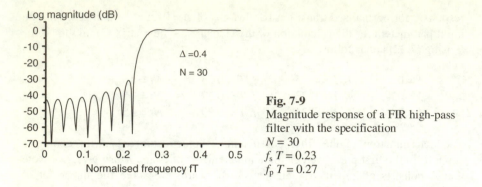

Fig. 7-9
Magnitude response of a FIR high-pass
filter with the specification
$N = 30$
$f_s T = 0.23$
$f_p T = 0.27$

The presented method, which minimises the mean-square deviation between desired and realised magnitude responses, offers some degree of freedom for the filter design, which goes beyond the possibilities of frequency sampling. Besides the filter order and the cutoff frequency, which can be arbitrarily chosen, we can additionally influence the sharpness of the cutoff, the passband ripple and the stopband attenuation by an appropriate choice of the width of the transition band. Passband ripple and stopband attenuation cannot be chosen independently of each other, since there is a fixed relationship between both parameters. Figure 7-10 shows the deviation $\delta(\omega)$ between desired and realised magnitude response for a 30th-order low-pass filter with $\Delta = 0.4$. The maximum deviations in the passband and stopband are equal in the vicinity of the transition band and amount to about $\delta_{max} = 0.073$. Passband ripple and stopband attenuation can be expressed in logarithmic scale as

$$D_p = 20 \left| \log(1 + \delta_{max}) \right| = 0.7 \text{dB}$$

$$D_s = -20 \log \left| \delta_{max} \right| = 22.7 \text{dB} \quad .$$

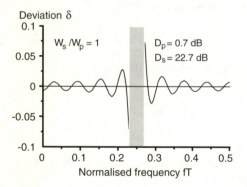

Fig. 7-10
Deviation from the ideal magnitude
response for a 30th-order low-pass filter
with $\Delta = 0.4$

It would be certainly desirable if, for a given filter degree N, the stopband attenuation could be increased at the expense of a higher ripple in the passband or vice versa. By introduction of a weighting function $W(\omega)$ into the error measure (7.13), it is possible to influence the design in this direction.

$$E_{LS} = \frac{T}{\pi} \int_{\mathcal{A}} \left(W(\omega) \left[H_0(\omega) - H_{0D}(\omega) \right] \right)^2 d\omega$$

(7.25)

$\mathcal{A} = $ interval $0 \ldots \pi/\mathrm{T}$, excluding the transition band

$W(\omega)$ determines at which degree the various frequency ranges are taken into account for the minimisation of the mean-square error. If $W(\omega)$ assumes higher values in the passband than in the stopband, for instance, the passband ripple will be reduced and the stopband attenuation becomes worse. If we choose a higher $W(\omega)$ in the stopband, we have the inverse behaviour. Only the ratio between the values of $W(\omega)$ in the passband and stopband is of importance; the absolute values can be chosen arbitrarily. Figure 7-11 shows again the error curve of a 30th-order low-pass filter with $\Delta = 0.4$, but the stopband is weighted 50 times higher than the passband. The maximum deviation in the stopband is now reduced to $\delta_{max} = 0.022$, corresponding to an attenuation of 33.2 dB. The ripple in the passband increases to 2.1 dB.

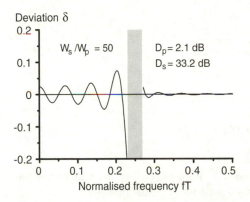

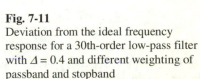

Fig. 7-11
Deviation from the ideal frequency response for a 30th-order low-pass filter with $\Delta = 0.4$ and different weighting of passband and stopband

There is no mathematical relation available in the literature that allows the calculation of the necessary filter order N, the width of the transition band Δ and the weighting function $W(\omega)$ from the data of a prescribed tolerance scheme. So some iteration steps are required, which could be best performed with computer support.

Finally, it must be noted that the described method can, in principle, also be used for bandpass and bandstop filters. Instead of two frequency intervals we have to optimise three intervals with respect to the least-squares error criterion: stopband–passband–stopband for bandpass filters or passband–stopband–passband for bandstop filters respectively. The transition bands, two in this case, are left out of consideration again.

7.4.2 The Window Method

We demonstrated in Sect. 7.4 how the frequency response is influenced by truncation (windowing) of the infinite unit-sample response of the ideal low-pass filter. A similar behaviour can be observed if the infinite unit-sample responses of ideal high-pass, bandpass and bandstop filters are truncated in the same way to obtain $N + 1$ samples ($n = 0 \dots N$). These filter types and the already considered low-pass filter have the following truncated unit-sample responses.

low-pass

$$h(n) = \frac{\sin \omega_c T(n - N/2)}{\pi(n - N/2)} \qquad n = 0 \dots N \tag{7.26a}$$

high-pass

$$h(n) = 1 - \frac{\sin \omega_c T(n - N/2)}{\pi(n - N/2)} \qquad n = 0 \dots N \tag{7.26b}$$

bandpass

$$h(n) = \frac{\sin \omega_u T(n - N/2) - \sin \omega_l T(n - N/2)}{\pi(n - N/2)} \qquad n = 0 \dots N \tag{7.26c}$$

bandstop

$$h(n) = 1 - \frac{\sin \omega_u T(n - N/2) - \sin \omega_l T(n - N/2)}{\pi(n - N/2)} \qquad n = 0 \dots N \tag{7.26d}$$

ω_c is the cutoff frequency of the low-pass or high-pass filter. ω_l and ω_u are the lower and upper cutoff frequencies of the bandpass or bandstop filter.

Figure 7-12 shows the magnitude responses of FIR filters with unit-sample responses according to (7.26a–d). The filter order is $N = 32$ in all cases. The first lobe in the stopband is about 20 dB below the maximum for all filter types. A better stopband attenuation than 20 dB is apparently not realisable if the $N + 1$ samples are simply cut out of the infinite unit-step response of the ideal filter. This behaviour is characteristic for the kind of windowing that we have performed, i.e. a rectangular window in which all samples are equally weighted.

A better filter performance can be achieved in the stopband, however, if the samples are progressively reduced in magnitude as n approaches 0 or N from the centre of the window. It can be observed that the stopband attenuation may increase considerably but at the expense of a less rapid transition between passband and stopband. A multitude of windows with this property have been extensively studied in the literature, which differ in terms of the sharpness of the

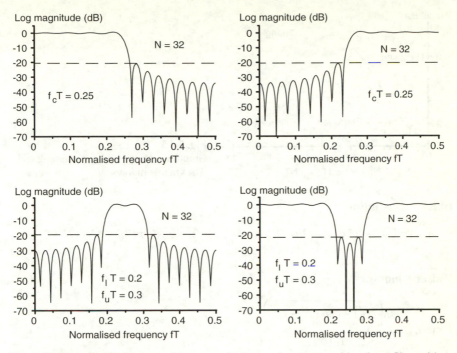

Fig. 7-12 Magnitude responses of low-pass, high-pass, bandpass and bandstop filter with
rectangular window and $N = 32$

cutoff, the attainable stopband attenuation and the evenness of the ripple.
Figure 7-13 depicts three examples of curves of such windows. In the following
we give the mathematical expressions for the most important windows [51] with
$0 \leq n \leq N$ in all cases.

Rectangular window

$$w(n) = 1$$

Triangular window

$$w(n) = 1 - |(2n - N) / (N + 2)|$$

Hamming window

$$w(n) = 0.54 + 0.46 \cos(\pi(2n - N) / N)$$

Hanning window

$$w(n) = 0.5 + 0.5 \cos(\pi(2n - N) / (N + 2))$$

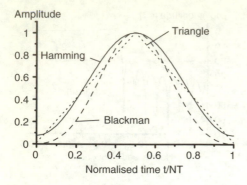

Fig. 7-13
Graphs of the triangular, Hamming and
Blackman windows

Blackman window

$$w(n) = 0.42 + 0.5 \cos\big(\pi(2n - N)/(N + 2)\big) + 0.08 \cos\big(2\pi(2n - N)/(N + 2)\big)$$

Kaiser window

$$w(n) = \frac{I_0\left(\alpha\sqrt{1 - (2n/N - 1)^2}\right)}{I_0(\alpha)}$$

$\alpha = 0.1102\,(ATT{-}8.7)$ $ATT > 50$ dB

$\alpha = 0.5842\,(ATT{-}21)^{0.4} + 0.07886\,(ATT{-}21)$ 21 dB $\leq ATT \leq 50$ dB

$\alpha = 0$ $ATT < 21$ dB

ATT is the desired stopband attenuation in dB. I_0 is the zeroth-order modified Bessel function of first kind.

Chebyshev window

$$W(f) = \delta_p\, T_{N/2}\big(\alpha\cos(2\pi fT) + \beta\big)$$

$$\alpha = (X_0 + 1)/2 \qquad\qquad \beta = (X_0 - 1)/2$$

$$X_0 = \big(3 - \cos(2\pi\Delta F)\big)/\big(1 + \cos(2\pi\Delta F)\big)$$

$$N = \cosh^{-1}\big((1 + \delta_p)/\delta_p\big)\big/\cosh^{-1}\big(1/\cos(\pi\Delta F)\big) \tag{7.27}$$

The Chebyshev window is specified in the frequency domain. $T_{N/2}(x)$ is the $N/2$th-order Chebyshev polynomial according to (2.17). Figure 7-14 shows the shape of the spectrum of the Chebyshev window and the parameters that specify the window. Filter order N, ripple δ_p and width of the main lobe ΔF cannot be chosen arbitrarily; these parameters have to satisfy equation (7.27). In order to obtain the window in the time domain, $W(f)$ is sampled at $N + 1$ equidistant frequency points. These samples are transformed into the discrete-time domain by inverse DFT using, for instance, the procedure CoeffIDFT in the Appendix.

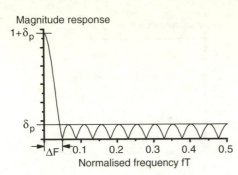

Magnitude response

Fig. 7-14
Spectrum of a Chebyshev window

The first step in the practical filter design is the calculation of the truncated unit-sample response of the ideal filter, choosing one of the relations (7.26a–d), depending on the filter type. These calculated samples are afterwards multiplied by one of the window functions, which yields the unit-sample response and hence the coefficients of the desired FIR filter. Each of the window functions results in characteristic curves of the magnitude response. Figure 7-15 depicts the magnitude responses of 50th-order low-pass filters with $f_cT = 0.25$ that have been designed using the various window functions introduced in this section. For the calculation of the magnitude response from the filter coefficients we can again make use of the procedure FirFreqResp.

The appropriate window is chosen according to the given specification with respect to the sharpness of the cutoff and the stopband attenuation of the filter under design. Five of the presented windows have fixed values for the stopband attenuation and are thus not very flexible. The width of the transition band is determined by the filter order. Table 7-1 summarises the main characteristics of the various windows. The triangular window has the special property that the magnitude response of the resulting filter has no ripple. The gain decreases more or less monotonously (refer to Fig. 7-15). The Hamming window leads to an almost even ripple which comes very close to the ideal case. Of all the windows with the transition bandwidth $4/NT$, the Hamming window possesses the highest stopband attenuation. The Blackman window has the highest stopband attenuation of all fixed windows but at the expense of a relatively wide transition band of $6/NT$, which is 3 times the width of the rectangular and 1.5 times the width of the Hamming window.

Window type	Transition width	Stopband attenuation
rectangular	2/NT	21 dB
triangular	4/NT	25 dB
Hamming	4/NT	53 dB
Hanning	4/NT	44 dB
Blackman	6/NT	74 dB
Kaiser	variable	variable
Chebyshev	variable	variable

Table 7-1
Characteristics of the window functions for the FIR filter design

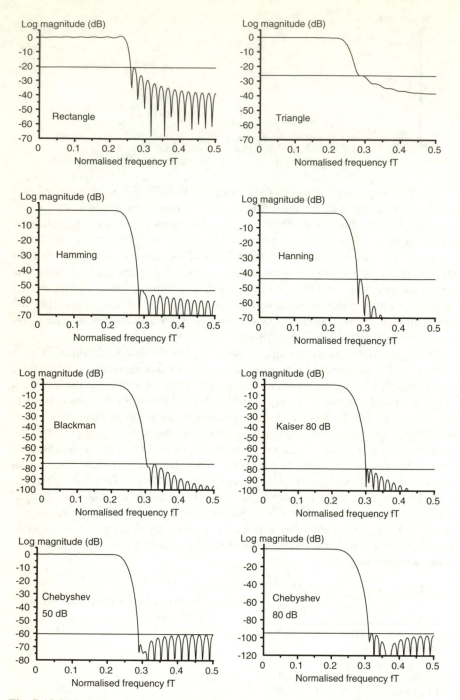

Fig. 7-15 Magnitude responses of 50th-order FIR low-pass filters, designed using the windows introduced in this section

The Kaiser window allows the explicit specification of the stopband attenuation. The width of the transition band Δ between passband and stopband can be adjusted through the filter order N. According to [33], the necessary filter order for a Kaiser window design can be approximately calculated as

$$N \approx \frac{ATT - 8}{14.36\,\Delta}\ . \tag{7.28}$$

The Chebyshev window enables the design of filters with relatively even ripple. The double width of the main lobe ΔF in Fig. 7-14 corresponds to about the width Δ of the transition band between passband and stopband of the realised filter. The ripple of the window is in general smaller than the ripple of the resulting transfer function so that we cannot exactly predict the achievable stopband attenuation. Relation (7.27) allows an approximate calculation of the required filter order. Choosing ΔF as half the width of the transition band and δ_p as the desired ripple δ of the filter is more likely to yield a too-pessimistic estimate of the filter order.

7.5 Chebyshev Approximation

Inspection of Table 7-1 shows that, of all the windows with a transition width of $4/NT$, the Hamming window exhibits the highest stopband attenuation and thus the best performance for a given filter order N. A special property of the Hamming window is, as mentioned earlier, the almost even ripple in the stopband. Both properties – high stopband attenuation and even ripple – are related to each other. The stopband attenuation is defined as the level of the lobe with the largest amplitude in the stopband with the gain in passband normalised to 0 dB (refer to Fig. 7-15). All other lobes with lower amplitude waste filter resources, in principle, since they lead in wide areas to a higher attenuation than required to realise the stopband attenuation as defined above in the whole stopband. The optimum utilisation of filter resources can apparently be achieved if all ripples in the stopband have equal amplitude. The same considerations also apply to the ripple in the passband. The achievement of magnitude responses with the named property is a Chebyshev approximation problem.

7.5.1 The Remez Exchange Algorithm

The starting point for an equiripple FIR filter design is the requirement that, by an appropriate choice of the coefficients $B(n)$, a frequency response $H(\omega)$ is achieved that minimises the maximum of the deviation $E(\omega)$ from a desired response $H_D(\omega)$. The target filter characteristic is specified in the form of a zero-phase frequency response $H_{0D}(\omega)$, which avoids any further consideration of the phase. In order to avoid any filter resources being wasted for optimisations in the transition band, we only include passband and stopband in the approximation process, as we have done previously in the context of the least-squares error approach. The frequency range $\omega = 0 \ldots \pi/T$, excluding the transition band, is denoted as $\mathcal{A}$ in the following. The error measure, the so-called Chebyshev error E_C, that has to be minimised in the approximation process can thus be mathematically expressed as

$$E_C = \max_{\omega \in \mathcal{A}} \left| H_0(\omega) - H_{0D}(\omega) \right| = \max_{\omega \in \mathcal{A}} \left| E(\omega) \right| . \tag{7.29}$$

For case 1 filters, which we consider for the time being, we obtain by substitution of (7.6a)

$$E_C[B(0), B(1), \ldots, B(Nt)] = \max_{\omega \in \mathcal{A}} \left| \sum_{n=0}^{Nt} B(n) \cos n\omega T - H_{0D}(\omega) \right| . \tag{7.30}$$

The minimisation of E_C cannot be performed analytically, since the right-hand side of (7.30) is not a differentiable function. The determination of the optimum coefficients by setting the derivative of E_C with respect to the coefficients $B(n)$ to zero is therefore not possible. In order to solve (7.30), a well-known property of the Chebyshev approximation problem may be used. The alternation theorem defines properties of the maxima and minima of the error function $E(\omega)$ [50]:

Alternation theorem: If $H_0(\omega)$ is a linear combination of r cosine functions of the form

$$H_0(\omega) = \sum_{n=0}^{r-1} B(n) \cos n\omega T , \tag{7.31}$$

then a necessary and sufficient condition that $H_0(\omega)$ be the unique, best Chebyshev approximation to a continuous function $H_{0D}(\omega)$ on $\mathcal{A}$, is that the error function $E(\omega)$ exhibit at least $r + 1$ extremal frequencies ω_i in $\mathcal{A}$ with the following properties: All extreme values have the same magnitude. The signs of the extreme values alternate:

$$E(\omega_i) = -E(\omega_{i+1}) \qquad\qquad \text{with } i = 1 \ldots r .$$

The magnitude of the extreme values $\left| E(\omega_i) \right|$ is denoted as δ in the following considerations.

For case 1 filters we have $r - 1 = Nt$. The number of alternating maxima and minima amounts to at least $N/2 + 2$ or $Nt + 2$. With this knowledge we can establish a set of $Nt + 2$ equations.

$$E(\omega_i) = H_0(\omega_i) - H_{0D}(\omega_i) = -(-1)^i \delta \qquad \text{or}$$

$$H_0(\omega_i) = H_{0D}(\omega_i) - (-1)^i \delta \tag{7.32}$$

$$i = 0 \ldots Nt + 1$$

Relation (7.32) expresses that $Nt + 2$ frequencies exist in $\mathcal{A}$ where the deviation between desired and realised frequency responses assumes the maximum value δ. The signs of the extrema alternate. Substitution of (7.6a) in (7.32) yields a set of equations that can be solved with respect to the desired filter coefficients.

$$\sum_{n=0}^{Nt} B(n) \cos n\omega_i T = H_{0D}(\omega_i) - (-1)^i \delta \qquad i = 0 \dots Nt + 1 \qquad (7.33)$$

By means of these $Nt + 2$ equations, we can determine the $Nt + 1$ unknown filter coefficients $B(n)$ as well as the maximum deviation δ. This only works, of course, if the extremal frequencies ω_i are known. For want of these values we start with a first rough estimation of the location of the extremal frequencies. A possible approach is to distribute the $Nt + 2$ frequencies regularly over the frequency interval $\mathcal{A}$. Figure 7-16 illustrates that the transition band between the edge frequencies ω_p and ω_s remains unconsidered for the allocation of the frequencies ω_i.

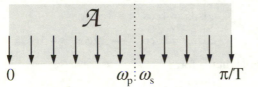
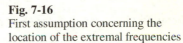

Fig. 7-16
First assumption concerning the location of the extremal frequencies

With these initial values as the first assumption, we solve the set of equations (7.33) with respect to the filter coefficients $B(n)$. These are used to calculate the zero-phase frequency response $H_0(\omega)$ of the filter and the error function $E(\omega) = H_0(\omega) - H_{0D}(\omega)$ with high resolution, which concludes the first iteration step. We have found a first approximation of the desired frequency response.

For the next iteration step, we search for the frequencies at which maxima and minima occur in the error curve $E(\omega)$. In the event that more than $Nt + 2$ extrema are found in $\mathcal{A}$, we choose those with the largest peaks. The extremal frequencies determined in this way are used to solve the set of equations (7.33). With the obtained coefficient $B(n)$ we again calculate the zero-phase frequency response $H_0(\omega)$ of the filter and the error function $E(\omega)$ with high resolution, which concludes the second iteration step. The next step starts again with the search of extremal frequencies in $E(\omega)$. The described iteration procedure to find the position of the extremal frequencies in a Chebyshev approximation problem is the basic Remez exchange algorithm.

The iteration loop can be terminated if the extremal frequencies do not significantly change compared to the results of the previous iteration step. A good termination criterion is also provided by the quantity δ, which is a result of each iteration step. δ increases with each step and converges to the actual maximum deviation of the frequency response from the desired characteristic. If the relative change of δ compared to the previous step is smaller than a given limit or if it even decreases, due to calculation inaccuracies, then the iteration is stopped.

The computational effort for the described algorithm is relatively high, since we have to solve a set of equations in each iteration step. The modification of the algorithm that we describe in the following section makes use of the fact that the knowledge of the filter coefficients is not necessarily required during the iteration process. What we need is the frequency response and derived from that the error function $E(\omega)$ in high resolution in order to determine the extremal frequencies for the next iteration step. These curves can be calculated in a different way, as we will see in the next section.

7.5.2 Polynomial Interpolation

The simplification of the algorithm is based on the possibility of expressing the zero-phase frequency response according to (7.31) by a polynomial. We can show this by means of trigonometric identities which express the cosine of integer multiples of the argument by powers of the cosine. Examples are:

$$\cos 2x = 2\cos^2 x - 1 \qquad\qquad \cos 3x = 4\cos^3 x - 3\cos x$$

$$\cos 4x = 8\cos^4 x - 8\cos^2 x + 1 \qquad\qquad \cos 5x = 16\cos^5 x - 20\cos^3 x + 5\cos x$$

Relation (7.31) can therefore be written in the form

$$H_0(\omega) = \sum_{n=0}^{Nt} C(n)\big(\cos\omega T\big)^n \;,$$

where the actual relationship between the $C(n)$ and $B(n)$ is not of direct interest. If we finally substitute $x = \cos\omega T$, we have the desired form, a polynomial.

$$H_0(\omega) = \sum_{n=0}^{Nt} C(n)\, x^n \qquad\qquad x = \cos\omega T \qquad\qquad (7.34)$$

The polynomial $H_0(\omega)$ of degree Nt can on the one hand be characterised by the $Nt + 1$ coefficients $C(0) \dots C(Nt)$ as in (7.34), and also by $Nt + 1$ distinct points through which the graph of $H_0(\omega)$ passes. The Lagrange interpolation formula, here in the barycentric form [25, 50], allows a complete reconstruction of the polynomial and hence of the frequency response from these $Nt + 1$ points.

$$H_0(\omega) = \frac{\displaystyle\sum_{k=0}^{Nt}\left[\frac{\beta_k}{x - x_k}\right] H_0(\omega_k)}{\displaystyle\sum_{k=0}^{Nt}\left[\frac{\beta_k}{x - x_k}\right]} \qquad\qquad \text{with } x = \cos\omega T \text{ and } x_k = \cos\omega_k T$$

The $H_0(\omega_k)$ are the $Nt + 1$ points which are interpolated by the Lagrange polynomial. The coefficients β_k are calculated as

$$\beta_k = \prod_{\substack{i=0 \\ i \neq k}}^{Nt} \frac{1}{x_k - x_i} \; .$$

The interpolation can be done with arbitrary resolution. As interpolation points we choose the alternating extremal values as expressed in (7.32), which are derived from the alternation theorem. $Nt + 1$ of the $Nt + 2$ extremal values are used to characterise the polynomial of Ntth degree.

$$H_0(\omega) = \frac{\displaystyle\sum_{k=0}^{Nt} \left[\frac{\beta_k}{x - x_k} \right] \left[H_{0D}(\omega_k) - (-1)^k \delta \right]}{\displaystyle\sum_{k=0}^{Nt} \left[\frac{\beta_k}{x - x_k} \right]} \qquad \text{with } x = \cos \omega T \qquad (7.35)$$
$$\text{and } x_k = \cos \omega_k T$$

The remaining extremal value is used to determine the ripple δ. The requirement that the polynomial (7.35) also passes through this last point yields a relation with δ as the only unknown.

$$H_{0D}(\omega_{Nt+1}) - (-1)^{Nt+1} \delta = \frac{\displaystyle\sum_{k=0}^{Nt} \left[\frac{\beta_k}{x_{Nt+1} - x_k} \right] \left[H_{0D}(\omega_k) - (-1)^k \delta \right]}{\displaystyle\sum_{k=0}^{Nt} \left[\frac{\beta_k}{x_{Nt+1} - x_k} \right]}$$
$$\text{with } x_k = \cos \omega_k T$$

Solving for δ results in

$$\delta = \frac{a_0 H_{0D}(\omega_0) + a_1 H_{0D}(\omega_1) + \; ..... \; + a_{Nt+1} H_{0D}(\omega_{Nt+1})}{a_0 - a_1 + a_2 - \; ..... \; + (-1)^{Nt+1} a_{Nt+1}} \; . \qquad (7.36)$$

The coefficients a_k in (7.36) are calculated as

$$a_k = \prod_{\substack{i=0 \\ i \neq k}}^{Nt+1} \frac{1}{x_k - x_i} \; .$$

Instead of solving the set of equations (7.33) in each iteration step, as described in the previous section, the polynomial interpolation approach reduces the mathematical effort to the determination of the ripple δ using (7.36) and the calculation of the frequency response with high resolution using (7.35).

If the iteration is terminated, we still have to determine the filter coefficients, which we avoided doing in the iteration process. We evaluate $H_0(\omega)$ at $N + 1 = 2Nt + 1$ equidistant frequencies in the interval $\omega = 0 \ldots 2\pi/T$. Inverse discrete Fourier transform (IDFT), according to (7.12a), yields the filter coefficients of the desired filter (refer to Sect. 7.3). Subroutine CoeffIDFT in the Appendix may be utilised for this purpose.

Example

The following example will illustrate how, by application of the Remez exchange algorithm, the maximum deviation is reduced step by step. At the same time, the ripple becomes more and more even.

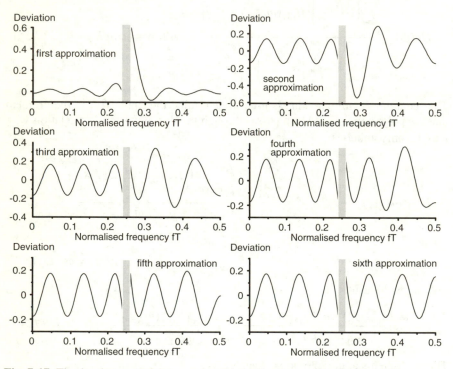

Fig. 7-17 The development of the error function in the course of the iteration for a 22nd-order low-pass filter

We consider a 22nd-order low-pass filter ($N = 22$, $Nt = 11$) with normalised edge frequencies at $f_p T = 0.24$ and $f_s T = 0.26$. According to the alternation theorem, the error curve $E(\omega)$ has at least 13 extremal frequencies with deviations of equal magnitude and alternating sign. As starting values for the iteration, we choose 13 frequencies that are equally distributed over the frequency range $fT = 0 \ldots 0.24$ and $0.26 \ldots 0.5$. The transition band $fT = 0.24 \ldots 0.26$ is ignored. The resulting frequencies can be found in the first column (step 1) of Table 7-2. With these 13 values, and using (7.36), we calculate a first approximation of the maximum error δ, which amounts to 0.01643 in this case. Fig. 7-18 shows the first approximation of the magnitude response which has been calculated using (7.35). The corresponding error curve can be found in Fig. 7-17. This curve has exactly 13 extremal frequencies, which are entered into the second column of Table 7-2. With this set of frequencies we start the next iteration. Figure 7-17 and Fig. 7-18 illustrate how the frequency response and the error curve continuously approach the ideal case of the Chebyshev behaviour with an even ripple in the passband and stopband.

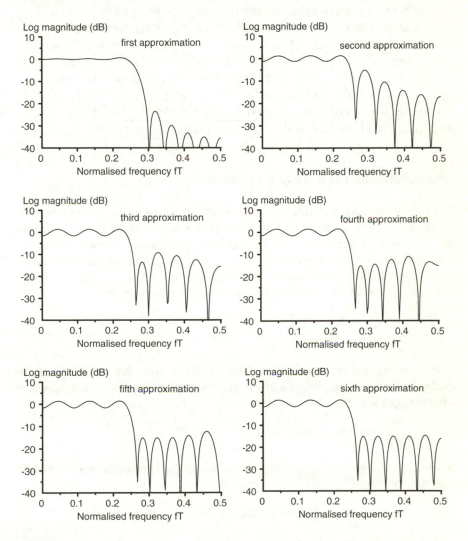

Fig. 7-18 The development of the magnitude response in the course of the iteration for a 22nd-order low-pass filter

Table 7-2 The extremal frequencies in the course of the iteration

	Step 1	Step 2	Step 3	Step 4	Step 5	Step 6	Step 7	Step 8
f_1T	0.00000	0.00000	0.00000	0.00000	0.00000	0.00000	0.00000	0.00000
f_2T	0.04000	0.04427	0.04427	0.04427	0.04427	0.04427	0.04427	0.04427
f_3T	0.08000	0.08854	0.09115	0.09115	0.08854	0.08854	0.08854	0.08854
f_4T	0.12000	0.13021	0.13542	0.13542	0.13281	0.13281	0.13281	0.13281
f_5T	0.16000	0.17708	0.17969	0.17708	0.17708	0.17708	0.17708	0.17708
f_6T	0.20000	0.22135	0.21875	0.21875	0.21875	0.21875	0.21875	0.21875
f_7T	0.24000	0.24000	0.24000	0.24000	0.24000	0.24000	0.24000	0.24000
f_8T	0.30000	0.26000	0.26000	0.26000	0.26000	0.26000	0.26000	0.26000
f_9T	0.34000	0.31729	0.29125	0.28344	0.28344	0.28344	0.28344	0.28344
$f_{10}T$	0.38000	0.36156	0.34594	0.32771	0.32250	0.32250	0.32250	0.32250
$f_{11}T$	0.42000	0.40844	0.39542	0.37979	0.36677	0.36677	0.36677	0.36677
$f_{12}T$	0.46000	0.45271	0.44750	0.43188	0.41625	0.41104	0.41104	0.41104
$f_{13}T$	0.50000	0.50000	0.50000	0.50000	0.47354	0.45792	0.45531	0.45531
δ	0.01643	0.13797	0.16549	0.17363	0.17448	0.17453	0.17453	0.17455

7.5.3 Introduction of a Weighting Function

The Remez exchange algorithm, as described in the previous section, leads to an equal ripple in the passband and stopband. Hence it follows that passband ripple and stopband attenuation cannot be chosen independently of each other. We encountered this problem already in Sect. 7.4.1 in the context of the least-squares error design. In the present example we obtain with $\delta = 0.17455$ (refer to column 8 in Table 7-2) for passband ripple and stopband attenuation:

$$D_p = 20 \log (1 + \delta) = 1.40 \text{ dB}$$
$$D_s = -20 \log \delta = 15.16 \text{ dB} .$$

By introduction of a weighting function $W(\omega)$ into the definition of the Chebyshev error (7.29), we can increase the stopband attenuation at the expense of a larger ripple in the passband and vice versa.

$$E_C = \max_{\omega \in A} \left| W(\omega) \left(H_0(\omega) - H_{0D}(\omega) \right) \right| = \max_{\omega \in A} \left| W(\omega) E(\omega) \right|$$

In this case, the product $W(\omega) E(\omega)$ has to satisfy the alternation theorem, which results in the following modification of (7.32):

$$W(\omega_i)E(\omega_i) = W(\omega_i)\left[H_0(\omega_i) - H_{0D}(\omega_i)\right] = -(-1)^i \delta$$

$$H_0(\omega_i) = H_{0D}(\omega_i) - (-1)^i \delta / W(\omega_i)$$

$$i = 0 \ ..... \ Nt + 1 \ .$$

(7.37)

Relation (7.37) leads to the following relations between the ripple in the passband δ_p and in the stopband δ_s:

$$\left.\begin{array}{l} \delta_p = \delta/W_p \\[1.5em] \delta_s = \delta/W_s \end{array}\right\} \quad \Rightarrow \quad \delta_p/\delta_s = W_s/W_p \; .$$

W_p and W_s are the values of the weighting function in the passband and stopband respectively. The relations (7.35) and (7.36), which are important for carrying out the Remez algorithm in practice, assume the following form after introduction of the weighting function:

$$H_0(\omega) = \frac{\displaystyle\sum_{k=0}^{Nt}\left[\frac{\beta_k}{x - x_k}\right]\left[H_{0D}(\omega_k) - (-1)^k \delta / W(\omega_k)\right]}{\displaystyle\sum_{k=0}^{Nt}\left[\frac{\beta_k}{x - x_k}\right]} \quad \begin{array}{l} \text{with } x = \cos\omega T \\[0.5em] \text{and } x_k = \cos\omega_k T \; . \end{array} \tag{7.38}$$

The coefficients β_k are calculated as

$$\beta_k = \prod_{\substack{i=0 \\ i \neq k}}^{Nt} \frac{1}{x_k - x_i} \; .$$

$$\delta = \frac{a_0 H_{0D}(\omega_0) + a_1 H_{0D}(\omega_1) + \; ..... \; + a_{Nt+1} H_{0D}(\omega_{Nt+1})}{a_0 / W(\omega_0) - a_1 / W(\omega_1) + \; ..... \; + (-1)^{Nt+1} a_{Nt+1} / W(\omega_{Nt+1})} \tag{7.39}$$

The coefficients a_k in (7.39) are calculated as

$$a_k = \prod_{\substack{i=0 \\ i \neq k}}^{Nt+1} \frac{1}{x_k - x_i} \; .$$

Example

We consider again the 22nd-order low-pass filter ($N = 22$, $Nt = 11$) with normalised edge frequencies at $f_p T = 0.24$ and $f_s T = 0.26$. The weighting factor in the passband is chosen as $W_p = 1$, and that in the stopband as $W_s = 10$. Compared to the example in Sect. 7.5.2, we can expect a larger ripple in the passband and a reduced ripple in the stopband. Figure 7-19 shows the frequency response and the error curve which we obtain using the relations (7.38) and (7.39). The ripple is 10 times larger in the passband than in the stopband, since $W_s / W_p = \delta_p / \delta_s = 10$.

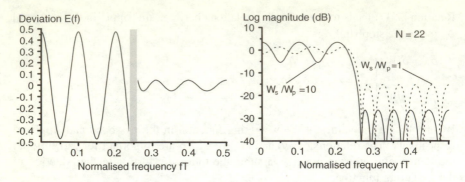

Fig. 7-19 Magnitude response and error curve of a 22nd-order low-pass filter with a
weighting ratio $W_s/W_p = 10$

The magnitude curve is plotted against the one with equal weighting in both
bands.

The mathematical implementation of the Remez exchange algorithm, which we
have introduced in this section, forms the core of the optimum equiripple design as
utilised in many commercial filter design programs. Rabiner, McClellan and Parks
published a very popular FORTRAN version in [43, 44].

The number of values that have to be calculated for the high resolution
representation of the frequency response $H_0(\omega)$ and of the error curve $E(\omega)$
depends on the filter order. A number of $16{\times}Nt$ in $\mathcal{A}$ is a good compromise
between achievable design accuracy and computational effort.

The edge frequencies of the filter, which determine the approximation region $\mathcal{A}$,
the filter order N and the gain and the weighting factors in the passband and
stopband, are the natural design parameters of the Remez algorithm. The quantities
δ_p and δ_s, which determine the ripple in the passband and the attenuation in the
stopband, are a result of the Remez iteration. Since filter specifications are in general
given in the form of tolerance schemes, it would be desirable to have a relation
available that allows the calculation of the required filter order in terms of the ripple
and the edge frequencies. Unfortunately there exists no analytical relation between
these quantities. Shen and Strang [55] derived the following semi-empirical formula
(7.40), which provides the most precise results that can be found in literature for the
unweighted case ($\delta = \delta_p = \delta_s$).

$$N \approx \frac{20 \log_{10}\left(\pi\delta\right)^{-1} - 10 \log_{10} \log_{10} \delta^{-1}}{4.343 \ln \cot\left(\pi\dfrac{1-2\Delta}{4}\right)} \tag{7.40}$$

For the weighted case ($\delta_p \neq \delta_s$), we utilise an approach by Kaiser [33] in which δ
in (7.40) is replaced with the geometric mean of δ_p and δ_s. The results are
acceptable for a wide range of Δ, δ_p and δ_s. Narrow transition widths Δ and at the
same time small ratios of δ_s to δ_p lead to estimates of N that are too pessimistic.
With increasing N, the results become more and more accurate. For a better

prediction, especially of lower filter orders in the weighted case, we can make use of the following relation [27].

$$N \approx \frac{D_\infty(\delta_p, \delta_s)}{\Delta F} - f(\delta_p, \delta_s)\Delta F \qquad (7.41)$$

$$D_\infty(\delta_p, \delta_s) = \left[a_1 \left(\log_{10}\delta_p\right)^2 + a_2 \log_{10}\delta_p + a_3 \right] \log_{10}\delta_s$$

$$+ \left[a_4 \left(\log_{10}\delta_p\right)^2 + a_5 \log_{10}\delta_p + a_6 \right]$$

with
$$\begin{aligned} a_1 &= 5.309 \cdot 10^{-3} & a_4 &= -2.66 \cdot 10^{-3} \\ a_2 &= 7.114 \cdot 10^{-2} & a_5 &= -5.941 \cdot 10^{-1} \\ a_3 &= -4.761 \cdot 10^{-1} & a_6 &= -4.278 \cdot 10^{-1} \end{aligned}$$

$$f(\delta_p, \delta_s) = b_1 + b_2 \log_{10}\left(\delta_p / \delta_s\right)$$

with
$$\begin{aligned} b_1 &= 11.01217 \\ b_2 &= 0.51244 \end{aligned}$$

7.5.4 Case 2, 3 and 4 Filters

The equiripple design algorithm that we have derived in this chapter is based on case 1 filters, which have a zero-phase frequency response of the form

$$H_0(\omega) = \sum_{n=0}^{Nt} B(n) \cos n\omega T \ .$$

For the practical implementation of filter design software, it would be desirable if this algorithmic kernel could also be used for case 2, case 3 and case 4 filters, which cover the variants of odd filter orders and antisymmetric unit-sample responses. This is made possible by the fact that the frequency responses for these symmetry cases can be expressed as the product of a function $Q(\omega)$ and a linear combination of cosine functions $P(\omega)$.

Case 2

$$H_0(\omega) = \sum_{n=1}^{Nt} B(n) \cos\left(n - 1/2\right)\omega T = \cos\omega T/2 \sum_{n=0}^{Nt_{mod}} C(n) \cos n\omega T \qquad (7.42a)$$

$$\text{with } Nt_{mod} = Nt - 1 = (N-1)/2$$

Case 3

$$H_0(\omega) = \sum_{n=1}^{Nt} \sin n\omega T = \sin \omega T \sum_{n=0}^{Nt_{\mathrm{mod}}} C(n) \cos n\omega T \tag{7.42b}$$

$$\text{with } Nt_{\mathrm{mod}} = Nt - 1 = N/2 - 1$$

Case 4

$$H_0(\omega) = \sum_{n=1}^{Nt} B(n) \sin \left(n - 1/2\right) \omega T = \sin \omega T/2 \sum_{n=0}^{Nt_{\mathrm{mod}}} C(n) \cos n\omega T \tag{7.42c}$$

$$\text{with } Nt_{\mathrm{mod}} = Nt - 1 = (N-1)/2$$

These relations can be generalised to the form

$$H_0(\omega) = Q(\omega) P(\omega) \ , \tag{7.43}$$

with

$$Q(\omega) = \cos \omega T/2 \qquad \text{for case 2}$$
$$Q(\omega) = \sin \omega T \qquad \text{for case 3}$$
$$Q(\omega) = \sin \omega T/2 \qquad \text{for case 4} \ .$$

Substitution of (7.43) in (7.29) results in the following expression for the Chebyshev error E_C:

$$E_C = \max_{\omega \in \mathcal{A}} \left| W(\omega) \left(Q(\omega) P(\omega) - H_{0D}(\omega)\right) \right|$$

$$E_C = \max_{\omega \in \mathcal{A}} \left| W(\omega) Q(\omega) \left(P(\omega) - H_{0D}(\omega)/Q(\omega)\right) \right| \ .$$

With the substitution

$$W_{\mathrm{mod}}(\omega) = W(\omega) Q(\omega) \qquad \text{and}$$
$$H_{0D\,\mathrm{mod}}(\omega) = H_{0D}(\omega)/Q(\omega) \ , \tag{7.44}$$

we finally arrive at a form of the Chebyshev error that can be treated completely by analogy with the case 1 design problem.

$$E_C = \max_{\omega \in \mathcal{A}} \left| W_{\mathrm{mod}}(\omega) \left(P(\omega) - H_{0D\,\mathrm{mod}}(\omega)\right) \right|$$

With regard to the quotient in (7.44), we merely have to make sure that the frequencies at which $Q(\omega)$ vanishes are excluded from the approximation interval $\mathcal{A}$.

As a result of the Remez algorithm, we obtain the frequency response $P(\omega)$, which, multiplied by $Q(\omega)$, yields the frequency response $H_0(\omega)$ of the desired filter. In order to determine the filter coefficients, we evaluate $H_0(\omega)$ at $N + 1$

equidistant frequencies in the interval $\omega = 0 \ldots 2\pi/T$ and apply to this set of values one of the variants of the IDFT (7.12a–d), which are tailored to the respective symmetry cases.

7.5.5 Minimum-Phase FIR Filters

All the filter design methods that we have introduced in this chapter up to now are related to linear-phase filters. Apart from the simple filter structure and the guaranteed stability, the possibility of realising strictly linear-phase filters, featuring excellent performance with respect to signal distortion, is one of the outstanding properties of FIR filters. By means of analog or discrete-time IIR filters, this behaviour can only be approximated by means of complex implementations. The design algorithms are simplified by the fact that we only have to deal with real-valued approximation problems. The constant group delay of $NT/2$ corresponding to half the register length of the FIR filter might be a disadvantage in some applications. If delay is a problem, minimum-phase filters, which exhibit in general a lower but frequency-dependent delay, become attractive. Minimum-phase FIR filters require fewer coefficients than FIR filters to obtain the same degree of approximation, however, a saving with respect to implementation complexity is in general not achievable, since the coefficients of linear-phase filters, on the other hand, occur in pairs, so that the number of multiplications can be halved.

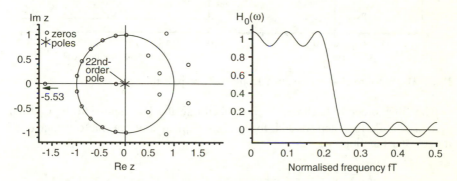

Fig. 7-20   Pole-zero plot and zero-phase frequency response of a 22nd-order low-pass filter

Figure 7-20 shows the pole-zero plot and the zero-phase frequency response of a 22nd-order linear-phase low-pass filter. This filter possesses zeros on the unit circle in the stopband and zero pairs with reciprocal distance from the origin in the passband. In Sect. 5.4.2 we explained that minimum-phase filters only have zeros on or within the unit circle. The fact that zeros with reciprocal distance from the origin have identical magnitude responses leads to the idea of decomposing the linear-phase frequency response into two partial frequency responses, each having a magnitude corresponding to the square root of the magnitude of the linear-phase filter. One comprises the zeros within, the other the zeros outside the unit circle.

The zeros on the unit circle and the poles in the origin are equally distributed over both partial filters. This would only work perfectly, however, if the zeros on the unit circle occurred in pairs, which is unfortunately not the case according to Fig. 7-20. A trick [28] helps to create the desired situation. If we add a small positive constant to the zero-phase frequency response shown in Fig. 7-20, the whole curve is shifted slightly upwards. The zeros on the unit circle move closer together in pairs. If the added constant equals exactly the stopband ripple δ_s, we obtain the desired zero pairs on the unit circle (Fig. 7-21). The starting point of the following mathematical considerations is the general form of the transfer function of a FIR filter.

$$H_1(z) = \sum_{n=0}^{N} b_n \, z^{-n} = z^{-N} \sum_{n=0}^{N} b_n \, z^{N-n}$$

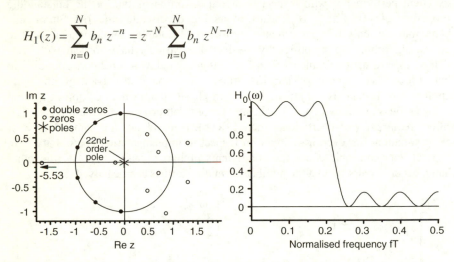

Fig. 7-21 Pole-zero plot of a modified 22nd-order low-pass filter

The filter order N is chosen even to facilitate the later split into two partial systems. In the first step we add, for reasons explained above, the constant δ_s to the transfer function.

$$H_2(z) = z^{-N} \sum_{n=0}^{N} b_n \, z^{N-n} + \delta_s$$

The polynomial in z can be expressed as the product of the zero terms. In addition we introduce a correction term $1/(1+\delta_s)$ that normalises the gain to unity again.

$$H_2(z) = \frac{b_0}{1+\delta_s} \frac{(z - z_{01})(z - z_{02}) \dots (z - z_{0N})}{z^N}$$

If z_0 is a zero within or on the unit circle, the corresponding zero with reciprocal radius can be expressed as $z_0' = 1/z_0{}^*$. The linear-phase transfer function $H_2(z)$ can therefore be grouped into two products.

$$H_2(z) = \frac{b_0}{(1+\delta_s)z^N} \prod_{i=1}^{N/2}(z - z_{0i}) \prod_{i=1}^{N/2}(z - 1/z_{0i}^*)$$

Equivalent rearrangement yields

$$H_2(z) = K\, z^{-N/2} \prod_{i=1}^{N/2}(z - z_{0i}) \prod_{i=1}^{N/2}\left(z^{-1} - z_{0i}^*\right) , \tag{7.45}$$

with

$$K = \frac{b_0\,(-1)^{N/2}}{(1+\delta_s)\displaystyle\prod_{i=1}^{N/2} z_{0i}} .$$

K is, in any case, real-valued, since the zeros in the product term are real or occur in complex-conjugate pairs. By the substitution of $z = e^{j\omega T}$ in (7.45), we obtain the corresponding frequency response.

$$H_2\left(e^{j\omega T}\right) = K\, e^{-j\omega NT/2} \prod_{i=1}^{N/2}\left(e^{j\omega T} - z_{0i}\right) \prod_{i=1}^{N/2}\left(e^{-j\omega T} - z_{0i}^*\right) \tag{7.46}$$

Both products in (7.46) have identical magnitude since the product terms are complex-conjugate with respect to each other. If we realise a new transfer function $H_3(z)$ from the left product in (7.45), we obtain a minimum-phase filter of order $N/2$ and a magnitude response according to the square root of the magnitude response of the modified linear-phase filter $H_2(z)$.

$$H_3(z) = \sqrt{K}\, z^{-N/2} \prod_{i=1}^{N/2}(z - z_{0i}) \tag{7.47}$$

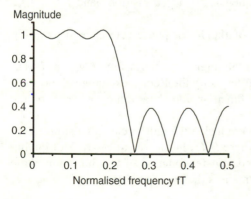

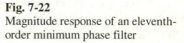

Fig. 7-22
Magnitude response of an eleventh-order minimum phase filter

Figure 7-22 depicts the magnitude response of the eleventh-order minimum-phase filter that we derived from the linear-phase filter according to Fig. 7-20. The ripple in the passband is smaller, and the one in the stopband is larger, compared to the original linear-phase filter with double filter order. The exact relationships between these ripple values can be expressed as

$$\delta_{p3} = \sqrt{1 + \frac{\delta_{p1}}{1 + \delta_{s1}}} - 1 \qquad (7.48a)$$

$$\delta_{s3} = \sqrt{\frac{2\,\delta_{s1}}{1 + \delta_{s1}}} \ . \qquad (7.48b)$$

Multiplying out the product in (7.47) yields the transfer function in form of a polynomial in z and thus the coefficients of the resulting eleventh-order FIR filter. These coefficients are equivalent to the unit-sample response of the filter, which is graphically represented in Fig. 7-23.

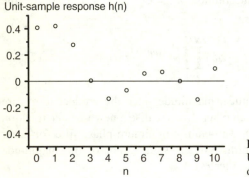

Unit-sample response h(n)

Fig. 7-23
Unit-sample response of the eleventh-order minimum-phase FIR filter

For the practical design of a minimum phase filter, we can summarise the following design steps:
- Specify the target filter in terms of the ripple in the passband δ_{p3}, the ripple in the stopband δ_{s3} and the edge frequencies of the transition band $f_p T$ and $f_s T$ with $\Delta = (f_s - f_p)T$.
- Calculate the ripples δ_{p1} and δ_{s1} of the linear-phase reference filter using (7.48).
- Estimate the required filter order of the reference filter using (7.40) or (7.41).
- Design the linear-phase reference filter using the Remez exchange algorithm.
- Add δ_{s1} to the obtained transfer function and determine the zeros of the polynomial.
- Choose those zeros of $H_2(z)$ which are inside the unit circle and a simple zero at those points on the unit circle where $H_2(z)$ has a pair of zeros. Multiplying out the zero terms yields the desired transfer function of the minimum-phase filter. The additional factor $\sqrt{K}$ normalises the gain to unity.

7.6 Maximally Flat (MAXFLAT) Design

According to (7.6a), we have the following zero-phase frequency response of a linear-phase filter with even filter order N:

$$H_0(\omega) = \sum_{n=0}^{Nt} B(n) \cos n\omega T \qquad \text{with } Nt = N/2 .$$

(7.48)

We can determine a set of coefficients $B(n)$ in closed form that leads to a monotonic decay of the magnitude in the frequency range $\omega = 0 \ldots \pi/T$ similar to the behaviour of a Butterworth low-pass filter. For the determination of the $Nt + 1$ coefficients $B(0) \ldots B(Nt)$, we need an equal number of equations. Some of these equations, say a number L, is used to optimise the frequency response in the vicinity of $\omega = 0$; the remaining equations, say a number K, dictates the behaviour around $\omega = \pi/T$. The design parameters L, K and Nt are related by

$$K + L = Nt + 1 .$$

(7.49)

One of the L equations is used to normalise the gain to unity at $\omega = 0$. The remaining $L - 1$ equations enforce $L - 1$ derivatives of $H_0(\omega)$ to vanish at $\omega = 0$.

$$\left| H_0(\omega) \right| \Big|_{\omega = 0} = 1$$

(7.50a)

$$\frac{d^\nu H_0(\omega)}{d\omega^\nu} \Big|_{\omega = 0} = 0 \qquad \nu = 1 \ldots L - 1$$

(7.50b)

K equations are used to enforce $H_0(\omega)$ and $K - 1$ derivatives of $H_0(\omega)$ to vanish at $\omega = \pi/T$.

$$\left| H_0(\omega) \right| \Big|_{\omega = \pi/T} = 0$$

(7.50c)

$$\frac{d^\mu H_0(\omega)}{d\omega^\mu} \Big|_{\omega = \pi/T} = 0 \qquad \mu = 1 \ldots K - 1$$

(7.50d)

For the time being, we solve the problem on the basis of a polynomial of degree Nt in the variable x.

$$P_{Nt,K}(x) = \sum_{\nu=0}^{Nt} a_\nu x^\nu$$

The coefficients a_ν of the polynomial are determined in such a way that $P_{Nt,K}(0) = 1$ and $P_{Nt,K}(1) = 0$. $L - 1$ derivatives of $P_{Nt,K}(x)$ at $x = 0$ and $K - 1$

derivatives at $x = 1$ are forced to vanish. The closed-form solution of this problem can be found in [26]:

$$P_{Nt,K}(x) = \left(1 - x\right)^K \sum_{v=0}^{Nt-K} \binom{K + v - 1}{v} x^v \ . \tag{7.51}$$

The term in parenthesis is a binomial coefficient [9], which is defined as

$$\binom{n}{m} = \frac{n!}{m!\left(n - m\right)!} \ . $$

Relation (7.51) can thus be written in the form

$$P_{Nt,K}(x) = \left(1 - x\right)^K \sum_{v=0}^{Nt-K} \frac{(K - 1 + v)!}{(K - 1)! \, v!} x^v \ . \tag{7.52}$$

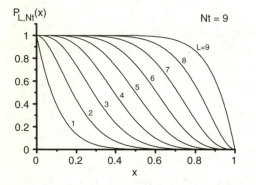

Fig. 7-24
Graphs of the polynomial family $P_{Nt,L}(x)$ for $Nt = 9$ and $L = 1 \dots 9$

Figure 7-24 shows plots of this polynomial family for $Nt = 9$ in the interval $x = 0 \dots 1$. For $Nt = 9$, equivalent to a filter order of $N = 18$, there are exactly 9 different curves and hence 9 possible cutoff frequencies. In order to obtain the frequency response of a realisable FIR filter, a transformation of the variable x to the variable ω has to be found that meets the following conditions:

- The transformation must be a trigonometric function.
- The range $x = 0 \dots 1$ has to be mapped to the frequency range $\omega = 0 \dots \pi/T$.
- The relation between x and ω must be a monotonic function in the interval $x = 0 \dots 1$ in order to maintain the monotonic characteristic of the magnitude response.

The relation

$$x = \left(1 - \cos \omega T\right)/2 \tag{7.53}$$

meets these conditions. Substitution of (7.53) in (7.52) yields the following representation:

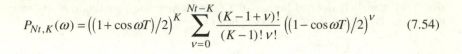

$$P_{Nt,K}(\omega) = \left((1 + \cos\omega T)/2\right)^K \sum_{v=0}^{Nt-K} \frac{(K-1+v)!}{(K-1)!\,v!} \left((1 - \cos\omega T)/2\right)^v \qquad (7.54)$$

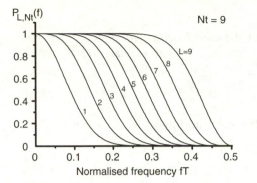

Fig. 7-25
Graphs of the polynomial family
$P_{Nt,L}(\cos\omega T)$ for $Nt = 9$ and $L = 1 \ldots 9$

Figure 7-25 shows the graphs of (7.54) again for the case $Nt = 9$. The monotonic curves are preserved by application of transformation (7.53); only the cutoff frequencies are slightly shifted.

Multiplying out (7.54) and rearrangement of the terms with respect to powers of $\cos\omega T$ leads to a polynomial of the form

$$P_{Nt,K}(\omega) = \sum_{v=0}^{Nt} c_v (\cos\omega T)^v \ .$$

Application of trigonometric identities finally yields a representation of the zero-phase transfer function according to (7.48).

$$H_0(\omega) = P_{Nt,K}(\omega) = \sum_{v=0}^{Nt} b_v \cos v\omega T$$

It is possible to derive a relation that allows the calculation of the coefficients b_v in closed form [32], but the resulting formula is rather unwieldy. A more elegant way to determine the filter coefficients [34] is to evaluate the frequency response $H_0(\omega_i) = P_{Nt,K}(\omega_i)$ at $N+1$ equidistant frequencies,

$$\omega_i = i\frac{2\pi}{(N+1)\,T} \qquad i = 0 \ldots N \ ,$$

using (7.54), and to calculate the coefficients by inverse discrete Fourier transform (7.12a) on this obtained set of values.

The actual filter design process consists of providing a set of parameters L, K and Nt that leads to an optimum approximation of the prescribed filter specification. Nt can be approximately calculated from the transition band of the filter [34], which is defined here as the range where the magnitude drops from 95% (passband edge) down to 5% (stopband edge).

$$Nt > \frac{1}{\left[2\left(f_{0.05} - f_{0.95}\right)T\right]^2} \tag{7.55}$$

Inspection of Fig. 7-24 shows that the cutoff points $x_{0.5}$ of the polynomial, defined as the values of x where the gain equals 0.5, are regularly distributed. $x_{0.5}$ can therefore be approximated as

$$x_{0.5} \approx L/(Nt + 1) \ . \tag{7.56}$$

Substitution of (7.53) involves the half-decay cutoff frequency $f_{0.5}$ instead of $x_{0.5}$ in this design relation.

$$L/(Nt + 1) \approx \left(1 - \cos 2\pi f_{0.5} T\right)/2 \tag{7.57a}$$

$$L = \text{int}\left[(Nt + 1)\left(1 - \cos 2\pi f_{0.5} T\right)/2\right] \tag{7.57b}$$

The practical filter design is thus performed in the following 4 steps:
• Calculation of the required filter order $N = 2 \times Nt$ from the width of the transition band (95%–5%) according to (7.55).
• Calculation of the design parameters L and $K = Nt + 1 - L$ using (7.57b). The input parameter is the half-decay cutoff frequency $f_{0.5}$. The result of this calculation has to be rounded to the nearest integer.
• Evaluation of the frequency response $H_0(\omega_i) = P_{Nt,K}(\omega_i)$ at $N + 1$ equidistant frequencies using (7.54).
• Calculation of the filter coefficients by inverse DFT on the $N + 1$ samples of the frequency response using (7.12a).

The presented design method is not very accurate. The calculations of the parameter Nt (7.55) and relation (7.56) are only approximations. The rounding of L in (7.57b) is a further reason for inaccuracies. The latter is especially critical in the case of low filter orders since, with Nt given, only Nt different cutoff frequencies $f_{0.5}$ can be realised. An improvement of the rounding problem can be achieved by allowing Nt to vary over a certain range, say from a value calculated using (7.55) to twice this value [34]. We finally choose that Nt which leads to the best match between the fraction $L/(Nt + 1)$ and the right-hand side of equation (7.57a).

The problem that (7.55) and (7.56) are only approximations is not completely solved by this procedure. The design results will still not be very accurate. A further drawback of the design procedure described above is the fact that the filter cannot be specified in terms of a tolerance scheme but only by the unusual parameters $f_{0.5}$ and a transition band defined by the edge frequencies $f_{0.95}$ and $f_{0.05}$.

More flexibility and, in general, more effective results with respect to the required filter order are obtained by the method described in [52], which is essentially based on a search algorithm. Specified are passband edge (v_p, ω_p) and stopband edge (v_s, ω_s) of a tolerance scheme. Starting with a value of 2, Nt is incremented step by step. At each step, the algorithm checks whether there exists an L with $L = 1 \ldots Nt$ and hence a frequency response $H_0(\omega)$ according to (7.54) that complies with the tolerance scheme. If a combination of L and Nt is found that meets the specification, the algorithm is stopped. Because of certain properties of the polynomial $P_{Nt,L}$, the search algorithm can be implemented very efficiently, since for each Nt only three values of L have to be checked.

Two Pascal subroutines are included in the Appendix that support the design of MAXFLAT filters. The function **MaxFlatPoly** is a very efficient implementation of the polynomial calculation according to (7.52). The procedure **MaxFlatDesign** determines the parameters Nt and hence the required filter order $N = 2 \times Nt$, as well as the parameter L, according to the algorithm introduced in [52] from a given tolerance scheme.

Example

A maximally flat FIR filter be specified as follows:

Passband edge: $f_p T = 0.2$	Max. passband attenuation:	1 dB
Stopband edge: $f_s T = 0.3$	Min. stopband attenuation:	40 dB

With this tolerance scheme, the search algorithm yields $Nt = 32$ and $L = 15$ as result. The attenuation amounts to 0.933 dB at the passband edge and to 40.285 dB at the stopband edge. The magnitude response fully complies with the prescribed tolerance scheme (Fig. 7-26).

By inverse DFT on $N + 1$ samples of the frequency response we obtain the following set of filter coefficients:

$b(0)$	=	$b(64)$	=	$1.38 \cdot 10^{-11}$		$b(17)$	=	$b(47)$	= $-8.12 \cdot 10^{-05}$
$b(1)$	=	$b(63)$	=	$8.04 \cdot 10^{-11}$		$b(18)$	=	$b(46)$	= $1.33 \cdot 10^{-03}$
$b(2)$	=	$b(62)$	=	$-2.47 \cdot 10^{-10}$		$b(19)$	=	$b(45)$	= $5.99 \cdot 10^{-04}$
$b(3)$	=	$b(61)$	=	$-2.43 \cdot 10^{-09}$		$b(20)$	=	$b(44)$	= $-2.83 \cdot 10^{-03}$
$b(4)$	=	$b(60)$	=	$-2.00 \cdot 10^{-10}$		$b(21)$	=	$b(43)$	= $-2.74 \cdot 10^{-03}$
$b(5)$	=	$b(59)$	=	$3.72 \cdot 10^{-08}$		$b(22)$	=	$b(42)$	= $5.17 \cdot 10^{-03}$
$b(6)$	=	$b(58)$	=	$3.42 \cdot 10^{-08}$		$b(23)$	=	$b(41)$	= $6.94 \cdot 10^{-03}$
$b(7)$	=	$b(57)$	=	$-2.72 \cdot 10^{-07}$		$b(24)$	=	$b(40)$	= $-1.16 \cdot 10^{-02}$
$b(8)$	=	$b(56)$	=	$-4.91 \cdot 10^{-07}$		$b(25)$	=	$b(39)$	= $-1.11 \cdot 10^{-02}$
$b(9)$	=	$b(55)$	=	$1.37 \cdot 10^{-06}$		$b(26)$	=	$b(38)$	= $1.23 \cdot 10^{-02}$
$b(10)$	=	$b(54)$	=	$3.29 \cdot 10^{-06}$		$b(27)$	=	$b(37)$	= $3.93 \cdot 10^{-02}$
$b(11)$	=	$b(53)$	=	$-5.83 \cdot 10^{-06}$		$b(28)$	=	$b(36)$	= $-2.61 \cdot 10^{-02}$
$b(12)$	=	$b(52)$	=	$-2.94 \cdot 10^{-05}$		$b(29)$	=	$b(35)$	= $-8.02 \cdot 10^{-02}$
$b(13)$	=	$b(51)$	=	$1.95 \cdot 10^{-05}$		$b(30)$	=	$b(34)$	= $2.55 \cdot 10^{-02}$
$b(14)$	=	$b(50)$	=	$9.29 \cdot 10^{-05}$		$b(31)$	=	$b(33)$	= $3.22 \cdot 10^{-01}$
$b(15)$	=	$b(49)$	=	$-1.61 \cdot 10^{-05}$		$b(32)$	=		$4.06 \cdot 10^{-01}$
$b(16)$	=	$b(48)$	=	$-3.83 \cdot 10^{-04}$					

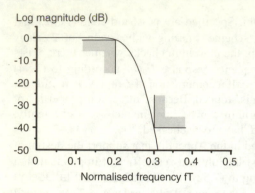
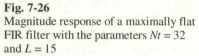

Log magnitude (dB)

Normalised frequency fT

Fig. 7-26
Magnitude response of a maximally flat
FIR filter with the parameters $Nt = 32$
and $L = 15$

Figure 7-27 shows a graphical representation of the obtained filter coefficients and hence of the unit-sample response of the filter. It is worthy of note that the values of the coefficients span an enormous dynamic range, which is a property typical of MAXFLAT filters. Additionally, the filter order is much larger than for a standard equiripple design with identical filter specification. However, maximally flat FIR filters have the advantage of having a monotonous frequency response, which is desirable in certain applications. An attenuation exceeding 100 dB almost everywhere in the stopband can be easily achieved.

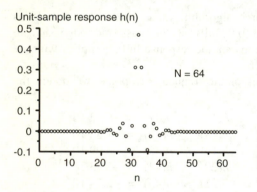
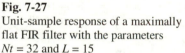

Unit-sample response h(n)

N = 64

n

Fig. 7-27
Unit-sample response of a maximally
flat FIR filter with the parameters
$Nt = 32$ and $L = 15$

The large dynamic range of the coefficients that would be needed in our example is certainly not realisable in fixed-point representation with usual wordlengths of 8, 16 or 24 bit. If we map the calculated filter coefficients to the named fixed-point formats, many coefficients will become zero, which reduces the filter order. While the filter has 65 coefficients in floating-point representation, the filter length is reduced to 51 coefficients with 24-bit, 41 coefficients with 16-bit and 21 coefficients with 8-bit coefficient wordlength. Figure 7-28 shows the magnitude responses for these coefficient resolutions.

If the magnitude response obtained with the limited coefficient wordlength is not satisfactory, other filter structures may be used that are less sensitive to coefficient inaccuracies. One possibility is the cascade structure, where the whole filter is decomposed into first- and second-order sections. For the design, one has to determine the zeros of the transfer function and combine complex-conjugate zero pairs to second-order sections with real coefficients. A further alternative is the structure presented in [56], which is especially tailored to the realisation of maximally flat filters. This structure features very low sensitivity with respect to coefficient deviations and has the additional advantage of requiring fewer multiplications than the direct form.

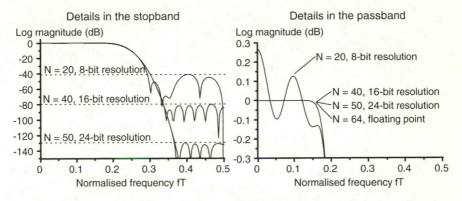

Fig. 7-28 Magnitude response of the example filter with different coefficient wordlengths

8 Effects of Finite Register Lengths

8.1 Classification of the Effects

All practical implementations of digital signal processing algorithms have to cope with finite register lengths. These lead to impairments of the filter performance that become visible in different ways.

The filter coefficients, which can be determined by means of the presented design algorithms with arbitrary precision, can only be approximated by numbers with a given limited wordlength. This implies that the characteristics of the resulting filter deviate more or less from the original specification.

Further effects such as the generation of noise and unstable behaviour, are caused by the truncation of bits which becomes necessary after multiplication. If we multiply, for instance, a signal sample with a resolution of a bits by a coefficient with a wordlength of b bits, we obtain a result with $a+b$ bits. In recursive filters without that truncation, the number of digits would increase by b after each clock cycle. Under certain circumstances, the effects caused by the truncation of digits can be treated mathematically like an additive noise signal. Into the same category falls, by the way, the noise generated in analog-to-digital converters by the quantisation of the analog signal. Significant differences can be observed between floating-point and fixed-point arithmetic with respect to the order of magnitude of the additive noise and the dependencies of these effects on the signal to be processed.

The occurrence of so-called limit cycles is another effect that can be observed in the context of recursive filters realised with fixed-point arithmetic. A limit cycle is a constant or periodic output signal which does not vanish even if a constant zero is applied to the input of the filter. The filter thus shows a certain unstable behaviour.

The size of this chapter, about 85 pages, points to the large complexity of this topic. Nevertheless, it is not possible to present all aspects of finite wordlengths in detail in this textbook. The explanations should be sufficient enough, however, to avoid the grossest design mistakes and to finally arrive at filters which exhibit the desired performance.

8.2 Number Representation

Fixed-point arithmetic offers a bounded range of numbers which requires careful scaling of the DSP algorithms in order to avoid overflow and achieve an optimum signal-to-noise ratio. The implementation of floating-point arithmetic is more complex but avoids some of the drawbacks of the fixed-point format. In the following sections we introduce in detail the properties of both number formats.

8.2.1 Fixed-Point Numbers

As in the case of decimal numbers, we can distinguish between integer and fractional numbers. Each position of a digit in a number has a certain weight. In case of binary numbers, these weights are powers of 2. On the left of the binary point we find the coefficients of the powers with positive exponent, including the zero; on the right are placed the coefficients of the powers with negative exponent, which represent the fractional part of the fixed point number.

$$Z = a_4\, a_3\, a_2\, a_1\, a_0 \,.\, a_{-1}\, a_{-2}\, a_{-3}$$

The a_i can assume the values 0 or 1. The value of this number is

$$Z = a_4\, 2^4 + a_3\, 2^3 + a_2\, 2^2 + a_1\, 2^1 + a_0\, 2^0 + a_{-1}\, 2^{-1} + a_{-2}\, 2^{-2} + a_{-3}\, 2^{-3}$$

Let us consider the binary number 0001101.01010000 as an example. One byte (8 bits) is used to represent the integer part and again one byte for the representation of the fractional part of the fixed-point number. The corresponding decimal number is 13.3125.

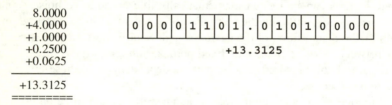

```
    8.0000
   +4.0000
   +1.0000
   +0.2500
   +0.0625
  ─────────
  +13.3125
  =========
```

+13.3125

The most significant bit of binary numbers is commonly used as a sign bit if signed numbers have to be represented. A "0" denotes positive, a "1" negative numbers. Three different representations of negative numbers are commonly used; of these, the one's complement is only of minor importance in practice.

Sign-magnitude

The binary number following the sign bit corresponds to the magnitude of the number.

Example

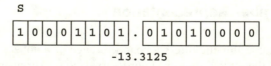

-13.3125

One's complement

All bits of the positive number are inverted to obtain the corresponding negative number.

Example

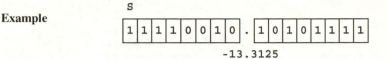

-13.3125

Two's complement

All bits of the positive number are inverted again. After that, a 1 is added to the least significant bit (LSB). A carry that may be generated by an overflow in the most significant bit (MSB) will be ignored.

Example

s
1 1 1 1 0 0 1 0 . 1 0 1 1 0 0 0 0

-13.3125

Table 8-1 Signed one-byte number representations

Decimal	-128	-127	-1	-0	+0	+1	+127
Sign-magnitude		11111111	10000001	10000000	00000000	00000001	01111111
One's complement		10000000	11111110	11111111	00000000	00000001	01111111
Two's complement	10000000	10000001	11111111	00000000		00000001	01111111

Table 8-1 shows the results of these rules for various exposed values of signed one-byte numbers. Both the sign-magnitude and the one's complement format have the special property that there are two representations of the zero (+0, −0). The two's complement arithmetic has only one representation of the zero, but the value ranges for positive and negative numbers are asymmetrical. Each of these representations has its pros and cons. Their choice depends on the hardware or software implementation of the arithmetical operations. Multiplications are preferably performed in sign-magnitude arithmetic. Additions and subtractions are more easily realised in two's complement form. Positive and negative numbers can be added in arbitrary sequence without the need for any pre-processing of the numbers. An advantageous property of the complement representations becomes effective if more than two numbers have to be added in a node of the flowgraph. If it is guaranteed that the total sum will remain within the amplitude limits of the respective number representation, intermediate results may lead to overflows without causing any error in the total sum. After the last addition we obtain in any case the correct result. Possible overflows occurring in these additions are ignored. The described behaviour will be illustrated by means of a simple example using an 8 bit two's complement representation, which has an amplitude range from −128 to +127.

Example

```
        100                    01100100
    +    73                +   01001001
    ─────────              ──────────────
        173                    10101101    (-83)

        173                    10101101
    +   -89                +   10100111
    ─────────              ──────────────
         84                 (1)01010100
    ═════════              ══════════════
```

After the first addition, the permissible number range is exceeded. A negative result is pretended. By addition of the third, a negative number, we return back into the valid number range and obtain a correct result. The occurring overflow is ignored. Care has to be taken in practical implementations that such "false" intermediate results never appear at the input of a multiplier since this leads to errors that cannot be corrected.

A significant difference exists between the signed number formats with respect to the overflow behaviour. This topic is discussed in detail in Sect. 8.7.1. Special cases of fixed-point numbers are integers and fractions.

Programming languages like C and Pascal offer various types of signed and unsigned integers with different wordlengths. In practice, number representations with 1, 2, 4 and 8 bytes are used. Table 8-2 shows the valid ranges of common integer types together with their names in C and Pascal terminology. Negative numbers are in all cases realised in two's complement arithmetic.

Table 8-2 The range of various integer types

Number representation	C	Pascal	Range
1 byte signed	char	ShortInt	−128 ... 127
1 byte unsigned	unsigned char	Byte	0 ... 255
2 bytes signed	int	Integer	−32768 ... 32767
2 bytes unsigned	unsigned int	Word	0 ... 65535
4 bytes signed	long int	LongInt	−2147483648 ... 2147483647
4 bytes unsigned	unsigned long		0 ... 4294967295
8 bytes signed		Comp	$-9.2\,10^{18}$ ... $9.2\,10^{18}$

In digital signal processing, it is advantageous to perform all calculations with pure fractions. The normalisation of signal sequences and coefficients to values less than unity has the advantage that the multiplication of signals by coefficients always results in fractions again, which makes overflows impossible. The return to the original wordlength which is necessary after multiplications can be easily performed by truncation or rounding of the least significant bits. This reduces the accuracy of the number representation, but the coarse value of the number remains unchanged.

Example

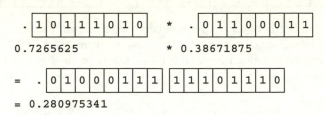

0.7265625   * 0.38671875

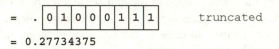

= 0.280975341

Truncation to the original wordlength of 8 bits yields

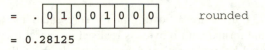

= 0.27734375

The error that we make by the truncation of the least significant bits can be reduced on an average by rounding. For rounding we consider the MSB in the block of bits that we want to remove. If it is 0, we simply truncate as before. If it is 1, we truncate and add a 1 to the LSB of the remaining number.

<table>
<tr><td>=</td><td>.</td><td>0</td><td>1</td><td>0</td><td>0</td><td>1</td><td>0</td><td>0</td><td>0</td><td>rounded</td></tr>
</table>

= 0.28125

8.2.2 Floating-Point Numbers

If the available range of integers is not sufficient in a given application, which is often the case in technical and scientific problems, we can switch to the floating point representation. A binary floating-point number is of the general form

$$F = Mantissa \times 2^{Exponent}$$

An important standard, which defines the coding of mantissa and exponent and details of the arithmetic operations with floating-point numbers, has been developed by the group P754 of the IEEE (Institute of Electrical and Electronical Engineers). The widely used floating-point support processors of Intel (80287, 80387, from the 80486 processor on directly integrated in the CPU) and Motorola (68881, 68882), as well as most of the floating-point signal processors, obey this standard.

According to IEEE P754, the exponent is chosen in such a way that the mantissa (M) is always in the range

$$2 > M \geq 1.$$

This normalisation implies that the digit on the left of the binary point is always set to 1. Applying this rule to the above example of the decimal number 13.3125 yields the following bit pattern for a representation with 2 bytes for the mantissa and one byte for the exponent:

1	.	1	0	1	0	1	0	1	0	0	0	0	0	0	0	0	*2^	0	0	0	0	0	0	1	1

We still need to clarify how negative numbers (negative mantissa) and numbers with magnitude less than unity (negative exponent) are represented.

For the exponent, a representation with offset has been chosen. In case of an one-byte exponent we obtain

$$Z = M \times 2^{E-127}$$

The mantissa is represented according to IEEE P754 in the sign-magnitude format.

Since we guarantee by the normalisation of the number that the bit on the left of the binary point is always set to 1, this digit needs not be represented explicitly. We thus gain a bit that can be used to increase the accuracy of the mantissa. The normalisation rule makes it impossible, however, to represent the number zero. A special agreement defines that $Z = 0$ is represented by setting all bits of the mantissa (M) and of the exponent (E) to zero. The exponent $E = 255$ (all bits of the exponent are set) plays another special role. This value identifies infinite and undefined numbers, which may result from divisions by zero or taking the square root of negative numbers.

According to IEEE P754, there are 3 floating-point representations, with 4, 8 and 10 bytes. An additional format with 6 bytes is realised in Pascal by software. This format is not supported by mathematical coprocessors. In the following, we give details of how the bits are distributed over sign bit, mantissa and exponent. Table 8-3 shows the valid ranges of these floating-point types together with their names in C and Pascal terminology.

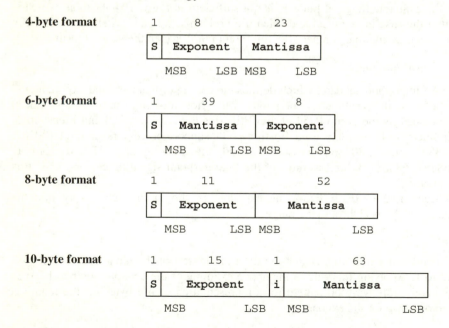

The coprocessor's internal calculations are based on the 10-byte format, which is the only one to explicitly show the bit (i) on the left of the binary point. All floating-point numbers are converted to this format if transferred to the coprocessor.

Table 8-3 The ranges of various floating-point types

Number representation	C	Pascal	Range
4 bytes, 8-bit exponent	float	Single	$1.5 \ 10^{-45} \dots 3.4 \ 10^{+38}$
6 bytes, 8-bit exponent		Real	$2.9 \ 10^{-39} \dots 1.7 \ 10^{+38}$
8 bytes, 11-bit exponent	double	Double	$5.0 \ 10^{-324} \dots 1.7 \ 10^{+308}$
10 bytes, 15-bit exponent	long double	Extended	$3.4 \ 10^{-4932} \dots 1.1 \ 10^{+4932}$

8.3 Quantisation

The multiplication of numbers yields results with a number of bits corresponding to the sum of the bits of the two factors. Also in the case of the addition of floating-point numbers, the number of bits increases depending on the difference between the exponents. In DSP algorithms, there is always a need to truncate or round off digits to return to the original wordlength. The errors introduced by this quantisation manifest themselves in many ways, as we will see later.

8.3.1 Fixed-Point Numbers

If $x_q(n)$ is a signal sample that results from $x(n)$ by quantisation, the quantisation error can be expressed as

$$e(n) = x_q(n) - x(n) \ . \tag{8.1}$$

If $b1$ is the number of bits before truncation or rounding and b the reduced number of bits, then

$$N = 2^{b1-b}$$

steps of the finer resolution before quantisation are reduced to one step of the new coarser resolution. The error $e(n)$ thus can assume N different values. The range of the error depends on the one hand on the chosen number representation (sign-magnitude or two's complement), on the other hand on the type of the quantiser (truncation or rounding).

In the case of truncation, the corresponding number of least significant bits is simply dropped.

Example

Figure 8-1a shows how the grid of possible values of the previous finer resolution is related to the coarser grid after quantisation. The details depend on the method of representation of negative numbers.

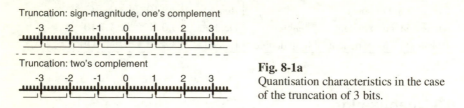

Fig. 8-1a

Quantisation characteristics in the case of the truncation of 3 bits.

Rounding is performed by substituting the nearest number that can be represented by the reduced number of bits. If the old value lies exactly between two possible rounded values, the number is always rounded up without regard of sign or number representation. This rule simplifies the implementation of the rounding algorithm, since the MSB of the dropped block of bits has to be taken into account only. In the following examples, the resolution is reduced by two bits.

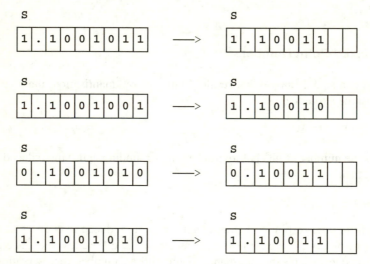

The relation between fine and coarse grid of values is depicted in Fig. 8-1b, where the resolution is again reduced by three bits. We omit the consideration of the one's complement representation, since rounding leads to large errors in the range of negative numbers.

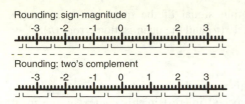

Rounding: sign-magnitude

Rounding: two's complement

Fig. 8-1b
Quantisation characteristics in the case
of rounding of 3 bits.

Mathematical rounding is similar. Rounding is also performed by substituting the nearest number that can be represented by the reduced number of bits. If the old value lies exactly between two possible rounded values, however, the number is always rounded in such a way that the LSB of the rounded value becomes zero. This procedure has the advantage that in this borderline case rounding up and rounding down is performed with equal incidence.

Example

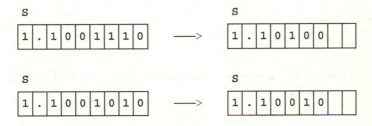

Figure 8-1c shows the quantisation characteristic, which is identical for sign-magnitude and two's complement representation.

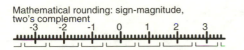

Mathematical rounding: sign-magnitude, two's complement

Fig. 8-1c
Quantisation characteristic in the case of mathematical rounding

There are further rounding methods with the same advantage, where rounding up and rounding down is performed randomly in the borderline case.

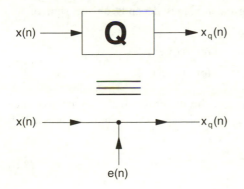

Fig. 8-2
Equivalent representation of a quantiser

By rearrangement of (8.1), the output signal of the quantiser $x_q(n)$ can be interpreted as the sum of the input signal $x(n)$ and the quantisation error $e(n)$.

$$x_q(n) = x(n) + e(n)$$

According to this relation, a quantiser can be represented by an equivalent flowgraph, as shown in Fig. 8-2. Of special interest in this context is the mean value of the error $e(n)$, which has the meaning of an offset at the output of the quantiser, as well as the mean-square error, which corresponds to the power of the superimposed error signal. Assuming a concrete sequence of samples, the error signal $e(n)$ can be determined exactly. If the signal sequence is sufficiently complex, as in the case of speech or music, we can characterise the error signal by statistical parameters such as the mean value or the variance. In these cases, $e(n)$ is a noiselike signal, for which all N possible error values occur with equal probability. If the latter is fulfilled at least approximately, the mean error can be calculated as

$$\overline{e(n)} = \frac{1}{N}\sum_{k=1}^{N} e_k \ .$$
(8.2)

The e_k represent the N possible error values. The mean-square error can be expressed accordingly.

$$\overline{e^2(n)} = \frac{1}{N}\sum_{k=1}^{N} e_k^2$$
(8.3)

The quantity

$$\sigma_e^2 = \overline{e^2(n)} - \overline{e(n)}^2$$
(8.4)

is commonly referred to as the variance of the error, which has the meaning of the mean power of the superimposed noise. As an example, we calculate offset and noise power for the case of sign-magnitude truncation. Inspection of Fig. 8-1a shows that the range of the occurring error values depends on the sign of the signal $x(n)$.

$$e_k = k\frac{\Delta x_q}{N} \qquad \begin{array}{ll} 0 \leq k \leq N-1 & \text{for } x(n) < 0 \\ -N+1 \leq k \leq 0 & \text{for } x(n) > 0 \ . \end{array}$$
(8.5)

Δx_q is the quantisation step at the output of the quantiser. If the signal sequence $x(n)$ is normalised to the range $-1 \leq x \leq +1$, then Δx_q can be expressed in dependence of the wordlength b as

$$\Delta x_q = 2 \times 2^{-b} \ .$$
(8.6)

We substitute (8.5) in (8.2) to calculate the mean error. For negative signal values we obtain

$$\overline{e(n)} = \frac{1}{N} \sum_{k=0}^{N-1} k \frac{\Delta x_q}{N}$$

$$\overline{e(n)} = \frac{\Delta x_q}{N^2} \sum_{k=0}^{N-1} k = \frac{\Delta x_q}{N^2} \frac{N(N-1)}{2}$$

$$\overline{e(n)} = \frac{\Delta x_q}{2} \left(1 - \frac{1}{N}\right) \qquad \text{for } x(n) < 0 .$$

Positive signal values accordingly yield

$$\overline{e(n)} = -\frac{\Delta x_q}{2} \left(1 - \frac{1}{N}\right) \qquad \text{for } x(n) > 0 .$$

For the calculation of the mean-square error, we substitute (8.5) in (8.3). In this case, the result is independent of the sign of the signal $x(n)$.

$$\overline{e^2(n)} = \frac{1}{N} \sum_{k=0}^{N-1} k^2 \frac{\Delta x_q^2}{N^2}$$

$$\overline{e^2(n)} = \frac{\Delta x_q^2}{N^3} \sum_{k=0}^{N-1} k^2 = \frac{\Delta x_q^2}{N^3} \left(\frac{N^3}{3} - \frac{N^2}{2} + \frac{N}{6}\right)$$

$$\overline{e^2(n)} = \frac{\Delta x_q^2}{3} \left(1 - \frac{3}{2N} + \frac{1}{2N^2}\right)$$

The variance, equivalent to the mean power of the superimposed noise, is calculated according to (8.4) as

$$\sigma_e^2 = \overline{e^2(n)} - \overline{e(n)}^2 = \frac{\Delta x_q^2}{12} \left(1 - \frac{1}{N^2}\right) .$$

The mean and mean-square values of the error for the other number representations and quantiser types are summarised in Table 8-4. A comparison of the various results shows that the terms to calculate the noise power are identical in all but one case. For large N, the variances of the error approach one common limiting value in all considered cases:

$$\sigma_e^2 = \frac{\Delta x_q^2}{12} \tag{8.7}$$

For a reduction in the wordlength of 3 or 4 bits, (8.7) is already an excellent approximation. A result of this reasoning is that the noise power is not primarily dependent on the number of truncated bits but rather on the resulting accuracy of the signal representation after truncation or rounding off. A special quantiser in this context is the A/D converter which realises the transition from the practically infinite resolution of the analog signal to a resolution according to the number of bits of the chosen number representation. Thus the conversion from analog to digital generates a noise power according to (8.7). Substitution of (8.6) in (8.7) yields the noise power in terms of the wordlength.

$$\sigma_e^2 = \frac{4 \times 2^{-2b}}{12} = \frac{2^{-2b}}{3} \tag{8.8}$$

Table 8-4 Mean values and variances for various quantiser types

Type of quantiser	$\overline{e(n)}$	$\overline{e^2(n)}$	σ_e^2
Truncation: Sign-magnitude One's complement	$\pm\dfrac{\Delta x_q}{2}\left(1-\dfrac{1}{N}\right)$	$\dfrac{\Delta x_q^2}{3}\left(1-\dfrac{3}{2N}+\dfrac{1}{2N^2}\right)$	$\dfrac{\Delta x_q^2}{12}\left(1-\dfrac{1}{N^2}\right)$
Truncation: Two's complement	$-\dfrac{\Delta x_q}{2}\left(1-\dfrac{1}{N}\right)$	$\dfrac{\Delta x_q^2}{3}\left(1-\dfrac{3}{2N}+\dfrac{1}{2N^2}\right)$	$\dfrac{\Delta x_q^2}{12}\left(1-\dfrac{1}{N^2}\right)$
Rounding: Sign-magnitude	$\pm\dfrac{\Delta x_q}{2}\dfrac{1}{N}$	$\dfrac{\Delta x_q^2}{12}\left(1+\dfrac{2}{N^2}\right)$	$\dfrac{\Delta x_q^2}{12}\left(1-\dfrac{1}{N^2}\right)$
Rounding: Two's complement	$+\dfrac{\Delta x_q}{2}\dfrac{1}{N}$	$\dfrac{\Delta x_q^2}{12}\left(1+\dfrac{2}{N^2}\right)$	$\dfrac{\Delta x_q^2}{12}\left(1-\dfrac{1}{N^2}\right)$
Mathematical rounding: Sign-magnitude Two's complement	0	$\dfrac{\Delta x_q^2}{12}\left(1+\dfrac{2}{N^2}\right)$	$\dfrac{\Delta x_q^2}{12}\left(1+\dfrac{2}{N^2}\right)$

8.3.2 Floating-Point Numbers

Floating point numbers behave in a different way. We consider a floating-point number x of the form

$$x = M \times 2^E ,$$

where the mantissa M assumes values between 1 and 2. After reduction of the resolution of the mantissa by truncation or rounding we obtain the quantised value

$$x_q = M_q \times 2^E .$$

The error with respect to the original value amounts to

$$e = x_q - x$$

$$e = \left(M_q - M\right) \times 2^E \; . \tag{8.9}$$

Because of the exponential term in (8.9), the quantisation error, caused by rounding or truncation of the mantissa, is proportional to the magnitude of the floating-point number. It makes therefore sense to consider the relative error

$$e_{rel} = \frac{e}{x} = \frac{M_q - M}{M}$$

$$e_{rel} = \frac{e_M}{M} \; .$$

The calculation of the mean and the mean-square error is more complex in this case because the relative error is the quotient of two random numbers, the error of the mantissa and the value of the mantissa itself. The mean-square error is calculated as

$$\overline{e_{rel}^2} = \sum_{k=1}^{N} \sum_{j=1}^{2^b} \frac{e_{Mk}^2}{M_j^2} p(j,k) \; . \tag{8.10}$$

In this double summation, the error of the mantissa e_M progresses with the parameter k through all possible values which can occur by rounding or truncation. The mantissa M progresses with the parameter j through all possible values that it can assume in the range between 1 and 2 with the given wordlength. $p(j,k)$ is a so-called joint probability which defines the probability that an error value e_{Mk} coincides with a certain value of the mantissa M_j. Assuming that the error and mantissa values are not correlated with each other and all possible values e_{Mk} and M_j occur with equal likelihood, (8.10) can be essentially simplified. These conditions are satisfied by sufficient complex signals like voice and music. Under these assumptions, the joint probability $p(j,k)$ can be approximated by a constant. The value corresponds to the reciprocal of the number of all possible combinations of the e_{Mk} and M_j.

$$p(j,k) = \frac{1}{N \, 2^b}$$

N is the number of possible different error values e_{Mk}, 2^b is the number of mantissa values M_j that can be represented by b bits.

$$\overline{e_{rel}^2} = \frac{1}{N \, 2^b} \sum_{k=1}^{N} \sum_{j=1}^{2^b} \frac{e_{Mk}^2}{M_j^2}$$

Since the joint probability is a constant, the two summations can be decoupled and calculated independently.

$$\overline{e_{\text{rel}}^2} = \frac{1}{N} \sum_{k=1}^{N} e_{Mk}^2 \; \frac{1}{2^b} \sum_{j=1}^{2^b} \frac{1}{M_j^2} \tag{8.11}$$

With increasing length of the mantissa b, the second sum in (8.11) approaches the value $1/2$. $b = 8$ for instance yields a value of 0.503. We can therefore express the relative mean-square error in a good approximation as

$$\overline{e_{\text{rel}}^2} = \frac{1}{2N} \sum_{k=1}^{N} e_{Mk}^2 \; .$$

The mean value of the relative error is also of interest and can be calculated as

$$\overline{e_{\text{rel}}} = \sum_{k=1}^{N} \sum_{j=1}^{2^b} \frac{e_{Mk}}{M_j} \, p(j,k) \; . \tag{8.12}$$

For uniformly distributed and statistically independent error and mantissa values, the mean value becomes zero in any case as the mantissa assumes positive and negative values with equal probability. According to (8.4), the variance of the relative error and hence the relative noise power can be calculated as

$$\sigma_{\text{rel}}^2 = \overline{e_{\text{rel}}^2} - \overline{e_{\text{rel}}}^2 \; .$$

Since the mean value vanishes, we have

$$\sigma_{\text{rel}}^2 = \overline{e_{\text{rel}}^2} = \frac{1}{2N} \sum_{k=1}^{N} e_{Mk}^2 \; .$$

The mean-square error for various number representations and quantiser types has been derived in the previous section and can be directly taken from Table 8-4. The variance of the relative error of floating-point numbers is thus simply obtained by multiplying the corresponding value of the mean-square error in Table 8-4 by $1/2$. For sufficient large N, these values approach two characteristic values:

$$\sigma_{\text{rel}}^2 = \frac{1}{6} \Delta x_{\text{q}}^2 = \frac{1}{6} 2^{-2(b-1)} = \frac{2}{3} 2^{-2b} \qquad \text{for truncation} \tag{8.13a}$$

$$\sigma_{\text{rel}}^2 = \frac{1}{24} \Delta x_{\text{q}}^2 = \frac{1}{24} 2^{-2(b-1)} = \frac{1}{6} 2^{-2b} \qquad \text{for rounding} \; . \tag{8.13b}$$

Truncation generates a quantisation noise power that is four times (6 dB) higher than the noise generated by rounding.

8.4 System Noise Behaviour

Before we analyse in detail the noise generated in the various filter structures, we discuss basic concepts in the context of noise. We introduce the signal-to-noise ratio, which is an important quality parameter of signal processing systems, and consider the filtering of noise in LSI systems. Finally we touch on the problem of signal scaling, which is necessary to obtain an optimum signal-to-noise ratio in fixed-point implementations.

8.4.1 Signal-to-Noise Ratio

The signal-to-noise ratio (SNR) is calculated as the ratio of the signal power, equivalent to the variance of $x(n)$, to the noise power, equivalent to the variance of the error signal $e(n)$.

$$SNR = \frac{\sigma_x^2}{\sigma_e^2} \tag{8.14}$$

The signal-to-noise ratio is often expressed on a logarithmic scale in dB.

$$10 \log SNR = 10 \log \frac{\sigma_x^2}{\sigma_e^2} = 10 \log \sigma_x^2 - 10 \log \sigma_e^2 \tag{8.15}$$

In order to calculate the SNR, we have to relate the noise power generated in the quantiser, as derived in Sect. 8.3, to the power of the signal. In this context we observe significant differences between the behaviour of fixed-point and floating-point arithmetic.

We start with the simpler case of floating-point representation. By rearrangement of (8.14), the SNR can be expressed as the reciprocal of the relative noise power σ^2_{erel}.

$$SNR = \frac{1}{\sigma_e^2 / \sigma_x^2} = \frac{1}{\sigma_{erel}^2}$$

Substitution of (8.13a) and (8.13b) yields the following logarithmic representations of the SNR:

$$10 \log SNR = 1.76 + 6.02b \quad \text{(dB)} \qquad \text{for truncation} \tag{8.16a}$$

$$10 \log SNR = 7.78 + 6.02b \quad \text{(dB)} \qquad \text{for rounding} \tag{8.16b}$$

Comparison of (8.16a) and (8.16b) shows that we gain about 6 dB, equivalent to an increase of the wordlength by one bit, if we apply the somewhat more complex rounding rather than simple truncation.

In contrast with fixed-point arithmetic, which we will consider subsequently, the signal-to-noise ratio is only dependent on the wordlength of the mantissa and not on the level of the signal. Floating-point arithmetic thus avoids the problem that care has to be taken regarding overload and scaling of the signals in all branches of the flowgraph in order to optimise the *SNR*. It is only necessary that we do not leave the allowed range of the number format used, which is very unlikely to happen in normal applications. A one byte exponent already covers the range from 10^{-38} to 10^{+38}, equivalent to a dynamic range of about 1500 dB.

In the case of fixed-point arithmetic, the noise power amounts to $\sigma_e^2 = 2^{-2b}/3$ according to (8.8). Substitution in (8.15) yields the relation:

$$10 \log SNR = 10 \log \sigma_x^2 - 10 \log \left(2^{-2b}/3 \right)$$

$$10 \log SNR = 10 \log \sigma_x^2 + 4.77 + 6.02\, b \qquad \text{(dB) .} \qquad (8.17)$$

According to (8.17), the variance and thus the power of a signal $x(n)$ must assume the largest possible value in order to obtain the best possible signal-to-noise ratio if this signal is quantised. This is accomplished if the available number range is optimally utilised and hence the system operates as closely as possible to the amplitude limits. The maximum signal power, which can be achieved within the amplitude limits of the used number representation, strongly depends on the kind of signal. In the following, we assume a purely fractional representation of the signals with amplitude limits of ±1. For sinusoids, the rms value is $1/\sqrt{2}$ times the peak value. The maximally achievable signal power or variance thus amounts to 1/2. An equally distributed signal, which features equal probability for the occurrence of all possible signal values within the amplitude limits, has a maximum variance of 1/3. For music or voice, the variance is approximately in the range from 1/10 to 1/15. For normally distributed (Gaussian) signals, it is not possible to specify a maximum variance, as this kind of signal has no defined peak value. If overload is allowed with a certain probability p_o, however, it is still possible to state a maximum signal power. The corresponding values are summarised in Table 8-5, which is derived using the Gaussian error function.

p_o	σ_x^2
10^{-1}	0.3696
10^{-2}	0.1507
10^{-3}	0.0924
10^{-4}	0.0661
10^{-5}	0.0513
10^{-6}	0.0418
10^{-7}	0.0352
10^{-8}	0.0305
10^{-9}	0.0268

Table 8-5
Maximum variance of the normally distributed signal for various probabilities of overload

Substitution of the named maximum variances for the various signal types in (8.17) yields the following signal-to-noise ratios that can be achieved with a given wordlength of b bits.

sinusoids	: $6.02\,b + 1.76$	(dB)	(8.18a)
equal distribution	: $6.02\,b$	(dB)	(8.18b)
audio signals	: $6.02\,b + (-5 \dots -7)$	(dB)	(8.18c)
Gaussian ($p_o = 10^{-5}$)	: $6.02\,b - 8.13$	(dB)	(8.18d)
Gaussian ($p_o = 10^{-7}$)	: $6.02\,b - 9.76$	(dB)	(8.18e)

The results obtained in this section can be directly applied to a special quantiser, the analog-to-digital converter. The A/D converter realises the transition from the practically infinite resolution of the analog signal to the chosen resolution of the digital signal. So even at the beginning of the signal processing chain, a noise power according to (8.8) is superimposed on the digital signal. Equation (8.18) can be used to determine the least required number of bits from the desired signal-to-noise ratio. If audio signals are to be processed with an *SNR* of 80 dB, for instance, a resolution of about 15 bits is required according to (8.18c). Each multiplication or floating-point addition in the further course of signal processing generates additional noise, which in the end increases the required wordlength beyond the results of (8.18).

The digital filter as a whole behaves like a quantiser which can be mathematically expressed by generalisation of the equations (8.16) and (8.17) and introduction of two new constants C_{fx} and C_{fl}. If we assume a noiseless input signal, the *SNR* at the output of the filter can be calculated in the case of fixed-point arithmetic as

$$10 \log SNR_{fx} = 10 \log \sigma_x^2 + 6.02\,b + C_{fx} \quad \text{(dB)} . \tag{8.19a}$$

In the case of floating-point arithmetic, the *SNR* is independent of the signal level, which leads to the following relation:

$$10 \log SNR_{fl} = 6.02\,b + C_{fl} \quad \text{(dB)} . \tag{8.19b}$$

The constants C_{fx} and C_{fl} are determined by the realised transfer function, the order of the filter and the chosen filter structure. In case of the cascade structure, the pairing up of poles and zeros as well as the sequence of second-order partial systems is of decisive importance. In general we can say that an increase of the wordlength by one bit leads to an improvement of the signal-to-noise ratio by 6 dB.

8.4.2 Noise Filtering by LSI Systems

The noise that is superimposed on the output signal of the filter originates from various sources. Besides the errors produced in the analog-to-digital conversion process, all quantisers in the filter required after multiplications and floating-point additions contribute to the noise at the output of the filter. These named noise sources do not have a direct effect on the output of the system. For the noise that is already superimposed on the input signal, the transfer function between the input and output of the filter is decisive. For the quantisation noise generated within the filter, the transfer functions between the locations of the quantisers and the filter output have to be considered. In this context, it is of interest to first of all establish general relations between the noise power at the input and output of LSI filters, which are characterised by their transfer functions $H(e^{j\omega T})$ or unit-sample responses $h(n)$.

We characterise the input signal $x(n)$ by its power-density spectrum $P_{xx}(\omega T)$, which is the square magnitude of the Fourier transform of $x(n)$ according to (3.10a). The term $P_{xx}(\omega T)\,d\omega T$ specifies the contribution of the frequency range $\omega T \ldots (\omega + d\omega)T$ to the total power of the signal. The total power is calculated by integration of the power-density spectrum.

$$\sigma_x^2 = \int_{-\pi}^{+\pi} P_{xx}(\omega T)\,d\omega T \tag{8.20}$$

The power-density spectrum $P_{yy}(\omega T)$ at the output of the filter is obtained by the relation

$$P_{yy}(\omega T) = P_{xx}(\omega T)\left|H\left(e^{j\omega T}\right)\right|^2 . \tag{8.21}$$

Combining (8.20) and (8.21) yields the following relationship to calculate the variance and thus the power of the output signal:

$$\sigma_y^2 = \int_{-\pi}^{+\pi} P_{xx}(\omega T)\left|H\left(e^{j\omega T}\right)\right|^2\,d\omega T . \tag{8.22}$$

For the estimation of the signal power at the output of the filter, the distribution of the signal power at the input must be known. If this is not the case, we can only give an upper limit which would be reached if the signal power is concentrated in the frequency range where the filter has its maximum gain.

$$\sigma_y^2 \leq \int_{-\pi}^{+\pi} P_{xx}(\omega T)\left|H_{\max}\right|^2\,d\omega T$$

$$\sigma_y^2 \leq \sigma_x^2\left|H_{\max}\right|^2 \tag{8.23}$$

If the input signal is a noise-like signal with a constant spectrum, as approximately generated in a quantiser, we also obtain relatively simple relationships. The total power $\sigma_x^2 = N_0$ is equally distributed over the frequency range $-\pi \leq \omega T \leq +\pi$, which results in the constant power density

$$P_{xx}(\omega T) = N_0/2\pi .$$

Substitution of this equation in (8.22) yields a relationship for the noise power at the output of the system.

$$\sigma_y^2 = \int_{-\pi}^{+\pi} \frac{N_0}{2\pi} \left| H\left(e^{j\omega T}\right)\right|^2 d\omega T$$

$$\sigma_y^2 = N_0 \frac{1}{2\pi} \int_{-\pi}^{+\pi} \left| H\left(e^{j\omega T}\right)\right|^2 d\omega T = \sigma_x^2 \frac{1}{2\pi} \int_{-\pi}^{+\pi} \left| H\left(e^{j\omega T}\right)\right|^2 d\omega T \qquad (8.24)$$

The noise power at the output is calculated in this case as the product of the input power and the mean-square gain of the filter.

$$\sigma_y^2 = \sigma_x^2 \overline{\left| H\left(e^{j\omega T}\right)\right|^2}$$

According to (3.26), the mean-square gain of the system is equal to the sum over the squared unit-sample response.

$$\sum_{n=-\infty}^{+\infty} h^2(n) = \frac{1}{2\pi} \int_{-\pi}^{+\pi} \left| H\left(e^{j\omega T}\right)\right|^2 d\omega T$$

Substitution in (8.24) yields

$$\sigma_y^2 = \sigma_x^2 \sum_{n=-\infty}^{+\infty} h^2(n) . \qquad (8.25)$$

If the noise power of a quantiser $N_0 = \Delta x_q^2/12$ is superimposed on the input signal of a filter, the superimposed noise power at the output amounts to

$$\sigma_y^2 = \frac{\Delta x_q^2}{12} \sum_{n=-\infty}^{+\infty} h^2(n) = \frac{\Delta x_q^2}{12} \overline{\left| H\left(e^{j\omega T}\right)\right|^2} . \qquad (8.26)$$

8.4.3 Optimum Use of the Dynamic Range

With floating-point arithmetic, overflow of the system is usually not a problem, since the number representations used in practice provide a sufficient dynamical range. The situation is different with fixed-point arithmetic. On the one hand, it is important to avoid overload, which as a rule leads to severe deterioration of the signal quality. On the other hand, the available dynamic range has to be used in an optimum way in order to achieve the best possible signal-to-noise ratio. This does not only apply to the output but also to each node inside the filter. The transfer functions from the input of the filter to each internal node and to the output are subject to certain constraints in the time or frequency domain, that avoid the occurrence of overflow. We first derive the most stringent condition, that definitely avoids overflow. The starting point is the convolution sum (3.3). The magnitude of the output signal does not exceed unity if the following relation applies:

$$\left| y(n) \right| = \left| \sum_{m=-\infty}^{+\infty} x(n-m)\,h(m) \right| \le 1 \ .$$

In the following, we derive an upper bound for the magnitude of the above convolution term. The magnitude of a sum is in any case smaller than or equal to the sum of the magnitudes (triangle inequality):

$$\left| \sum_{m=-\infty}^{+\infty} x(n-m)\,h(m) \right| \le \sum_{m=-\infty}^{+\infty} \left| x(n-m) \right| \left| h(m) \right| \le 1 \ .$$

If we replace in the right sum the magnitude of the signal by the maximally possible value, which is unity, the above inequality certainly still holds.

$$\left| \sum_{m=-\infty}^{+\infty} x(n-m)\,h(m) \right| \le \sum_{m=-\infty}^{+\infty} \left| h(m) \right| \le 1$$

If the magnitudes of input and output signal are limited to unity, we thus obtain the following condition, which guarantees the absence of overflow.

$$\sum_{m=-\infty}^{+\infty} \left| h(m) \right| \le 1 \tag{8.27}$$

For an optimum use of the dynamic range, the equals sign has to be aimed at. If a given filter does not satisfy this condition, the input signal of the filter has to be multiplied by a scaling factor S, which equals the reciprocal of (8.27).

$$S = \frac{1}{\displaystyle\sum_{m=-\infty}^{+\infty} \left| h(m) \right|} \tag{8.28}$$

A scaling according to (8.28) is too pessimistic in practice, and leads in most cases to a bad utilisation of the available dynamic range. Other scalings are therefore applied, that on the one hand allow occasional overload, but on the other hand lead to far better signal-to-noise ratios. Relation (8.29) defines a scaling factor which, as a rule, is too optimistic.

$$S = \frac{1}{\sqrt{\displaystyle\sum_{m=-\infty}^{+\infty} h^2(m)}} \tag{8.29}$$

Relation (8.29) is a good choice, especially for broadband signals. A compromise often used in practice follows from the inequality

$$\sqrt{\sum_{m=-\infty}^{+\infty} h^2(m)} \leq \max_{\omega T} \left| H(\omega T) \right| \leq \sum_{m=-\infty}^{+\infty} \left| h(m) \right| \ .$$

The resulting scaling sees to it that the maximum gain of the filter is restricted to unity. This guarantees absence of overflow in the case of sinusoidal input signals.

$$S = \frac{1}{\displaystyle\max_{\omega T} \left| H(\omega T) \right|} \tag{8.30}$$

Multiplication of the input signal of a filter by one of the scaling factors (8.28–8.30) reduces or avoids the probability of overflow. In the case of a cascade structure, it may be more advantageous to distribute the scaling factor over the partial filters. This is done in such a way that overflow is avoided at the output of each partial filter block.

8.5 Noise Optimisation with Fixed-Point Arithmetic

The noise at the output of a cascade filter realisation depends on various factors. The pairing of poles and zeros to form second-order filter blocks has a decisive influence. Furthermore, the sequence of the filter blocks plays an important role. In order to give quantitative estimations of the noise power at the output of the filter, we first have to determine the noise generated by the individual first- and second-order filter sections. In the next step, we will optimise the pole/zero pairing and the sequence of the partial filters in such a way that the noise of the overall filter is minimised.

8.5.1 Noise Behaviour of First- and Second-Order Filter Blocks

The noise power generated in first- and second-order filter sections and its spectral distribution mainly depends on three factors:
- the location of the poles and zeros of the filter,
- the filter structure and
- the number and positioning of the quantisers.

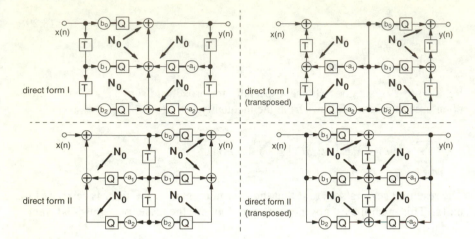

Fig. 8-3 Noise sources in the second-order direct-form structures

Figure 8-3 shows the four possible block diagrams of second-order direct-form filters with each multiplier followed by a quantiser. The arrows, labelled with N_0, indicate the locations where a noise signal with the power $N_0 = \Delta x_q^2/12$ is fed into the circuit. With the results of section 8.4.2, it is easy to determine the noise power at the output of the filter. For that purpose, me must calculate the mean-square gain of the paths from the noise sources to the output. If $H(e^{j\omega T})$ denotes the frequency response of the filter and $H_R(e^{j\omega T})$ the frequency response of the recursive part, the noise power can be calculated by means of the following relations.

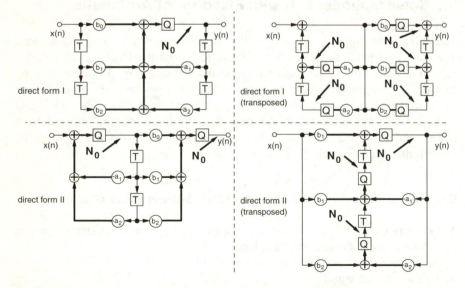

Fig. 8-4 Direct-form block diagrams with reduced number of quantisers

direct form I and
direct form II (transposed)

$$N_{\text{qout}} = 5N_0 \left| \overline{H_R\left(e^{j\omega T}\right)} \right|^2 \qquad (8.30a)$$

direct form I (transposed) and
direct form II

$$N_{\text{qout}} = 3N_0 + 2N_0 \left| \overline{H\left(e^{j\omega T}\right)} \right|^2 \qquad (8.30b)$$

The number of quantisers and hence the number of noise sources can be reduced if not all multipliers are directly followed by a quantiser. If adders or accumulators with double wordlength are available in the respective implementation, the quantisers can be shifted behind the adders as depicted in Fig. 8-4. The biggest saving can be achieved in the case of the direct form I, where five quantisers can be replaced by one. The paths in the block diagrams that carry signals with double wordlength are printed in bold. The general strategy for the placement of the quantisers is that a signal with double wordlength is not allowed at the input of a coefficient multiplier. The noise power at the output of the various direct-form variants is now calculated as follows:

direct form I

$$N_{\text{qout}} = N_0 \left| \overline{H_R\left(e^{j\omega T}\right)} \right|^2 \qquad (8.31a)$$

direct form II

$$N_{\text{qout}} = N_0 + N_0 \left| \overline{H\left(e^{j\omega T}\right)} \right|^2 \qquad (8.31b)$$

direct form I (transposed)

$$N_{\text{qout}} = 3N_0 + 2N_0 \left| \overline{H\left(e^{j\omega T}\right)} \right|^2 \qquad (8.31c)$$

direct form II (transposed)

$$N_{\text{qout}} = 3N_0 \left| \overline{H_R\left(e^{j\omega T}\right)} \right|^2 . \qquad (8.31d)$$

With direct form I and direct form II, we have found the configurations with the lowest possible number of noise sources. The transposed structures can only be further optimised if registers with double wordlength are made available for an interim storage of state variables. The big advantage of the non-transposed structures becomes obvious, as most of the mathematical operations take place in the accumulator. The transposed structures, on the other hand, require more frequent memory accesses, which must happen even with double wordlength if noise is to be minimised. Fig. 8-5 shows the corresponding block diagrams with only one or two quantisers respectively.

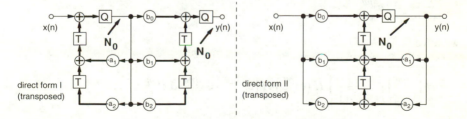

Fig. 8-5 Reduction of the quantisers in the case of the transposed structures

With double precision storage of state variables, the noise power at the output is calculated as follows:

direct form I (transposed) $\qquad N_{\text{qout}} = N_0 + N_0 \left| \overline{H\left(e^{j\omega T}\right)} \right|^2$ $\qquad$ (8.32a)

direct form II (transposed) $\qquad N_{\text{qout}} = N_0 \left| \overline{H_R\left(e^{j\omega T}\right)} \right|^2$. $\qquad$ (8.32b)

Direct form I and direct form II look quite different if considered separately. In a cascade arrangement of direct form I filter blocks, however, the various parts of the block diagram can be newly combined in such a way that we obtain cascaded direct form II blocks and vice versa. The direct form II has the disadvantage that an additional scaling coefficient has to be placed in front of the first block in order to optimally exploit the dynamic range of the filter chain (Fig. 5-18). In direct form I, this scaling can be done by an appropriate choice of the coefficients b_ν.

According to (8.30), (8.31) and (8.32), the mean-square gain of the path between the noise source and the output of the filter has to be determined in order to explicitly calculate the total noise power at the output. The most elegant way to evaluate the occurring integrals of the form

$$\overline{\left| H\left(e^{j\omega T}\right) \right|^2} = \frac{1}{2\pi/T} \int\limits_0^{2\pi/T} \left| H\left(e^{j\omega T}\right) \right|^2 d\omega = \frac{1}{2\pi} \int\limits_0^{2\pi} \left| H\left(e^{j\omega T}\right) \right|^2 d\omega T \qquad (8.33)$$

is to use the residue method, a powerful tool of functions theory. For that purpose, the integral (8.33) has to be converted into an equivalent integral over the complex variable z. First of all, we make use of the fact that the square magnitude can be expressed as the product of the frequency response and its complex-conjugate counterpart.

$$\overline{\left| H\left(e^{j\omega T}\right) \right|^2} = \frac{1}{2\pi} \int\limits_0^{2\pi} H\left(e^{j\omega T}\right) H\left(e^{-j\omega T}\right) d\omega T$$

Using the differential relationship

$$\frac{d\,e^{j\omega T}}{d\omega T} = j\,e^{j\omega T} ,$$

we have

$$\overline{\left| H\left(e^{j\omega T}\right) \right|^2} = \frac{1}{2\pi j} \int\limits_0^{2\pi} H\left(e^{j\omega T}\right) H\left(e^{-j\omega T}\right) e^{-j\omega T} \, d\,e^{j\omega T} .$$

The substitution $e^{j\omega T} = z$ results in the integral

$$\overline{\left| H\left(e^{j\omega T}\right)\right|^2} = \frac{1}{2\pi j} \oint_{z=e^{j\omega T}} H(z)\, H(1/z)\, z^{-1}\, dz .$$ (8.34)

Because of the relation $z = e^{j\omega T}$ and $0 \le \omega \le 2\pi/T$, the integration path is a full unit circle in the z-plane. All the filter structures that we consider in the following have noise transfer functions of the general form

$$H(z) = \frac{1 + k_1 z^{-1} + k_2 z^{-2}}{1 + a_1 z^{-1} + a_2 z^{-2}} .$$ (8.35)

Substitution of (8.35) in (8.34) yields:

$$\frac{1}{2\pi j} \oint_{z=e^{j\omega T}} \frac{1 + k_1 z^{-1} + k_2 z^{-2}}{1 + a_1 z^{-1} + a_2 z^{-2}} \frac{1 + k_1 z + k_2 z^2}{1 + a_1 z + a_2 z^2} \frac{1}{z}\, dz .$$ (8.36)

Within the integration contour, which is the unit circle, the integrand possesses three singularities: two at the complex-conjugate poles of the transfer function and one at $z = 0$. Integral (8.36) can therefore be represented by three residues [38]. We obtain the following expression as a result [12]:

$$\sigma_y^2 = N_0 \frac{2 k_2 \left(a_1^2 - a_2^2 - a_2\right) - 2 a_1 (k_1 + k_1 k_2) + \left(1 + k_1^2 + k_2^2\right)(1 + a_2)}{(1 + a_1 + a_2)(1 - a_1 + a_2)(1 - a_2)} .$$ (8.37)

In order to calculate the noise power at the output of a direct form I filter block, we have to determine the mean-square gain of the recursive filter part according to (8.31a).

$$H_R(z) = \frac{1}{1 + a_1 z^{-1} + a_2 z^{-2}}$$

So we have to set $k_1 = 0$ and $k_2 = 0$ in (8.35). With (8.37), we obtain the following variance of the noise:

$$\sigma_y^2 = N_0 \frac{1 + a_2}{1 - a_2} \frac{1}{(1 + a_2)^2 - a_1^2} .$$ (8.38a)

The coefficients a_1 and a_2 can also be represented by the polar coordinates of the complex-conjugate pole pair

$$\left(z - r \cdot e^{j\varphi}\right)\left(z - r \cdot e^{-j\varphi}\right) = z^2 - 2 r \cos\varphi \; z + r^2$$

$$a_1 = -2 r \cos\varphi \qquad\qquad a_2 = r^2 .$$

The noise power can thus be expressed as a function of the radius and the angle of the pole pair.

$$\sigma_y^2 = N_0 \frac{1+r^2}{1-r^2} \frac{1}{\left(1+r^2\right)^2 - 4r^2 \cos^2 \varphi} \tag{8.38b}$$

The noise power increases more the closer the pole pair comes to the unit circle ($r \to 1$). Additionally, there is a dependence on the angle and hence on the frequency of the pole. The minimum noise occurs when $\varphi = \pm \pi/2$. With respect to noise, the direct form is apparently especially appropriate for filters whose poles are located around a quarter of the sampling frequency.

First-order filter blocks can be treated in the same way. In the block diagrams in Fig. 8-4, the branches with the coefficients a_2 and b_2 simply have to be deleted. In order to calculate the noise power, the coefficients in (8.37) have to be set as follows:

$$k_1 = 0 \qquad\qquad k_2 = 0 \qquad\qquad a_2 = 0 \ .$$

The resulting noise power can be expressed as

$$\sigma_y^2 = N_0 \frac{1}{1-a_1^2} \ . \tag{8.39}$$

Equation (8.39) shows that the noise power increases also for first-order filter blocks if the pole approaches the unit circle.

Normal-form and wave digital filter have been introduced in Chap. 5 as alternatives to the direct form. Under the idealised assumption, that signals and coefficients can be represented with arbitrary precision, all these filter structures exhibit completely identical behaviour. They only differ with respect to the complexity. In practical implementations with finite wordlengths, however, significant differences arise with respect to noise and coefficient sensitivity.

Figure 8-6 shows the block diagram of a second-order normal form filter including a nonrecursive section to realise the numerator polynomial of the transfer function. The structure requires at least two quantisers. The transfer functions between the noise sources and the output of the filter can be expressed as

$$H_1(z) = \frac{\beta z^{-2}}{1-2\alpha z^{-1} + \left(\alpha^2 + \beta^2\right) z^{-2}} = \frac{\beta z^{-2}}{1+a_1 z^{-1} + a_2 z^{-2}}$$

$$H_2(z) = \frac{z^{-1} - \alpha z^{-2}}{1-2\alpha z^{-1} + \left(\alpha^2 + \beta^2\right) z^{-2}} = \frac{z^{-1}\left(1-\alpha z^{-1}\right)}{1+a_1 z^{-1} + a_2 z^{-2}} \ .$$

The delay terms in the numerator can be neglected as they do not influence the noise power at the output. The filter coefficients and the coefficients of the transfer function are related as follows:

$$\alpha = r\cos\varphi = -a_1/2$$

$$\beta = r\sin\varphi = \sqrt{a_2 - a_1^2/4} \ .$$

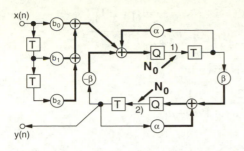

Fig. 8-6
Noise sources in a second-order normal-form structure

The two contributions to the overall noise at the output can be calculated using relation (8.37).

$$\sigma_y^2 = N_0 \frac{\beta^2}{\left(1+a_2\right)^2 - a_1^2} + N_0 \frac{2a_1\alpha + \left(1+\alpha^2\right)\left(1+a_2\right)}{\left(1+a_2\right)^2 - a_1^2}$$

$$\sigma_y^2 = N_0 \frac{1}{1-a_2} = N_0 \frac{1}{1-r^2} \tag{8.40}$$

The spectral distribution of the noise power, generated by the two sources, complement each other at the output of the filter in such a way that the total noise power becomes independent of the angle and hence of the frequency of the pole. Again, the noise power increases if the pole approaches the unit circle.

Finally, we consider the second-order wave digital filter block. Figure 8-7 shows that two quantisers are required if an accumulator with double wordlength is available.

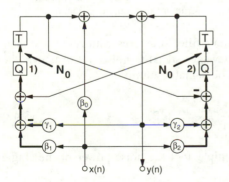

Fig. 8-7
Noise sources in a second-order wave digital filter block

In this case, the transfer functions between the noise sources and the output of the filter can be expressed as

$$H_1(z) = \frac{z^{-1}\left(1-z^{-1}\right)}{1+\left(\gamma_1-\gamma_2\right)z^{-1} + \left(1-\gamma_1-\gamma_2\right)z^{-2}} = \frac{z^{-1}\left(1-z^{-1}\right)}{1+a_1 z^{-1} + a_2 z^{-2}}$$

$$H_2(z) = \frac{z^{-1}(1+z^{-1})}{1+(\gamma_1-\gamma_2)\,z^{-1}+(1-\gamma_1-\gamma_2)\,z^{-2}} = \frac{z^{-1}(1+z^{-1})}{1+a_1z^{-1}+a_2z^{-2}} \; .$$

Using (8.37), the resulting noise power is given by

$$\sigma_y^2 = N_0 \frac{2}{1-a_2}\frac{1}{1+a_2-a_1} + N_0 \frac{2}{1-a_2}\frac{1}{1+a_2+a_1}$$

$$\sigma_y^2 = N_0 \frac{1+a_2}{1-a_2}\frac{4}{(1+a_2)^2 - a_1^2} = N_0 \frac{1+r^2}{1-r^2}\frac{4}{(1+r^2)^2 - 4r^2\cos^2\varphi} \; . \tag{8.41}$$

The filter structure according to Fig. 8-7 thus generates four times the quantisation noise power of the direct form I structure. The dependence on radius and angle of the poles is the same.

If the filter implementation allows the storage of the state variables with double wordlength, a filter with only one quantiser could be realised (Fig. 8-8). This structure tends to demonstrate unstable behaviour, however, as will be shown in Sect. 8.7. Therefore, we will not investigate this variant any further.

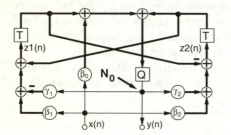

Fig. 8-8
Realisation of the wave digital filter structure with only one quantiser

In the following, we compare the various filter structures with respect to their noise behaviour. We derive the achievable signal-to-noise ratio under the assumption of a sinusoid signal and a 16-bit signal resolution.

$$10 \log SNR = 10 \log \frac{\sigma_{\text{signal}}^2}{\sigma_e^2} = 10 \log \frac{\sigma_{\text{signal}}^2}{\nu N_0} \qquad \text{(dB)} \tag{8.42}$$

The output noise is represented as a multiple ν of the noise power of the single quantiser N_0.

$$10 \log SNR = 10 \log \sigma_{\text{signal}}^2 - 10 \log \nu N_0 = 10 \log \sigma_{\text{signal}}^2 - 10 \log \nu - 10 \log N_0$$

$$10 \log SNR = 10 \log \sigma_{\text{signal}}^2 - 10 \log \nu - 10 \log 2^{-2b}/3$$

$$10 \log SNR = 10 \log \sigma_{\text{signal}}^2 - 10 \log \nu + 4.77 + 6.02b \qquad \text{(dB)} \tag{8.43}$$

With $\sigma_{\text{signal}}^2 = 1/2$ and $b = 16$, it follows that

$$10 \log SNR = 98.1 - 10 \log v \quad \text{(dB)} .$$

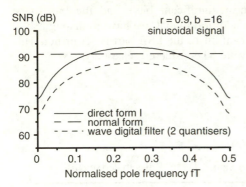

direct form I
normal form
wave digital filter (2 quantisers)

Fig. 8-9
Comparison of the quantisation noise of various filter structures with $r = 0.9$, $b = 16$, sinusoid signal

Figure 8-9 compares the achievable signal-to-noise ratios of the direct-form, normal-form and wave digital filter structure for a pole radius of $r = 0.9$. In the frequency range around $f_s/4$, the direct form exhibits the best performance. The normal form represents a compromise which features constant noise power over the whole frequency range.

8.5.2 Shaping of the Noise Spectrum

The explanations in the previous section have shown that the noise power at the output decisively depends on the filter structure. The transfer function between the location of the quantiser and the output of the filter is the determining factor. The quantisation noise passes through the same paths as the signal to be filtered. For the shaping of the noise spectrum, additional paths are introduced into the filter structure that affect the noise but leave the transfer function for the signal unchanged. For that purpose, it is necessary to get access to the error signal which develops from the quantisation process. So the surplus digits in the case of rounding or truncation must not be dropped but made available for further processing. Fig. 8-10 shows the equivalent block diagram of a quantiser with access to the error signal.

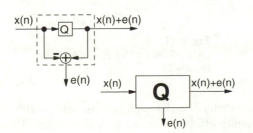

Fig. 8-10
Quantiser with access to the error signal

The relation between the dropped bits and the error signal $e(n)$ depends on the number representation (sign-magnitude or two's complement), the type of quantiser (truncation or rounding) and the sign of the signal value. The value $x(n)$ to be quantised is composed of three parts: the sign (S), the part that is to be preserved after quantisation (MSB, most significant bits), and the bits to be dropped (LSB, least significant bits). In addition to that, we have to distinguish in the case of rounding if the most significant bit of the LSB part is set or not. This leads to the two following cases that have to be considered separately:

Case 1:   S | MSB | 0 . . . . .

Case 2:   S | MSB | 1 . . . . . .

Table 8-6 Rules to derive the error signal $e(n)$

Number representation	Quantiser	Sign	Rules to derive $e(n)$
Sign-magnitude	Truncation	positive	1 | 00 ... 00 | LSB
		negative	0 | 00 ... 00 | LSB
	Rounding	positive	Case1: 1 | 00 ... 00 | LSB Case2: 0 | 00 ... 00 | LSB/2C
		negative	Case1: 0 | 00 ... 00 | LSB Case2: 1 | 00 ... 00 | LSB/2C
Two's complement	Truncation		1 | 11 ... 11 | LSB
	Rounding		Case1: 1 | 11 ... 11 | LSB/2C Case2: 0 | 00 ... 00 | LSB/2C

The rules to derive $e(n)$ from the LSB are summarised in Table 8-6. LSB/2C means the two's complement of the dropped bits.

We apply the method of noise shaping to the direct form I according to Fig. 8-4. The simplest way to influence the quantisation noise is to delay and feed back $e(n)$ to the summation node. This can be done with positive or negative sign as shown in Fig. 8-11a.

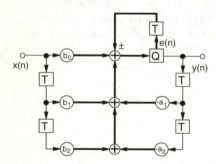

Fig. 8-11a
Direct form I with first-order noise feedback

The noise is now fed in twice, once directly into the quantiser and additionally in delayed form (Fig. 8-11b). The resulting effect is the one of a first-order FIR filter with the transfer function $H(z) = 1 \pm z^{-1}$. The noise is weighted by this transfer function before it is fed into the recursive part of the direct form structure. If the sign is negative, the transfer function has a zero at $\omega = 0$. This configuration is thus able to compensate more or less for the amplification of quantisation noise in the case of low frequency poles. In the reverse case, we obtain a filter structure that can be advantageously used for poles close to $\omega = \pi/T$. The transfer function between the point where the noise is fed in and the output of the filter can be expressed as

$$H(z) = \frac{1 \pm z^{-1}}{1 + a_1 z^{-1} + a_2 z^{-2}} \cdot \qquad (8.44)$$

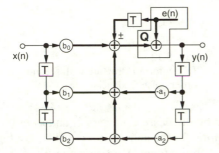

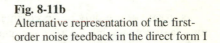

Fig. 8-11b
Alternative representation of the first-order noise feedback in the direct form I

Using (8.37), we obtain the following expressions for the noise power as a function of the radius and the angle of the pole pair. For the negative sign we have

$$\sigma_y^2 = N_0 \frac{2}{1 - a_2} \frac{1}{1 + a_2 - a_1} = N_0 \frac{2}{1 - r^2} \frac{1}{1 + r^2 + 2r\cos\varphi} \cdot \qquad (8.45)$$

The noise power becomes minimum at $\varphi = 0$. For the positive sign, the noise assumes a minimum for $\varphi = \pi$ (8.46):

$$\sigma_y^2 = N_0 \frac{2}{1 - a_2} \frac{1}{1 + a_2 - a_1} = N_0 \frac{2}{1 - r^2} \frac{1}{1 + r^2 - 2r\cos\varphi} \cdot \qquad (8.46)$$

The next logical step is to introduce another delay in Fig. 8-11b and to weight the delayed noise by coefficients k_1 and k_2 (Fig. 8-12). This leads to a noise transfer function between $e(n)$ and the output $y(n)$ of the following form:

$$H(z) = \frac{1 + k_1 z^{-1} + k_2 z^{-2}}{1 + a_1 z^{-1} + a_2 z^{-2}} \cdot \qquad (8.47)$$

If we choose $k_1 = a_1$ and $k_2 = a_2$, the effect of the pole is completely compensated. The noise signal $e(n)$ would reach the output without any spectral boost. k_1 and k_2 cannot be chosen arbitrarily, however. Both coefficients can only assume integer values, as otherwise quantisers would be needed in this noise feedback branch, generating noise again. It must be noted that $e(n)$ must be represented in double precision already. After another multiplication by a fractional coefficient, triple precision would be required.

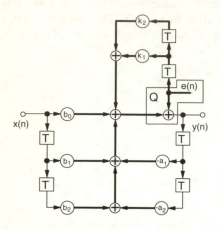

Fig. 8-12
Direct form I structure with second-order noise feedback

As a_1 and a_2 can only be approximated by the nearest integers k_1 and k_2 in the numerator polynomial, a perfect compensation for the poles is not possible. Useful combinations of k_1 and k_2 within or on the stability triangle are shown in Fig. 8-13. Simple feedback of the quantisation error, which results in the noise transfer function (8.44), corresponds to the cases A and B. Second order feedback is required in the cases C to H. The point O inside the triangle represents the case that no noise shaping takes place.

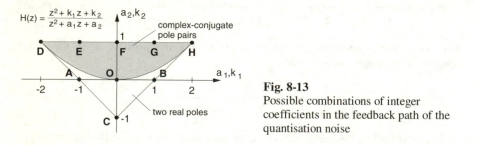

Fig. 8-13
Possible combinations of integer coefficients in the feedback path of the quantisation noise

Figure 8-14 shows the achievable signal to noise ratio for the nine possible coefficient combinations (k_1, k_2) as a function of the pole frequency. We assume a sinusoid signal with 16-bit resolution. The nine cases can be roughly subdivided in three groups:

- The case C yields a signal-to-noise ratio that is independent of the frequency of the pole pair. The values are in general below those that we can achieve in the other configurations. A look at Fig. 8-13 shows that the coefficient pair C will be more suitable to optimise real poles.

- The cases A and B correspond to the first order error feedback. These lead at high and low frequencies to signal-to-noise ratios that are comparable to those that we achieve in the case O around $f_s/4$ without any noise shaping.
- The best signal-to-noise ratios are obtained by second-order error feedback (cases D to H). The five pole frequencies for which the maximum is reached are $fT = 0$, 1/6, 1/4, 1/3 and 1/2.

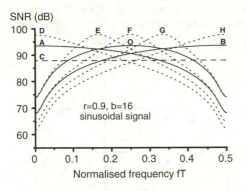

Fig. 8-14
Achievable signal-to-noise ratio for the coefficient pairs (k_1, k_2), sinusoid signal, $r = 0.9$, $b = 16$

Figure 8-15a shows a transposed direct form II with second-order error feedback. The access to the error signal $e(n)$ is explicitly realised by taking the difference between the quantised and unquantised signals. This structure is fully equivalent to Fig. 8-12. Additionally, the dependencies of the signal-to-noise ratio on the coefficients k_1 and k_2 are identical. Redrawing leads to the alternative block diagram Fig. 8-15b. The access to the error signal and the subsequent weighting by integer coefficients is automatically realised in this structure, even if it is not immediately visible. The complexity is not significantly higher compared to the corresponding structure without noise shaping. The coefficients k_1 and k_2 assume the values -2, -1, 0, 1 or 2, which in the worst case means a shifting of the signal value to the right by one digit. Furthermore, two extra additions are required.

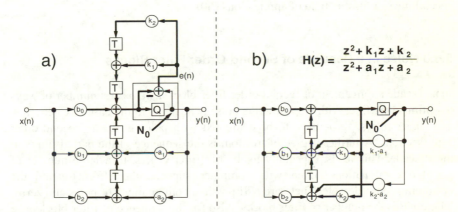

Fig. 8-15 Transposed direct form II with second-order error feedback

A particularly simple structure is obtained in the cases A and B, which yield decisive improvements for low and high frequency poles. The coefficients k_1 and k_2 assume the values

$$k_1 = \pm 1 \qquad\qquad k_2 = 0 \;.$$

One single extra addition or subtraction is required compared to the original transposed direct form II. Fig. 8-16 shows the corresponding block diagram. In addition, we show the modified direct form I which, beside the extra addition/sub-traction, requires an additional buffering of the sum in the accumulator.

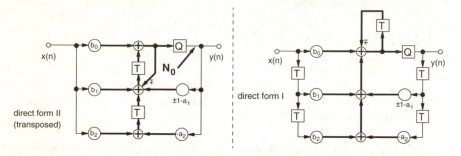

Fig. 8-16 Optimised structures for high or low frequency poles

Filters with poles at low frequencies are often needed for the processing of audio signals. The spectrum of voice and many musical instruments is mainly concentrated in the frequency range from 20 Hz up to about 4 ... 6 kHz. If we want to influence the sound using filters, poles and zeros also have to be placed into this frequency range. The standardised sampling frequencies are much higher, however. Signals are stored on compact discs at a sampling rate of 44.1 kHz, and DAT recorders use a sampling rate of 48 kHz. In order to achieve a good signal-to-noise ratio, filter structures according to Fig. 8-15 or Fig. 8-16 are therefore of special interest for this field of applications [60].

8.5.3 Noise in a Cascade of Second-Order Filter Blocks

In a cascade arrangement of second-order filter blocks, we have a number of ways to optimise the filter with respect to the noise behaviour. In the case of Chebyshev and Cauer filters, zeros have to be combined with poles to create second-order filter sections. A further degree of freedom is the sequence of the partial filters in the cascade. This leads, especially for higher filter orders, to a large number of possible combinations. Even with computer support, the effort to find the optimum is considerable. There are $NB!$ possible combinations of poles and zeros, where NB is the number of filter blocks. Also for the sequence of filter blocks, we have $NB!$ possibilities again. For higher filter orders, dynamic programming [42] is an effective way to find the optimum, but application of simple rules of thumb leads to usable suboptimal results as well.

Partially competing requirements have to be met in order to arrive at the optimum pole/zero pairing and sequence of filter blocks:

1. In the previous section, we have demonstrated that the generated noise power increases more, the closer the pole pair approaches the unit circle, and thus the higher the quality factor Q of the pole is. Obviously, it seems to be advisable to place the sections with high Q at the beginning of the cascade since the noise is filtered in the subsequent filter blocks and the contribution of the high-Q sections to the overall noise is reduced.

2. The sequence of the filter blocks has to be chosen in such a way that we obtain the flattest possible magnitude response in the passband between the input and each output of the individual filter blocks. A large ripple in the front stages would lead to small scaling coefficients between these stages in order to avoid overload. This would have to be compensated by a large gain in the back stages, with the consequence that the noise generated in the front stages would be amplified.

In the case of filters with finite zeros (inverse Chebyshev and Cauer), an appropriate combination of poles and zeros will strongly support the second requirement. A large overshot of the magnitude response is a distinctive feature of poles with high quality factor. This overshot can be reduced by combining these poles with nearby zeros. Since the most critical case is the pole with the largest quality factor, we start with assigning to this pole the nearest zero. We continue this procedure with decreasing Q by assigning the respective nearest of the remaining zeros.

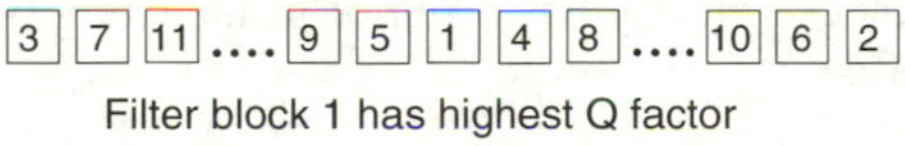

| 3 | 7 | 11 | .... | 9 | 5 | 1 | 4 | 8 | .... | 10 | 6 | 2 |

Filter block 1 has highest Q factor

Fig. 8-17
Rule of thumb for the optimum sequence of filter blocks

In the next step, the sequence of the partial filter blocks has to be optimised. For low-pass filters, a rule of thumb is given in [41], leading to suboptimal arrangements which only deviate by a few per cent from the optimum signal-to-noise ratio. The pole with the highest quality factor is placed in the middle of the filter chain, which is a compromise between the first and second requirement. The pole with the second highest Q is placed at the end, the one with the third highest Q at the beginning of the chain. Fig. 8-17 shows how the further filter blocks are arranged.

Table 8-7 shows examples for the noise power at the output of some concrete low-pass implementations. We deal in all cases with tenth-order filters realised by the first canonical structure. The cutoff frequency corresponds to a quarter of the sampling rate. The figures in the table give the output noise power as multiples v of the power N_0 of a single quantiser. Using (8.43), it is easy to calculate the achievable signal-to-noise ratio at the output of the filter. We compare Butterworth, Chebyshev and Cauer filters.

Table 8-7 Output noise power for various filter types (tenth-order, first canonical structure, $f_cT = 0.25$), normalised to N_0

Sequence	Butterworth	Chebyshev 1-dB	Cauer 1-dB/40-dB, pairing optimum	Cauer 1-dB/40-dB, pairing wrong
Optimum	②③④①⑤ 4.9	③④①⑤② 47	③④①⑤② 258	③④①⑤② 368
Worst case	⑤④③②① 10.9	①②③④⑤ 2407	⑤④①②③ 696	①②③④⑤ $9.6 \cdot 10^6$
③⑤①④② Rule of thumb	5.1	55	275	455
①②③④⑤	10.2	2407	474	$9.6 \cdot 10^6$
⑤④③②①	10.9	1189	376	$6.0 \cdot 10^6$
①⑤④③②	6.8	178	351	20050
②④③⑤①	6.2	53	270	660
③④①②⑤	5.2	77	261	13028
③④②⑤①	6.8	62	306	439
③④⑤①②	7.2	272	542	103057

If the optimum sequence is assumed, the Butterworth filter generates a noise power that exactly corresponds to the noise of the five coefficient multipliers ($\nu = 5$). A larger noise can be expected for Chebyshev filters since the Q-factor of the poles is higher for this filter type. For the Cauer filter, we show the results for both, the optimum pole/zero pairing and the contrary case, where the pole with the highest Q-factor is combined with the most distant zero etc. The noise power is again increased compared to the Chebyshev filter. The wrong pole/zero pairing may yield completely useless arrangements. If poles and zeros are well combined, the choice of the sequence of the filter blocks clearly has a minor influence only on the noise performance of the overall filter.

Table 8-8 Noise power of a tenth-order 1-dB Chebyshev filter ($f_cT = 0.1$) for various filter structures, normalised to N_0

	Direct form I	Normal form	Noise shaping (first-order)	Noise shaping (second-order)
Maximum	25995	6424	2909	972
Optimum	502	119	55	16
Rule of thumb	603	144	95	28

The noise figures in Table 8-7 are calculated for filters with the normalised cutoff frequency $f_cT = 0.25$, for which direct form I filters have optimum noise behaviour. As an example, application of the rule of thumb yields a noise power of $55\,N_0$ for a tenth-order 1-dB Chebyshev filter. A cutoff frequency of $f_cT = 0.1$, however, leads to a value about ten times higher. Transition to other filter structures is a way out of this situation. In Table 8-8, we compare the noise power of direct form I, normal form and first- and second-order error feedback for a

tenth-order 1-dB Chebyshev filter with $f_cT = 0.1$. With the help of Fig. 8-13 and Fig. 8-14, we choose the following integer coefficients for noise shaping:
– first-order error feedback: case A ($k_1 = 1, k_2 = 0$)
– second-order error feedback: case D ($k_1 = -2, k_2 = 1$) .

It turns out that, in all cases, matched against the most unfavourable configuration, the rule of thumb leads to fairly good results. The deviation from the optimum is in general well below 3 dB. By analogy, all the previous considerations apply to high-pass filters, too. For bandpass and bandstop filters, however, the rule has to be modified.

The rule for the sequence of filter blocks according to Fig. 8-17 leads to disappointing results if applied to bandpass and bandstop filters. For the following considerations, we choose a twelfth-order bandpass filter with mid-band frequency $f_mT = 0.25$ and bandwidth $(f_u–f_l)T = 0.1$ as an example. For a filter with Chebyshev characteristics, the sequence ⑤④①⑥②③ produces minimum noise. No obvious systematics can be identified from this sequence. We gain more informative results if the same calculations are performed for a corresponding Cauer filter with optimum combination of poles and zeros. Using direct form filter blocks yields the following noise figures:

maximum noise:	4123 N_0	(sequence ①③⑤②④⑥)
minimum noise:	12 N_0	(sequence ⑥⑤①②④③)
rule according to Fig. 8-17:	1322 N_0	(sequence ③⑤①④⑥②) .

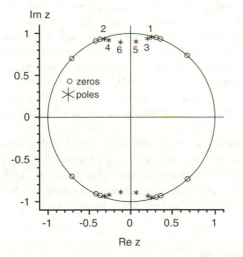

Fig. 8-18
Pole/zero plot for a twelfth-order band-pass filter with Cauer characteristics

From the optimum sequence ⑥⑤①②④③, the following rule can be derived:

The filter blocks are combined in pairs with decreasing Q-factor: highest with second highest, third highest with fourth highest Q-factor etc. These pairs are arranged again according to the well known rule as depicted in Fig. 8-17.

Figure 8-18 shows the location of the poles and zeros of our exemplary Cauer bandpass filter. These are arranged symmetrically about the mid-band frequency. The named rule thus combines in pairs the poles that are located symmetrically about the mid-band frequency and have about the same distance from the unit circle and consequently the same Q-factor. Table 8-9 shows how successfully this rule can be applied to other filter types.

Table 8-9 Noise power of twelfth-order low-pass filters (direct form I), $f_m T = 0.25$, $(f_u - f_l)T = 0.1$, normalised to N_0

Sequence	Butterworth	Chebyshev 1-dB	Cauer 1-dB/40-dB
Optimal	⑤④①②③⑥ 13	⑤④①⑥②③ 77	⑥⑤①②④③ 12
Worst case	①③⑤⑥④② 204	①③⑤⑥④② 17702	①③⑤②④⑥ 4123
⑤⑥①②③④ Bandpass rule	15	102	12
⑥⑤②①④③ Pairs in reverse order	15	103	12
③⑤①④⑥② Low-pass rule	95	6109	1322
⑤①③⑥②④ HP-LP	139	10595	260
⑥②④④①③ LP-HP	137	10458	260

Matched against the most unfavourable and the most optimum case, the modified rule for bandpass filters yields excellent results. The sequence within the pairs plays practically no role. Direct application of the rule for low- and high-pass filters leads to useless results. Bandpass filters can also be interpreted as the cascade of a low-pass and a high-pass filter. A logical procedure would be to optimise the low-pass (LP) and high-pass (HP) pole/zero pairs separately according to the low-pass/high-pass rule. Table 8-9 shows, however, that this approach goes completely wrong. By analogy, all the previous considerations apply to bandstop filters, too.

The considerations in this section show that observing some simple rules with respect to the pairing of poles and zeros and the sequence of partial second-order filter blocks leads to quite low-noise filter implementations. The deviation from the optimum is, in general, well below 3 dB. Furthermore, depending on the location of the poles, an appropriate filter structure should be chosen.

8.6 Finite Coefficient Wordlength

The coefficients of numerator and denominator polynomial or the poles and zeros of the transfer function (8.48) can be determined with arbitrary accuracy using the design procedures that we derived in Chap. 6 and Chap. 7.

$$H(z) = \frac{\displaystyle\sum_{r=0}^{M} b_r z^{-r}}{\displaystyle\sum_{i=0}^{N} a_i z^{-i}} = \frac{b_0}{a_0} \frac{\displaystyle\prod_{m=1}^{M}\left(1 - z_{0m} z^{-1}\right)}{\displaystyle\prod_{n=1}^{N}\left(1 - z_{\infty n} z^{-1}\right)} = \frac{N(z)}{D(z)} \tag{8.48}$$

Because of the finite register lengths in practical implementations, the filter coefficients can only be represented with limited accuracy, leading to deviations from the filter specification. Poles and zeros lie on a grid of discrete realisable values.

8.6.1 Coefficient Sensitivity

We first look at the sensitivity of the zero and pole location with respect to changes of the filter coefficients. We investigate in detail the behaviour of poles, but the obtained results completely apply to zeros, too.

The denominator polynomial $D(z)$, which determines the location of the poles, can be expressed by the filter coefficients a_i or by the poles $z_{\infty n}$:

$$D(z) = 1 + \sum_{i=1}^{N} a_i z^{-i} = \prod_{n=1}^{N}\left(1 - z_{\infty n} z^{-1}\right) . \tag{8.49}$$

The starting point for the determination of the sensitivity of the ith pole with respect to the inaccuracy of the kth filter parameter a_k is the partial derivative of the denominator polynomial, taken at the pole location $z_{\infty i}$.

$$\left. \frac{\partial D(z)}{\partial a_k} \right|_{z = z_{\infty i}}$$

Expansion of this term leads to a relation that includes the desired sensitivity of the poles to the coefficient quantisation.

$$\left. \frac{\partial D(z)}{\partial a_k} \right|_{z = z_{\infty i}} = \left. \frac{\partial D(z)}{\partial z_{\infty i}} \right|_{z = z_{\infty i}} \frac{\partial z_{\infty i}}{\partial a_k} \tag{8.50}$$

Equation (8.49) can be differentiated with respect to the filter coefficients a_k and the poles $z_{\infty i}$.

$$\left. \frac{\partial D(z)}{\partial a_k} \right|_{z = z_{\infty i}} = \left. z^{-k} \right|_{z = z_{\infty i}} = z_{\infty i}^{-k}$$

$$\left. \frac{\partial D(z)}{\partial z_{\infty i}} \right|_{z = z_{\infty i}} = \left. -z^{-1} \prod_{\substack{n=1 \\ n \neq i}}^{N}\left(1 - z_{\infty n} z^{-1}\right) \right|_{z = z_{\infty i}}$$

$$\left.\frac{\partial D(z)}{\partial z_{\infty i}}\right|_{z=z_{\infty i}} = -z_{\infty i}^{-1}\prod_{\substack{n=1\\n\neq i}}^{N}\left(1-z_{\infty n}\,z_{\infty i}^{-1}\right)$$

$$\left.\frac{\partial D(z)}{\partial z_{\infty i}}\right|_{z=z_{\infty i}} = -z_{\infty i}^{-N}\prod_{\substack{n=1\\n\neq i}}^{N}\left(z_{\infty i}-z_{\infty n}\right)$$

Substitution of these differential quotients in (8.50) yields an equation that gives a measure of the sensitivity of the ith pole to a change of the kth coefficient in the denominator polynomial of $H(z)$:

$$\frac{\partial z_{\infty i}}{\partial a_k} = -\frac{z_{\infty i}^{N-k}}{\displaystyle\prod_{\substack{n=1\\n\neq i}}^{N}\left(z_{\infty i}-z_{\infty n}\right)}\;. \tag{8.51}$$

Relation (8.51) shows that, if the poles (or zeros) are tightly clustered, it is possible that small errors in the coefficients can cause large shifts of the poles (or zeros). Furthermore, it is evident that the sensitivity increases with the number of poles (or zeros). Clusters of poles occur in the case of sharp cutoff or narrow bandpass/bandstop filters. For these filter types, it has to be carefully considered whether a chosen coefficient representation is sufficient in a given application. An improvement of the situation can be achieved in a cascade arrangement of second-order filter blocks which leads to clusters consisting of two (complex-conjugate) poles only.

The deviations of the poles from the exact value leads to consequences for the transfer function in three respects. Affected are:
- the frequency of the pole,
- the Q-factor or the 3-dB bandwidth of the resonant curve respectively (e.g. Fig. 6-11),
- the gain of the pole.

From the transfer function of a simple pole

$$H(z) = \frac{1}{z-z_\infty} = \frac{1}{z-r_\infty\,e^{j\varphi_\infty}}\;,$$

we can derive the three named quantities as

frequency of the pole:	$\omega_\infty T = \varphi_\infty$
3 - dB bandwidth:	$\Delta f_{3\text{dB}} T \approx \left(1-r_\infty\right)/\pi$
gain:	$\left|H\left(e^{j\omega_\infty T}\right)\right| = 1/\left(1-r_\infty\right)\;.$

Gain and bandwidth are especially critical, if the magnitude of the pole r_∞ approaches unity. The smallest deviations from the target value may lead to considerable deviations in the frequency response. At least the unpleasant distortion of the gain can be avoided by the choice of an appropriate filter structure. The boundedness of wave digital filters prevents an unlimited growth of the gain if the pole approaches the unit circle. We demonstrate this fact by means of a simple first-order low-pass filter. For the direct form, the transfer function is

$$H(z) = \frac{1}{1 + a_1 z^{-1}} \qquad\qquad H(z = 1, \omega = 0) = \frac{1}{1 + a_1} .$$

Figure 8-19 shows the analog reference filter for a wave digital filter implementation which only requires a symmetrical three-port parallel adapter and a delay.

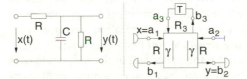

Fig. 8-19
Wave digital filter implementation of a first-order low-pass filter

The transfer function of the wave digital filter implementation according to Fig. 8-19 can be expressed as

$$H(z) = \frac{\gamma\left(1 + z^{-1}\right)}{1 - \left(1 - 2\gamma\right) z^{-1}} \qquad\qquad H(z = 1, \omega = 0) = 1 .$$

For $\omega = 0$ or $z = 1$, the gain amounts exactly to unity, independent of the choice of the parameter γ. By way of contrast, the gain approaches infinity as $a_1 \to -1$ in the case of the direct form.

Deviations of the pole frequency and the Q-factor cannot be avoided, however, even if wave digital filters are employed. In the following, we investigate the various types of second-order filter blocks with respect to the coefficient sensitivity. Criterion is the distribution of allowed pole locations in the z-plane.

8.6.2 Graphical Representation of the Pole Density

The coefficient sensitivity can be graphically illustrated by plotting all possible poles in the z-plane that can be represented with the given coefficient wordlength. In regions where poles are clustered, the sensitivity is low as many candidates are available in the vicinity to approximate a given exact pole location. In the reverse case, large deviations from the target frequency response are possible as the distance to the nearest realisable pole location may be considerable.

The resolution of the coefficients can be defined in different ways depending on the characteristics of the implemented arithmetical operations:

- by the quantisation step q of the coefficients which is determined by the number of available fractional digits
- by the total number of digits available for the coefficient representation

In the second case, the quantisation step q depends on the value range of the coefficients. If the magnitude is less than unity, all bits but the sign bit are used for the fractional digits. There are filter structures, however, whose coefficients may assume values up to 4. These require up to two integer digits which reduces the number of available fractional digits accordingly. In the case of pure fractional arithmetic, coefficients larger than 1 are realised by shifting the result of the multiplication by an appropriate number of bits to the left.

In the following graphical representations of possible discrete pole locations, we assume in all cases a coefficient format with four fractional bits. This corresponds to a quantisation step of $q = 1/16$. If we compare the various filter structures with respect to coefficient sensitivity with the aforementioned assumption, structures with a large coefficient range come off better than those with a small range. The reason is that, with a given quantisation step q, a large coefficient range yields more possible discrete pole locations than a small range. The following graphical representations show the upper half of the unit circle only because the pattern of allowed complex-conjugate pole pairs is symmetrical about the real axis.

Second-order direct form

For this filter type, the coefficients a_k of the denominator polynomial of the transfer function are identical with the filter coefficients in the recursive part of the block diagram. The direct form I structure (Fig. 8-3) is an example. The real and imaginary parts of the poles z_∞ and the filter coefficients a_1 and a_2 are related as follows:

$$\operatorname{Re} z_\infty = -a_1/2 \tag{8.52a}$$

$$\operatorname{Im} z_\infty = \sqrt{a_2 - a_1^2/4} \ . \tag{8.52b}$$

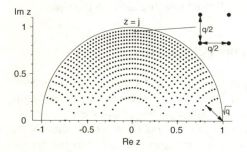

Fig. 8-20
Pattern of possible pole locations for direct form filters (coefficients with four fractional bits, $q = 1/16$)

The stability range of the coefficient pair (a_1, a_2) is the interior of the triangle according to Fig. 5-20. The coefficient a_1 assumes values in the range -2 to $+2$

while a_2 lies in the range -1 to $+1$. If we plot all possible complex poles in the z-plane, using (8.52), that lie within the unit circle and result from coefficients a_1 and a_2 with 4 fractional bits, we obtain the pattern according to Fig. 8-20. The possible pole locations are unevenly distributed over the unit circle. At low frequencies ($z = 1$) and high frequencies ($z = -1$), the pole density is rather low, and the poles lie on circular arcs. In the vicinity of $f_s/4$, the density is highest. The pole pattern is almost quadratic.

The distance of a pole from the point $z = 1$ can be calculated as

$$D_{z=1} = \sqrt{\left(1 - \operatorname{Re} z_\infty\right)^2 + \left(\operatorname{Im} z_\infty\right)^2} = \sqrt{\left(1 + a_1/2\right)^2 + a_2 - a_1^2/4}$$

$$D_{z=1} = \sqrt{1 + a_1 + a_2} \ . \tag{8.53}$$

For $z_\infty = 1$ we have $a_1 = -2$ and $a_2 = 1$. About this point, a_1 and a_2 take on the following discrete values:

$$a_1 = -2 + l_1\, q$$
$$a_2 = 1 - l_2\, q \ .$$

q is the quantisation step of the coefficients. l_1 und l_2 are integers that assume small positive values in the region about $z = 1$.

$$D_{z=1} = \sqrt{1 - 2 + l_1\, q + 1 - l_2\, q} = \sqrt{\left(l_1 - l_2\right) q}$$

The smallest possible distance of a pole from the point $z = 1$ thus amounts to $\sqrt{q}$.

For $z = j$ we have $a_1 = 0$ und $a_2 = 1$. About this point, a_1 und a_2 take on the following discrete values:

$$a_1 = l_1\, q$$
$$a_2 = 1 - l_2\, q \ .$$

Using (8.52a), the real part of the resulting realisable poles can be expressed as

$$\operatorname{Re} z_\infty = -a_1/2 = -l_1\, q/2 \ .$$

Use of (8.53b) yields the following expression for the imaginary part:

$$\operatorname{Im} z_\infty = \sqrt{a_2 - a_1^2/4} = \sqrt{1 - l_2\, q - l_1^2\, q^2/4}$$
$$\operatorname{Im} z_\infty \approx \sqrt{1 - l_2\, q} \approx 1 - l_2\, q/2 \ .$$

In the region about $z = j$, the distance between adjacent poles amounts to $q/2$ in the horizontal and vertical direction. It becomes obvious that approximately double the coefficient wordlength is needed to achieve the same pole density in the vicinity of $z = 1$ as we find it about the point $z = j$ with the original wordlength.

As an example, we consider the case of eight fractional bits for the coefficient representation, where q assumes the value $1/256 = 0.0039$. In the region about

$z = j$, the distance between adjacent poles amounts to $q/2 = 0.002$, wheras the smallest distance of a pole from the point $z = 1$ is $\sqrt{q} = 0.0625$. Fig. 8-20 shows the result of the corresponding example with $q = 1/16$.

The normal form

The normal-form structure according to Fig. 5-27 features an even distribution of the poles. This is due to the fact that the filter coefficients α and β are identical with the real and imaginary part of the complex-conjugate pole pair. A disadvantage of the normal form is the double number of required multiplications compared to the direct form. Based on the filter coefficients α and β, the transfer function can be expressed as

$$H(z) = \frac{1}{1 - 2\alpha\,z^{-1} + \left(\alpha^2 + \beta^2\right) z^{-2}} \ .$$

The stability region of the coefficient pair (α, β) is the interior of the unit circle. Both coefficients thus assume values in the range -1 to $+1$. If we plot all possible complex poles in the z-plane that lie within the unit circle and result from coefficients α and β with four fractional bits, we obtain the pattern according to Fig. 8-21. Compared to the direct form, we observe a higher pole density about $z = 1$ and $z = -1$ and a lower one about $z = \pm j$. For wide-band filters whose poles are spread over a large range of frequencies, the normal form is a good compromise.

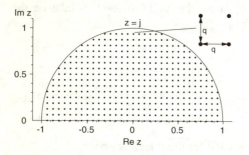
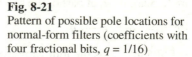

Fig. 8-21
Pattern of possible pole locations for normal-form filters (coefficients with four fractional bits, $q = 1/16$)

Second-order wave digital filter

Second-order wave digital filter sections are of interest due to their excellent stability properties. In Sect. 8.7 we will show that, under certain circumstances, unstable behaviour such as limit cycles can be avoided completely. At the beginning of this section, we also demonstrated that variations in the coefficients do not influence the maximum gain of the pole. The coefficients γ_1 and γ_2 in the block diagram (Fig. 5-61) are responsible for the location of the poles. The stability range of the coefficient paires (γ_1, γ_2) is again the interior of a triangle, as depicted in Fig. 8-22.

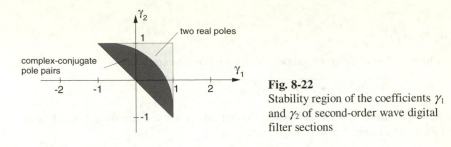

Fig. 8-22
Stability region of the coefficients γ_1
and γ_2 of second-order wave digital
filter sections

Based on the coefficients γ_1 and γ_2, the transfer function of the wave digital filter section can be expressed as

$$H(z) = \frac{1}{1 + (\gamma_1 - \gamma_2)\, z^{-1} + (1 - \gamma_1 - \gamma_2)\, z^{-2}} \ . \tag{8.54}$$

By making use of (8.52) again, the poles can be calculated as

$$\mathrm{Re}\, z_\infty = -a_1/2 = (\gamma_1 - \gamma_2)/2 \tag{8.55a}$$

$$\mathrm{Im}\, z_\infty = \sqrt{a_2 - a_1^2/4} = \sqrt{1 - \gamma_1 - \gamma_2 - (\gamma_1 - \gamma_2)^2/4} \ . \tag{8.55b}$$

According to Fig. 8-22, both coefficients γ_1 and γ_2 have a value range of -1 to $+1$. If we plot all possible complex poles in the z-plane that lie within the unit circle and result from coefficients γ_1 and γ_2 with four fractional bits, we obtain the pattern according to Fig. 8-23. As with the direct form, the possible pole locations are unevenly distributed over the unit circle. At low frequencies ($z = 1$) and high frequencies ($z = -1$), the pole density is rather low, and the poles lie on circular arcs. In the vicinity of $f_s/4$, the density is highest. The pole pattern is almost diamond shaped.

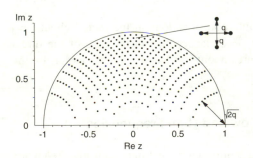

Fig. 8-23
Pattern of possible pole locations for
second-order wave digital filter sections
(coefficients with four fractional bits,
$q = 1/16$)

According to (8.53), the distance of a pole from the point $z = 1$ can be calculated as

$$D_{z=1} = \sqrt{1 + a_1 + a_2} \ .$$

From (8.54), we can derive the following relations between the coefficients a_ν and γ_ν:

$$a_1 = \gamma_1 - \gamma_2$$
$$a_2 = 1 - \gamma_1 - \gamma_2 \ .$$

The radius of the circular arcs about $z = 1$ can thus be expressed as

$$D_{z=1} = \sqrt{1 + \gamma_1 - \gamma_2 + 1 - \gamma_1 - \gamma_2} = \sqrt{2 - 2\gamma_2} \ .$$

For $z_\infty = 1$ we have $\gamma_1 = -1$ and $\gamma_2 = 1$. About this point, the coefficient γ_2 takes on the following discrete values:

$$\gamma_2 = 1 - l_2\, q \ .$$

$$D_{z=1} = \sqrt{2 - 2\left(1 - l_2\, q\right)} = \sqrt{2 l_2\, q}$$

The smallest possible distance of a pole from the point $z = 1$ thus amounts to $\sqrt{(2q)}$.

For $z = j$ we have $\gamma_1 = 0$ und $\gamma_2 = 0$. About this point, the coefficients γ_1 und γ_2 assume the following discrete values:

$$\gamma_1 = l_1\, q$$
$$\gamma_2 = l_2\, q \ .$$

Substitution in (8.55) yields the following discrete pole locations in the vicinity of $z = j$.

$$\mathrm{Re}\, z_\infty = \left(\gamma_1 - \gamma_2\right)/2 = \left(l_2 - l_1\right) q/2 \tag{8.56a}$$

$$\mathrm{Im}\, z_\infty = \sqrt{1 - \gamma_1 - \gamma_2 - \left(\gamma_1 - \gamma_2\right)^2 / 4} = \sqrt{1 - \left(l_1 + l_2\right) q - \left(l_1 - l_2\right)^2 q^2 / 4}$$

$$\mathrm{Im}\, z_\infty \approx \sqrt{1 - \left(l_1 + l_2\right) q} \approx 1 - \left(l_1 + l_2\right) q/2 \tag{8.56b}$$

The minimum distance between possible pole locations in the region about $z = j$ amounts to $\sqrt{2} \times q/2$, which is a factor of $\sqrt{2}$ larger than in the case of the direct form. Also the radius of the circular arcs about $z = 1$ and $z = -1$ is larger, by the same factor.

8.6.3 Increased Pole Density at Low and High Frequencies

In Sect. 8.5.2, we already pointed to the importance of filters with low-frequency poles for the processing of audio signals. But we are still missing a filter structure with increased pole density about $z = 1$. The elementary structures discussed above do not get us any further. Therefore, we have to construct specialised structures with the desired property.

A primary drawback of the direct form with respect to low-frequency poles is the fact that the radius of the pole locations about the point $z_\infty = 1$ is proportional to the square root of the coefficient quantisation q. Due to this dependence, the pole density is rather low in the vicinity of $z_\infty = 1$. We will overcome this situation by choosing a new pair of filter coefficients e_1/e_2. We try the following approach:

- The coefficient e_1 determines the radius of the circular arcs about the point $z_\infty = 1$ on which the poles lie. Using (8.53), e_1 can be expressed as

$$e_1 = \sqrt{1 + a_1 + a_2} \ . \tag{8.57}$$

- The coefficient e_2 is chosen equal to a_2 and thus determines the radius of the pole location about the origin.

The filter coefficients e_1 and e_2 and the coefficients of the transfer function a_1 and a_2 are related as follows:

$$e_1 = \sqrt{1 + a_1 + a_2} \qquad\qquad a_1 = e_1^2 - 1 - e_2$$
$$e_2 = a_2 \qquad\qquad\qquad\quad a_2 = e_2 \ .$$

By substitution of these relationships, the transfer function of the pole can be rewritten with the new coefficients e_1 and e_2 as parameters.

$$H(z) = \frac{1}{1 + a_1 z^{-1} + a_2 z^{-2}} = \frac{1}{1 + \left(e_1^2 - 1 - e_2\right) z^{-1} + e_2 z^{-2}} \tag{8.58}$$

Using (8.52), the real and imaginary parts of the pole location can also be expressed in terms of e_1 and e_2.

$$\operatorname{Re} z_\infty = -a_1/2 = \left(1 + e_2 - e_1^2\right)\big/2$$

$$\operatorname{Im} z_\infty = \sqrt{a_2 - a_1^2\big/4} = \sqrt{e2 - \left(1 + e_2 - e_1^2\right)^2 \big/ 4} \tag{8.59}$$

Figure 8-24 depicts the range of the new coefficients leading to stable poles.

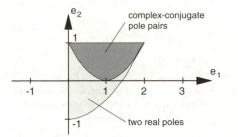

Fig. 8-24
Stability region of the coefficients e_1 and e_2 of second-order filter sections with increased pole density about $z = 1$

If we plot all possible complex poles in the z-plane that lie within the unit circle and result from coefficients e_1 and e_2 with four fractional bits, we obtain the pattern according to Fig. 8-25. The pole pattern shows the desired property. The highest pole density appears in the vicinity of $z = 1$. The poles lie on circular arcs about $z = 1$ with equidistant radii, which are determined by the coefficient e_1.

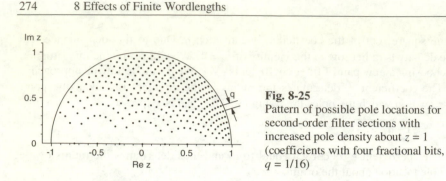

Fig. 8-25
Pattern of possible pole locations for second-order filter sections with increased pole density about $z = 1$ (coefficients with four fractional bits, $q = 1/16$)

Figure 8-26 shows the prototype of a block diagram which is especially suited to the implementation of filters with low coefficient sensitivity in the case of high and low pole frequencies. By an appropriate choice of the filter coefficients s, t, u and v, we can realise a variety of transfer functions from the literature which optimise the coefficient wordlength requirements for the placement of poles near $z = 1$ and $z = -1$. The general transfer function of this structure is given by

$$H(z) = \frac{d_0 - (2d_0 + d_1 + d_2)\, z^{-1} + (d_0 + d_1)\, z^{-2}}{1 - (2 - vs - uvt)\, z^{-1} + (1 - vs)\, z^{-2}} \ . \tag{8.60}$$

For the realisation of the transfer function (8.58), the filter coefficients take on the following values:

$$s = 1 - e_2 \qquad\qquad u = e_1 \qquad\qquad t = e_1 \qquad\qquad v = 1 \ .$$

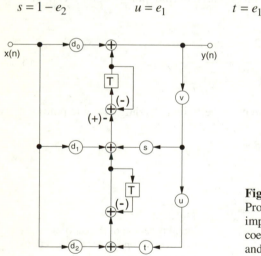

Fig. 8-26
Prototype block diagram for the implementation of filters with reduced coefficient sensitivity in the case of low and high pole frequencies

In the following, we introduce further filter structures with low coefficient sensitivity for low and high frequency poles. These feature in general a higher concentration of poles near $z = 1$ and $z = -1$ than we could achieve with the above approach. For all these structures, we show the valid range of filter coefficients and the respective pole distribution pattern in the z-plane.

Avenhaus [1]

$$H(z) = \frac{d_0 - \left(2d_0 + d_1 + d_2\right) z^{-1} + \left(d_0 + d_1\right) z^{-2}}{1 - \left(2 - c_1\right) z^{-1} + \left(1 - c_1 + c_1 \, c_2\right) z^{-2}}$$

$c_1 = a_1 + 2$ $d_0 = b_0$

                                   $d_1 = b_2 - b_0$

$c_2 = \dfrac{1 + a_1 + a_2}{2 + a_1}$ $d_2 = -b_0 - b_1 - b_2$

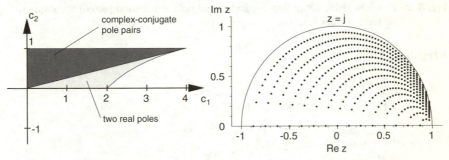

Fig. 8-27 Range of stable coefficients c_1 and c_2 and pole distribution pattern according to Avenhaus (four fractional bits, $q = 1/16$)

The coefficients in block diagram Fig. 8-26 assume the values

$s = 1 - c_2$ $u = 1$

$t = c_2$ $v = c_1$.

In [1], a block diagram is given that is especially tailored to this special case, saving one multiplication.

An increased pole density at high frequencies about $z = -1$ can be achieved if we choose the signs in parenthesis in Fig. 8-26. With these signs, the pole pattern in Fig. 8-27 is mirrored about the imaginary axis (Fig. 8-28). The same is valid by analogy for the two structures introduced in the following.

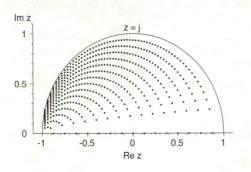

Fig. 8-28
Increased pole density about $z = -1$

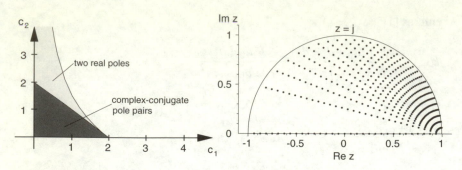

Fig. 8-29 Range of stable coefficients c_1 and c_2 and pole distribution pattern according to Kingsbury (four fractional bits, $q = 1/16$)

Kingsbury [36]

$$H(z) = \frac{d_0 - (2d_0 + d_1 + d_2)\, z^{-1} + (d_0 + d_1)\, z^{-2}}{1 - (2 - c_1\, c_2 - c_1^2)\, z^{-1} + (1 - c_1\, c_2)\, z^{-2}}$$

$$c_1 = \sqrt{1 + a_1 + a_2}$$
$$c_2 = (1 - a_2)/c_1$$

$$d_0 = b_0$$
$$d_1 = b_2 - b_0$$
$$d_2 = -b_0 - b_1 - b_2$$

The coefficients in block diagram Fig. 8-26 assume the values

$$s = c_2 \qquad\qquad u = 1$$
$$t = c_1 \qquad\qquad v = c_1 \; .$$

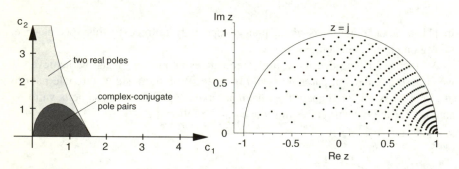

Fig. 8-30 Range of stable coefficients c_1 and c_2 and pole distribution pattern according to Zölzer (four fractional bits, $q = 1/16$)

Zölzer [60]

$$H(z) = \frac{d_0 - (2d_0 + d_1 + d_2)\, z^{-1} + (d_0 + d_1)\, z^{-2}}{1 - (2 - c_1\, c_2 - c_1^3)\, z^{-1} + (1 - c_1\, c_2)\, z^{-2}}$$

$$c_1 = \sqrt[3]{1 + a_1 + a_2}$$
$$c_2 = (1 - a_2)/c_1$$

$$d_0 = b_0$$
$$d_1 = b_2 - b_0$$
$$d_2 = -b_0 - b_1 - b_2$$

The coefficients in block diagram Fig. 8-26 assume the values

$$s = c_2 \qquad\qquad u = c_1$$
$$t = c_1 \qquad\qquad v = c_1 \; .$$

8.6.4 Further Remarks Concerning Coefficient Sensitivity

There is a relationship between coefficient sensitivity and sampling frequency, which is not so obvious at first glance. Let us imagine the pole and zero distribution of an arbitrary filter. If we increase the sampling frequency without changing the frequency response, poles and zeros migrate towards $z = 1$ and concentrate on an ever-decreasing area. In order to still provide a sufficient pole density in this situation, the wordlength of the filter parameters has to be increased accordingly. The choice of a filter structure with low coefficient sensitivity about $z = 1$ would keep the required wordlength within reasonable limits.

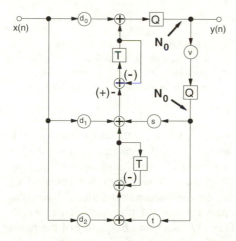

Fig. 8-31
Noise sources in the structure according to Kingsbury

A further interesting relationship exists between coefficient sensitivity and quantisation noise. It is striking that the noise of a normal-form filter is independent of the angle of the pole, and the poles are evenly distributed over the unit circle. Direct-form and wave digital filter structure exhibit the largest coefficient sensitivity at high ($z = -1$) and low ($z = 1$) pole frequencies, which are the pole locations that create the largest quantisation noise. In order to confirm a general rule, we consider in the following a filter with low coefficient sensitivity at low frequencies, such as the Kingsbury structure. Figure 8-31 shows a block diagram with the relevant noise sources. The quantisation after the multiplication by v is needed to return back to the original wordlength before the signal in this path is multiplied by the coefficients s and t. The transfer functions between the noise sources and the output of the filter are

$$H_1(z) = \frac{\left(1 - z^{-1}\right)^2}{1 + a_1 z^{-1} + a_2 z^{-2}}$$

$$H_2(z) = -(s+t)\, z^{-1} \frac{1 - s/(s+t)\, z^{-1}}{1 + a_1 z^{-1} + a_2 z^{-2}} \quad .$$

Using the relations that we derived in Sect. 8.5.1, the generated quantisation noise power can be calculated as a function of the angle of the pole. Figure 8-32 shows the result together with the curves for the noise power of the direct and normal forms. It proves true that the smallest quantisation noise is created for those pole locations where the filter exhibits the lowest coefficient sensitivity.

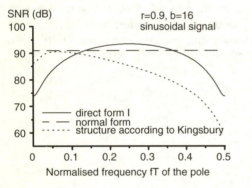

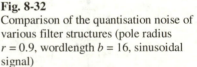

Fig. 8-32

Comparison of the quantisation noise of various filter structures (pole radius $r = 0.9$, wordlength $b = 16$, sinusoidal signal)

The coincidence of low quantisation noise and low coefficient sensitivity can be easily understood if the process of quantisation is interpreted in a somewhat different way. The effect of rounding or truncation could also be achieved by modifying the filter coefficients slightly in such a way that the rounded or truncated values appear as a result of the coefficient multiplications. Noiselike variations are superimposed onto the nominal values of the coefficients, which finally results in a shift-variable system. Gain, Q-factor and frequency of the poles and zeros change irregularly within certain limits and thus modulate the quantisation noise onto the signal. The lower the coefficient sensitivity is, the slighter are the variations of the transfer function and the smaller is the resulting quantisation noise.

8.6.5 A Concluding Example

The decomposition of transfer functions into second-order sections is a proven measure to reduce coefficient sensitivity. The mathematical background of this fact has been given in Sect. 8.6.1. A further possibility to reduce coefficient sensitivity is the transition to wave digital filter structures (refer to Sect. 5.7.4), whose properties are based on the choice of analog lossless networks inserted between resistive terminations as prototypes for the digital filter design. The following example is to show what benefits there are in practice from choosing a cascade or wave digital filter structure. Figure 8-33a shows the magnitude response of a third-order Chebyshev filter with a ripple of 1.256 dB and a cutoff frequency at $f_c T = 0.2$. The transfer function of the filter is

$$H(z) = 0.0502 \frac{\left(1 + z^{-1}\right)^3}{1 - 1.3078\, z^{-1} + 1.0762\, z^{-2} - 0.3667\, z^{-3}} .$$

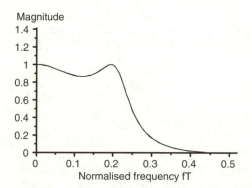

Fig. 8-33a
Third-order Chebyshev filter (1.256 dB ripple, $f_c T = 0.2$), ideal curve

In the case of the direct form, the filter coefficients are equal to the coefficients of the transfer function.

$$H(z) = V \frac{\left(1 + z^{-1}\right)^3}{1 + a_1 z^{-1} + a_2 z^{-2} + a_3 z^{-3}}$$

with
$$V = 0.0502$$
$$a_1 = -1.3078$$
$$a_2 = 1.0762$$
$$a_3 = -0.3667$$

We will check in the following how the magnitude curve is influenced by variations of the coefficients. Figure 8-33b illustrates the result for the case that each of the three decisive filter coefficients a_1, a_2 and a_3 assumes its nominal as well as a 3% higher or lower value, which leads to 27 different curves in total. We observe considerable variations in the curves and large deviations from the ideal Chebyshev (equiripple) characteristic.

We decompose now the third-order direct-form structure into the series arrangement of a first- and a second-order filter block, resulting in the following expression for the transfer function:

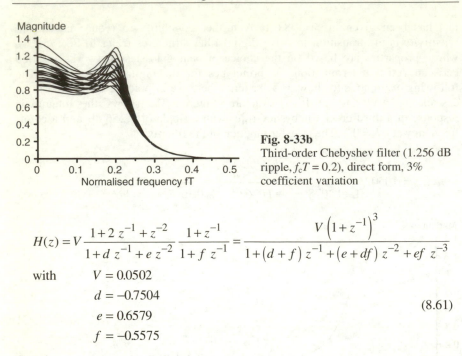

Fig. 8-33b
Third-order Chebyshev filter (1.256 dB ripple, $f_c T = 0.2$), direct form, 3% coefficient variation

$$H(z) = V \frac{1 + 2\,z^{-1} + z^{-2}}{1 + d\,z^{-1} + e\,z^{-2}} \frac{1 + z^{-1}}{1 + f\,z^{-1}} = \frac{V\left(1 + z^{-1}\right)^3}{1 + (d + f)\,z^{-1} + (e + df)\,z^{-2} + ef\,z^{-3}}$$

with
$$V = 0.0502$$
$$d = -0.7504$$
$$e = 0.6579$$
$$f = -0.5575$$

(8.61)

If the filter coefficients d, e and f vary by up to ±3%, we obtain magnitude curves as shown in Fig. 8-33c. It turns out that, by simply splitting off the first-order section from the third-order filter, we achieve a visible reduction of the variation range by more than a factor of two. The coefficient V, which determines the overall gain of the filter, was not included in these considerations since it has no direct influence on the shape of the magnitude curve. It must not be forgotten, however, that the gain can only be adjusted with an accuracy according to the wordlength of the coefficient V.

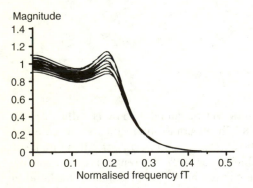

Fig. 8-33c
Third-order Chebyshev filter (1.256 dB ripple, $f_c T = 0.2$), cascade structure, 3% coefficient variation

In the case of filter structures with low coefficient sensitivity, the coefficients of the transfer function are calculated as products and sums of the filter coefficients, as (8.61) shows for the cascade structure. A more extreme behaviour in this respect is shown by the wave digital filter. The relations between the coefficients of the transfer function a_1, a_2 and a_3 and the filter coefficients γ_1, γ_2 and γ_3 according to the block diagram Fig. 5-55 is far more complex.

$$H(z) = \frac{\gamma_1 \, \gamma_3 \left(1 + z^{-1}\right)^3}{1 + a_1 z^{-1} + a_2 z^{-2} + a_3 z^{-3}}$$

$$a_1 = 2\gamma_1 + 2\gamma_2 + 4\gamma_3 - \gamma_1\gamma_3 - \gamma_2\gamma_3 - 3$$

$$a_2 = 6\gamma_1\gamma_3 + 6\gamma_2\gamma_3 + 4\gamma_1\gamma_2 - 4\gamma_1\gamma_2\gamma_3 - 4\gamma_1 - 4\gamma_2 - 4\gamma_3 + 3$$

$$a_3 = 4\gamma_1\gamma_2\gamma_3 - 4\gamma_1\gamma_2 - \gamma_2\gamma_3 - \gamma_1\gamma_3 + 2\gamma_1 + 2\gamma_2 - 1$$

$$\text{with} \qquad \gamma_1 = 0.2213$$

$$\gamma_2 = 0.2213$$

$$\gamma_3 = 0.2269$$

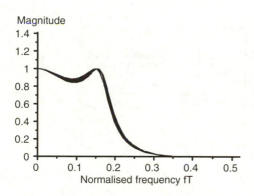

Fig. 8-33d
Third-order Chebyshev filter (1.256 dB ripple, $f_c T = 0.2$), wave digital filter, 3% coefficient variation

Variations of the coefficients by up to ±3% yield the magnitude curves in Fig. 8-33d. We find here nine different curves only, since the coefficients γ_1 and γ_2 assume identical values. The maximum gain is strictly limited to unity. The low variation range of the magnitude response proves the superiority of the wave digital filter structure with respect to coefficient sensitivity.

8.7 Limit Cycles

In the previous sections, we already identified roundoff noise as one of the deviations from the behaviour of an ideal discrete-time system which are due to the finite precision of signal representation. In recursive filters we can observe another phenomenon, which manifests itself by constant or periodical output signals which remain still present even when the input signal vanishes. These oscillations are called limit cycles and have their origin in nonlinearities in the feedback branches of recursive filter structures. Two different nonlinear mechanisms can be distinguished in this context:

- Rounding or truncation that is required to reduce the number of digits to the original register length after multiplications or after additions in the case of floating-point arithmetic.
- Overflow which occurs if the results of arithmetic operations exceed the representable number range.

Overflow cycles are especially unpleasant since the output signal may oscillate between the maximum amplitude limits. Quantisation limit cycles normally exhibit amplitudes of few quantisation steps only. But they may assume considerable values if the poles of the filter are located in close vicinity to the unit circle. The named unstable behaviour is mainly influenced by the choice of the filter structure and by the quantisation and overflow characteristics of the implemented arithmetic. As a consequence, the stable coefficient regions of the various filter structures as depicted in Fig. 8-38 are in most cases reduced compared to the ideal linear case.

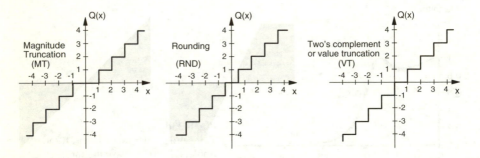

Fig. 8-34 Quantisation characteristics

8.7.1 Nonlinearities in Actual Implementations

The most common quantisation characteristics are:
- Magnitude truncation (MT)
- Two's complement truncation, also called value truncation (VT)
- Rounding (RND)

Fig. 8-34 shows graphical illustrations of the nonlinear characteristics of the three named types of quantisers. For all types of number representations, rounding is performed by substituting the nearest possible number that can be represented by the reduced number of bits. If binary digits are truncated in case of the two's complement format, the value is always rounded down independently of the sign of the number (value truncation). In the case of the sign-magnitude representation, the magnitude of the number is always rounded down (magnitude truncation), which may be in effect a rounding up or a rounding down depending on the sign.

For rounding and magnitude truncation we can define sectors, the shaded regions in Fig. 8-34, that are bounded by two straight lines and that fully enclose the steplike quantisation curves. The slopes of the bounding lines amount to 0 and k_q respectively, where k_q assumes the value 1 for magnitude truncation and 2 for rounding. The quantisation characteristics $Q(x)$ thus satisfy the following sector conditions:

$$0 \leq Q(x)/x \leq k_q \qquad \text{with} \qquad \begin{cases} k_q = 1 & \text{for magnitude truncation} \\ k_q = 2 & \text{for rounding} \end{cases} \qquad (8.62)$$

The sector condition (8.62) is a simple means to characterise the quantiser. The smaller the aperture angle of the sector is, the less the nonlinearity deviates from the ideal linear behaviour. For the truncation in the two's complement number format, we cannot define a corresponding sector that is bounded by straight lines with finite slope. This fact makes the further mathematical treatment of this format more difficult.

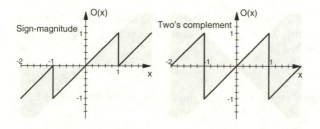

Fig. 8-35a
Natural overflow characteristics of two's complement and sign-magnitude arithmetic

Overflow nonlinearities can be treated in the same way. They may also be characterised by an overflow characteristic $O(x)$ and a certain sector condition. We start with considering the natural overflow characteristics of the sign-magnitude and the two's complement representation. Overflow in the case of two's complement arithmetic leads to a behaviour where the result of a summation jumps over the whole representable number range to the value with the opposite sign (Fig. 8-35a). For the sign-magnitude arithmetic, the magnitude jumps to zero in the case of an overflow. The sector, which comprises the nonlinear characteristic, is bounded by straight lines with the slopes 1 and k_o, where k_o assumes the value -1 for the two's complement and 0 for the sign-magnitude representation.

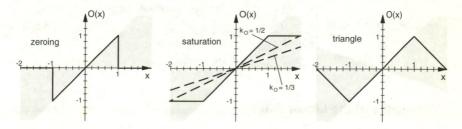

Fig. 8-35b Special overflow characteristics: zeroing, saturation, triangle

Special measures taken in the adders have the goal of narrowing down the sectors of the nonlinear characteristics in order to minimise the deviation from the ideal linear case as far as possible. Common measures are zeroing, saturation and triangle characteristic as depicted in Fig. 8-35b. All three methods have advantages in the context of two's complement arithmetic since the sector is clearly narrowed down. The slope of the bounding line amounts to $k_o = 0$ for zeroing and saturation; for the triangle characteristic we have $k_o = -1/3$. The behaviour of the saturation and triangle characteristics can be further improved by making special assumptions concerning the degree of overflow occurring in concrete implementations.

If we guarantee in the case of the triangle characteristic that the sum is always less than 2, the slope k_o will always be positive. This applies for instance to normal-form filters. The signal amplitude is limited to unity. Also the filter coefficients α and β have magnitudes of less than 1 for stable filters. In case of a vanishing input signal, only two signal values, weighted by these coefficients, are added up in the normal-form structure resulting in a sum whose magnitude will always be less than 2.

In case of saturation, k_o is always positive as long as the sum is finite. Denoting the maximum possible magnitude of the sum that can occur in a given filter structure as S_{max}, k_o is calculated as

$$k_o = 1/S_{max} \ .$$

Assuming that the input signal vanishes, we can thus estimate for the normal-form structure

$$k_o = 1/\left(|\alpha| + |\beta|\right) > 1/2 \ .$$

As for the direct form, the magnitude of the coefficient a_1 is limited to 2, the magnitude of a_2 to 1. If the input signal vanishes, we can thus give a lower limit for k_o:

$$k_o = 1/\left(|a_1| + |a_2|\right) > 1/3 \ .$$

Equation (6.83) summarises all the overflow sector conditions that we have discussed above.

$$k_o \le O(x)/x \le 1 \ \text{with} \ \begin{cases} k_o = 0 & \text{for sign - magnitude} \\ k_o = -1 & \text{for two's complement} \\ k_o = 0 & \text{for zeroing} \\ k_o = 1/S_{max} & \text{for saturation} \\ k_o = 2/S_{max} - 1 & \text{for triangle} \end{cases} \tag{8.63}$$

8.7.2 Stability of the Linear Filter

Up to now, we have considered stability from the point of view that a bounded input signal should always result in a bounded output signal. This so-called BIBO stability requires in the time domain that the unit-sample response is absolutely summable (refer to Sect. 3.4). We have derived a related condition in Sect. 5.4.1

for the frequency domain which states that all poles of the system must lie within the unit circle. For the discussion of limit cycles we need a different definition of stability, however, since we investigate the behaviour of filters under the assumption of vanishing input signals.

In the state-space representation, a general second-order system is described by the following vector relationship (5.29):

$$z(n+1) = A\ z(n) + b\ x(n)$$

$$y(n) = c^{T}\ z(n) + d\ x(n)\ .$$

The vector $z(n)$ combines the state variables which correspond to the contents of the delay memories of the system. If the input signal $x(n)$ vanishes, the state variables of the respective next clock cycle $z(n+1)$ are simply calculated by multiplication of the current state vector $z(n)$ by the system matrix A.

$$z(n+1) = A\ z(n) \tag{8.64}$$

Since the output signal $y(n)$ is obtained by linear combination of the state variables (multiplication of the state vector $z(n)$ by c^{T}), it is sufficient to prove stability for $z(n)$. In this context it is of interest to investigate the behaviour of the system if a starting vector z_0 is continuously multiplied by the system matrix A. The curve that the arrowhead of the state vectors passes in the state space by these multiplications is called a trajectory. The following considerations are restricted to second-order filter blocks but can be easily extended to higher-order systems. The state vector consists of the two components $z_1(n)$ and $z_2(n)$, which fully describe the state of the system in the time domain. Fig. 8-36 shows four examples of possible trajectories in the z_1/z_2-plane.

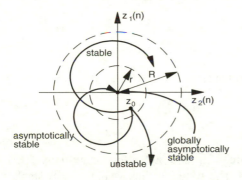

Fig. 8-36
Stability in the sense of Lyapunov

The stability criterion according to Lyapunov [16, 37] considers the behaviour of systems in the neighbourhood of equilibrium points which can be characterised in the linear case by the relation

$$z(n+1) = A\ z(n) = z(n)\ . \tag{8.65}$$

The origin $z = 0$ that should be reached with a vanishing input signal $x(n)$ is such an equilibrium point, which can be easily verified using (8.65). Stability can be defined in this context as follows:

For a neighbourhood of the origin with the radius r there exists another neighbourhood with radius R such that the following applies: If the starting point of a trajectory $z_0 = z(0)$ lies within the neighbourhood with radius r, then the entire trajectory $z(n)$ as $n \rightarrow \infty$ remains within the neighbourhood with radius R (Fig. 8-36).

A system is asymptotically stable if it is stable in the above sense and the trajectory converges to the origin as $n \rightarrow \infty$.

A system is globally asymptotically stable if it is asymptotically stable and z_0 can be chosen arbitrarily in the state space.

The eigenvalues λ_i of the matrix A, which are identical to the poles of the transfer function, have a decisive influence on the stability behaviour of the system. If z_0 is the state of the system for $n = 0$, we obtain the state after N clock cycles by means of N times multiplying z_0 by the Matrix A.

$$z(N) = A^N z_0 \tag{8.66}$$

Each nonsingular matrix can be written in the form

$$A = P^{-1} L P \;,$$

where L is a diagonal matrix with the eigenvalues λ_i of A appearing as diagonal elements. The matrix P contains the eigenvectors of A as columns. Relation (8.66) can thus be expressed as follows:

$$z(N) = \left(P^{-1} L P \right)\left(P^{-1} L P \right)\left(P^{-1} L P \right) \ldots \left(P^{-1} L P \right) z_0$$

$$z(N) = P^{-1} L L L \ldots L P z_0$$

$$z(N) = P^{-1} L^N P z_0 \;.$$

For second-order systems, L^N can be written as

$$L^N = \begin{pmatrix} \lambda_1^N & 0 \\ 0 & \lambda_2^N \end{pmatrix} \;.$$

If the magnitude of both eigenvalues λ_1 and λ_2 is less than unity, $z(n)$ approaches the origin with increasing N. The system is thus asymptotically stable. Matrices whose eigenvalues have magnitudes of less than unity are therefore called stable matrices. If single eigenvalues lie on the unit circle, $z(n)$ remains bounded. The system is stable. Multiple eigenvalues on the unit circle and eigenvalues with a magnitude larger than unity lead to unstable systems.

Stability as discussed above is closely related to the existence of a Lyapunov function for the considered system. A Lyapunov function v is a scalar function of the state vector $z(n)$ or of the two components $z_1(n)$ and $z_2(n)$ respectively. In a certain neighbourhood of the origin, this function has the following properties:

A. v is continuous.
B. $v(0) = 0$.
C. v is positive outside the origin.
D. v does not increase, mathematically expressed as $v[z(n+1)] \le v[z(n)]$.

If one succeeds in finding such a function, then the region in the neighbourhood of the origin is stable. For a given system, the fact that no Lyapunov has been found yet does not allow us to draw any conclusions concerning the stability of the system. If the more stringent condition D′ applies instead of D, then the origin is asymptotically stable.

D′. v decreases monotonically, mathematically expressed as $v[z(n+1)] < v[z(n)]$.

Condition D′ guarantees that the trajectories $z(n)$ which originate in the neighbourhood of the origin will converge to the origin.

Lyapunow functions are an extension of the energy concept. It is evident that the equilibrium state of a physical system is stable if the energy decreases continuously in the neighbourhood of this equilibrium state. The Lyapunov theory allows us to draw conclusions concerning the stability of a system without any knowledge of the solutions of the system equations. The concept of Lyapunov functions will be illustrated by means of the example of a simple electrical resonant circuit (Fig. 8-37). The reader is presumably more familiar with energy in the context of electrical circuits than with the energy of number sequences processed by a mathematical algorithm.

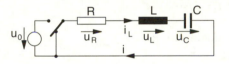

Fig. 8-37
Electrical resonant circuit

The energy stored in the circuit according to Fig. 8-37 depends on the state quantities i_L and u_C, which can be combined to the state vector $z(t)$.

$$z(t) = \begin{pmatrix} u_C(t) \\ i_L(t) \end{pmatrix}$$

We choose the total energy stored in the inductor and in the capacitor as candidate for a Lyapunov function:

$$v(t) = 1/2\,C\,u_C^2(t) + 1/2\,L\,i_L^2(t) \ ,$$

or in quadratic vector form

$$v(t) = \begin{pmatrix} u_C(t) & i_L(t) \end{pmatrix} \cdot \begin{pmatrix} 1/2C & 0 \\ 0 & 1/2L \end{pmatrix} \cdot \begin{pmatrix} u_C(t) \\ i_L(t) \end{pmatrix} . \tag{8.67}$$

The conditions A to C for the existence of a Lyapunov function are surely met by (8.67). v is continuous, positive and vanishes in the origin. Condition D requires that v must not increase. In order to check the observance of this condition, we form the derivative of (8.67) with respect to the time.

$$
\begin{aligned}
\mathrm{d}v/\mathrm{d}t &= C\,u_C\,\mathrm{d}u_C/\mathrm{d}t + L\,i_L\,\mathrm{d}i_L/\mathrm{d}t \\
&= u_C\,i + i_L\,u_L = u_C\,i + u_L\,i = i\left(u_C + u_L\right) \\
&= i\left(-u_R\right) = -i^2\,R
\end{aligned}
$$

For $R \geq 0$, v does not increase, which proves (8.67) to be a Lyapunov function. The stability analysis by means of the chosen Lyapunov function correctly predicts that the system according to Fig. 8-37 is stable for $0 \leq R \leq \infty$. It is worth mentioning that we did not have to solve a differential equation to arrive at this result.

Also for linear discrete-time systems, quadratic forms are a promising approach for the choice of Lyapunov functions:

$$
v(n) = z(n)^{\mathrm{T}}\,V\,z(n) \geq 0 \ . \tag{8.68}
$$

If the matrix V is positive definite, $v(n)$ will be positive for any arbitrary state vector $z(n) \neq 0$. A general 2x2 matrix

$$
\begin{pmatrix} A & C \\ D & B \end{pmatrix}
$$

possesses this property if

$$
A > 0, \qquad B > 0 \qquad \text{and} \qquad 4AB > \left(C + D\right)^2 \ .
$$

We still have to make sure that $v(n)$ does not increase:

$$
v(n+1) \leq v(n)
$$

$$
z(n+1)^{\mathrm{T}}\,V\,z(n+1) \leq z(n)^{\mathrm{T}}\,V\,z(n)
$$

$$
z(n)^{\mathrm{T}}\,A^{\mathrm{T}}\,V\,A\,z(n) \leq z(n)^{\mathrm{T}}\,V\,z(n)
$$

$$
z(n)^{\mathrm{T}}\,V\,z(n) - z(n)^{\mathrm{T}}\,A^{\mathrm{T}}\,V\,A\,z(n) \geq 0
$$

$$
z(n)^{\mathrm{T}}\left(V - A^{\mathrm{T}}\,V\,A\right)z(n) \geq 0 \ . \tag{8.69}
$$

Relation (8.68) is thus a Lyapunov function if both matrices V and $V - A^{\mathrm{T}}\,V\,A$ are positive definite. From (8.68) and (8.69) we can derive a number of relations that allow us to check whether a matrix V leads to a Lyapunov function, if the system matrix A is given. With a given matrix V, conversely, we can determine the ranges of the coefficients of the system matrix A in which stability is guaranteed. For the most important second-order structures (direct-form, normal-form, wave digital filter), appropriate V matrices are given in [10].

$$
V_{\text{direct}} = \begin{pmatrix} 1 + a_2 & a_1 \\ a_1 & 1 + a_2 \end{pmatrix} \qquad\qquad A_{\text{direct}} = \begin{pmatrix} -a_1 & -a_2 \\ 1 & 0 \end{pmatrix}
$$

$$V_{\text{normal}} = \begin{pmatrix} 1 & 0 \\ 0 & 1 \end{pmatrix} \qquad\qquad A_{\text{normal}} = \begin{pmatrix} \alpha & -\beta \\ \beta & \alpha \end{pmatrix}$$

$$V_{\text{WDF}} = \begin{pmatrix} 1-\gamma_2 & 0 \\ 0 & 1-\gamma_1 \end{pmatrix} \qquad A_{\text{WDF}} = \begin{pmatrix} -\gamma_1 & 1-\gamma_1 \\ \gamma_2-1 & \gamma_2 \end{pmatrix}$$

With these V matrices, (8.68) is always a Lyapunov function if the coefficients of the system matrix A lie in the stable regions as shown in Fig. 8-38.

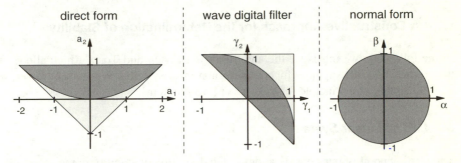

Fig. 8-38 Regions of stable coefficient pairs as derived in Chap. 5 and 8

8.7.3 Stability with Quantisation and Overflow Correction

Lyapunow functions that are based on the quadratic form can also be constructed for nonlinear systems. However, these are usually too pessimistic with regards to the predicted stable range of filter coefficients. In the following we introduce two algorithms which allow the investigation of stability in the nonlinear case: the "constructive algorithm" by Brayton and Tong [7, 8] and an "exhaustive search algorithm" in the state space [3, 49].

The constructive algorithm is based on the search for an appropriate Lyapunov function. Global asymptotical stability can be proved for those ranges of the filter coefficients for which a Lyapunov function can be found. For the remaining ranges of the coefficients we cannot make any statement, as pointed out in Sect. 8.7.2. A prerequisite for applying this method is the possibility of specifying sectors which comprise the respective quantisation and overflow characteristics as depicted in Fig. 8-34 and Fig. 8-35. The case of two's complement truncation can therefore not be treated by this method. The discussion of the constructive algorithm will give the reader a good insight into the dynamic behaviour of second-order filter blocks.

The exhaustive search algorithm is only suitable for the investigation of quantisation limit cycles. The case of two's complement truncation is covered by this method. Unlike the constructive algorithm which leaves some coefficient ranges undefined with respect to stability, the exhaustive search method provides clear results concerning stable and unstable regions in the coefficient plane. In the context of the introduction of this algorithm, we will also present formulas for the estimation of the amplitude of limit cycles.

The two aforementioned algorithms are presented in this chapter since they are universally valid in a certain sense. By way of contrast, numerous publications on this topic such as [2, 5, 6, 13, 14, 15, 17, 29, 30, 39] only consider partial aspects such as special quantisation or overflow characteristics, number representations or filter structures. They generally do not provide any new information. In most cases, the results of theses papers are more conservative with respect to stable ranges of filter coefficients than those of the constructive or search algorithm.

8.7.4 A Constructive Approach for the Determination of Stability

In the following, we introduce the algorithm by Brayton and Tong, which allows the construction of Lyapunov functions for nonlinear systems. We start, however, with illustrating the principle for the case of linear systems.

8.7.4.1 The Linear System

The starting point of our considerations is the definition of stability in the sense of Lyapunov, as introduced in a previous section. We choose a diamond-shaped region W_0 in the state space x_1/x_2 according to Fig. 8-39 as the initial neighbourhood in which all trajectories start. Our goal is to determine the neighbourhood W^* of the origin which all trajectories that start in W_0 pass through. If the system is stable, this neighbourhood W^* must be finite. In the first step we had to multiply, in principle, each point (x_1/x_2) of W_0 by the system matrix A. Since W_0 possesses the property of being convex, only a few points out of the infinite number of points of W_0 actually have to be calculated.

A set K is convex if the following applies to any arbitrary pair of points a and b in K: If a and b are in K, then all points on the straight connecting line between a and b are also in K.

With this property, only the corner points or extremal points of the convex set have to be multiplied by the matrix A. If we connect in our case the four resulting points to a new quadrangle, we obtain the region into which all trajectories move from W_0 after the first clock cycle. The four new extremal points are again multiplied by A which yields the region into which all trajectories move after the second clock cycle. This procedure is repeated until certain termination criteria are fulfilled which identify the system as stable or unstable.

The system is stable if the nth quadrangle, calculated by n successive multiplications by the system matrix A, does not increase the "total area" of all quadrangles calculated up to the $(n-1)$th step. "Total area" in this context means the convex hull which can be imagined as the area formed by a rubber band spanning all these quadrangles as indicated by the dotted lines in Fig. 8-39. If the new calculated quadrangle (drawn in grey) lies completely within the convex hull over all previous quadrangles (drawn in black), the algorithm can be terminated. The system is stable. Fig. 8-39 illustrates this algorithm by means of a system, that turns out to be stable after the 5th step because the new grey quadrangle lies completely within the convex hull W_5. With W_5 we have also found the neighbourhood W^* which all trajectories that start in W_0. pass through.

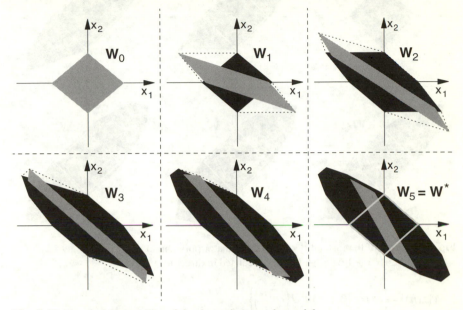

Fig. 8-39 Proof of the stability of the denominator polynomial
$N(z) = 1 + 1.67\, z^{-1} + 0.9\, z^{-2}$, realised in direct form

Figure 8-39 also shows that the convex hull W^* remains in touch with the initial region W_0, which is a general property of stable systems. This means illustratively that the W_k must not expand in all directions. If W_0 lies after the kth step completely within the convex hull W_k, the algorithm can be stopped with the result "unstable". Figure 8-40 shows an example in which instability is proved after the 5th iteration step. After the first step, W_0 touches the convex hull with the upper and lower corner only. After the 5th step, W_0 lies completely within W_5. The complex-conjugate pole pair, realised by $N(z)$, is unstable.

If the system is stable, the size of the region W^* is finite. W^* possesses an interesting property: All trajectories that start within W^* will completely run within or on the boundary of W^*. It is true that we have constructed W^* by trajectories which all have their origin in W_0, but each point that a trajectory passes can be used as starting point for a new trajectory which will follow the same curve as the original one that passed through this point. So all trajectories that start outside W_0 but within W^* will never leave W^*. The exact proof is given in [7].

With this property of W^*, we can define a Lyapunov function which can be later applied to nonlinear systems, too. We choose a positive constant α as Lyapunov function $v[z(n)]$ that scales the region W^* in such a way that $z(n)$ lies on the boundary of αW^*, or otherwise expressed as

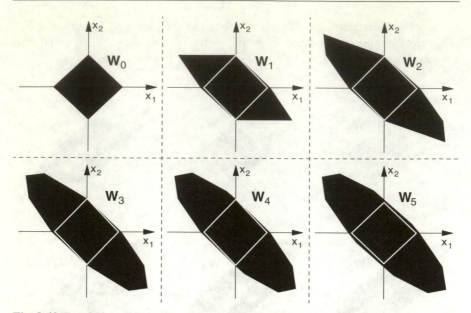

Fig. 8-40 Proof of the instability of the denominator polynomial
$N(z) = 1 + 1.67\,z^{-1} + 1.02\,z^{-2}$, realised in direct form

$$v[z(n)] = \left\{ \alpha : \alpha > 0, z(n) \in \partial\left(\alpha W^*\right) \right\},$$

$$\partial \text{ stands for the boundary of the region.}$$

It is easy to realise that v is a Lyapunov function. v is always positive and vanishes for $z(n) = 0$. Since the boundary of W^* is a continuous curve, v will also be continuous. As a state $z(n)$ on the boundary αW^* cannot migrate outwards, v cannot increase. Hence all criteria of a Lyapunov function are met.

By means of the presented algorithm, we can only prove stability and not asymptotic stability as we do not further investigate whether the trajectories converge into the origin. If we want to exclude, however, any occurrence of limit or overflow cycles, the system has to be asymptotically stable. A trick helps in this case. All coefficients of the system matrix A are multiplied by a factor a little larger than 1 such as 1.00001. The system becomes a little more unstable, since the eigenvalues of the matrix increase accordingly. If the system matrix, modified in this way, still turns out to be stable, then the original matrix is asymptotically stable [8].

8.7.4.2 Stability of the Nonlinear System

The method to prove stability that we illustrated in the previous section for linear systems is also suitable to investigate limit and overflow cycles in real implementations. The constructive approach enables us to consider any combination of the following features:

- structure of the filter
- type of quantiser (rounding or truncation)
- placing of the quantisers (after each coefficient multiplication or after summation)
- inclusion of arbitrary overflow characteristics

According to (8.63), the behaviour of autonomous systems (systems in the situation that no signal is applied to the input), can be described in vector form by the state equation

$$z(n+1) = A\, z(n)\ .$$

In component form we can write for a second-order system

$$z_1(n+1) = a_{11}\, z_1(n) + a_{12}\, z_2(n)$$
$$z_2(n+1) = a_{21}\, z_1(n) + a_{22}\, z_2(n)\ .$$

Overflows, which result in overflow corrections $O(x)$ as illustrated in Fig. 8-35, can occur in practice after the summation of several signals.

$$z_1(n+1) = O\big(a_{11}\, z_1(n) + a_{12}\, z_2(n)\big)$$
$$z_2(n+1) = O\big(a_{21}\, z_1(n) + a_{22}\, z_2(n)\big)$$

For quantisation, characterised by $Q(x)$ according to Fig. 8-34, we have two possibilities: the quantisation may take place after each coefficient multiplication or only after summation of the signals from several branches:

$$z_1(n+1) = O\big[Q\big(a_{11}\, z_1(n) + a_{12}\, z_2(n)\big)\big]$$
$$z_2(n+1) = O\big[Q\big(a_{21}\, z_1(n) + a_{22}\, z_2(n)\big)\big]$$

or

$$z_1(n+1) = O\big[Q\big(a_{11}\, z_1(n)\big) + Q\big(a_{12}\, z_2(n)\big)\big]$$
$$z_2(n+1) = O\big[Q\big(a_{21}\, z_1(n)\big) + Q\big(a_{22}\, z_2(n)\big)\big]\ .$$

In the following, we will interpret the influence of the quantisation characteristic $Q(x)$ and overflow characteristic $O(x)$ in a different way: we assume that the filter coefficients a_{ij} are modified by $Q(x)$ and $O(x)$ in such a way that we obtain the respective quantised or overflow corrected values of $z_1(n+1)$ and $z_2(n+1)$. This means in particular that the coefficients a_{ij} vary from cycle to cycle and hence become a function of n.

$$z_1(n+1) = a'_{11}(n)\, z_1(n) + a'_{12}(n)\, z_2(n)$$
$$z_2(n+1) = a'_{21}(n)\, z_1(n) + a'_{22}(n)\, z_2(n)$$

The nonlinear system is thus converted into a shift-variant system with the state equation

$$z(n+1) = A'(n)\, z(n)\ .$$

The range of values that the coefficients of $A'(n)$ may assume depends immediately on the parameters k_q as specified in (8.62) and k_o as specified in (8.63).

Examples

Quantisation after a coefficient multiplication

$$y = Q(ax) = \frac{Q(ax)}{ax} ax$$

The quotient $Q(ax)/ax$ assumes values between 0 and k_q according to (8.62).

$$y = \left(0 \ldots k_q\right) ax = a'x$$

In the case of magnitude truncation, the coefficient a' takes on values between 0 and a, for rounding between 0 and $2a$.

Overflow correction after an addition

$$y = O(ax_1 + bx_2) = \frac{O(ax_1 + bx_2)}{ax_1 + bx_2}(ax_1 + bx_2)$$

The quotient $O(ax_1 + bx_2)/(ax_1 + bx_2)$ assumes values between k_o and 1 according to (8.63).

$$y = \left(k_o \ldots 1\right)(ax_1 + bx_2) = a'x_1 + b'x_2$$

This leads to the following ranges for a' and b':

$$k_o a \le a' \le a \qquad \text{and} \qquad k_o b \le b' \le b \quad .$$

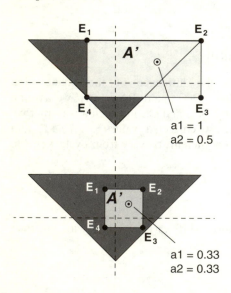

a1 = 1
a2 = 0.5

a1 = 0.33
a2 = 0.33

Fig. 8-41
Two examples for the scattering range of the coefficients of a direct form filter with triangular overflow characteristic and rounding after each coefficient multiplication

A stable filter may become unstable by these modifications of the coefficients a_{ij}, since eigenvalues of $A'(n)$ with magnitude larger than unity may occur. Figure 8-41 shows the example of a direct-form filter with two possible stable coefficient pairs. The applied overflow correction is triangle, the quantisation characteristic is rounding after each coefficient multiplication. In the one case, the scattering range of the coefficients extends beyond the stability triangle. In the other case, the coefficients remain completely within the stability triangle.

As for linear systems, the trajectories are simply calculated by continued multiplication by the system matrix A. In the case of nonlinear systems with rounding and possibly overflow, we obtain arbitrary sequences of matrices in the set A'. In our example, the set of matrices A' is defined as

$$A' = \begin{pmatrix} -a_1' & -a_2' \\ 1 & 0 \end{pmatrix},$$

where the pair a'_1/a'_2 may assume any value within the grey regions in Fig. 8-41. In order to guarantee stability, each possible sequence of matrices in A' has to satisfy the stability criterion that we introduced in the previous section. This requires that, from some finite iteration step, the neighbourhood of the origin passed by the trajectories does not grow any further. If some sequence satisfies the instability criterion (complete enclosure of W_0), stability is no longer guaranteed.

Our problem seems to be insoluble at first glance since an infinite number of possible matrix sequences of A' would have to be checked. Brayton and Tong [7] showed, however, that stability can be proven in a finite number of steps:

1. For proof of stability, it is sufficient to consider the extreme matrices E_i only, which form the corner points of the scattering region of the set of matrices A' as illustrated in Fig. 8-41. The proof is given in [7] that if the set of extreme matrices of A' is stable, then also A' is stable.

2. We start the algorithm with the extreme matrix E_1. The initial neighbourhood W_0 represented by its extremal points is continuously multiplied by the matrix E_1 until one of the known termination criteria is fulfilled. If the result is "unstable", the whole procedure can be terminated with the result "unstable". If the result is "stable", the algorithm is continued. We denote the convex hull that we obtain as a result of the described first step of the algorithm as W_1. In the second step, we multiply W_1, represented by its extremal points, continuously by the extreme matrix E_2 until one of the termination criteria is fulfilled. If the result is "unstable", we terminate. If the result is "stable", we obtain a new convex hull that we denote as W_2. We continue with the matrix E_3 accordingly. In the following, all m extreme matrices E_i are applied cyclically (E_1, E_2, E_3, ..., E_m, E_1, E_2, ...) in the described manner. The algorithm is terminated with the result "unstable" if the criterion for instability is fulfilled for the first time. We terminate with the result "stable" if we do not observe any further growth of the convex hull after a complete pass through all m extreme matrices. This termination criterion can be expressed as $W_k = W_{k+m} = W^*$.

In the same way as we did for the linear system in Sect. 8.7.4.2, the finite neighbourhood W^* of the origin, which we finally obtain as a result of the algorithm, can be used to define a Lyapunov function for the nonlinear system.

In the context of the described algorithm, instability means that at least one sequence of the E_i leads to the result "unstable". Since these unstable sequences may not be run through in practical operation, the actual behaviour of the filter must be considered undetermined. If the result is "stable", global stability is guaranteed in any case.

The constructive algorithm allows us to determine the range of coefficients in a given implementation for which the filter is globally asymptotically stable. The involved nonlinearities are characterised by the parameters k_o and k_q and hence by the sectors which enclose the quantisation or overflow characteristics. The details of the mechanisms that lead to limit or overflow cycles are not considered with this algorithm. Since the sectors for quantisation and overflow overlap to a great extend, it is not possible to conclude whether limit or overflow cycles or both will occur in the filter under consideration when the algorithm says "unstable". If we investigate both effects separately in the following, then we do that under the assumption that in each case only one of the nonlinearities – quantisation or overflow – is implemented.

The constructive algorithm yields in most cases comparable or even larger coefficient ranges with guaranteed stability than have been previously seen in the literature. In a few cases, the results are more pessimistic, however, which is a consequence of the fact that the algorithm is only based on the sectors of the nonlinear characteristics, and the detailed mechanisms of quantisation and overflow are not considered [18, 19].

In the following, we test direct form, normal form and wave digital filters for stability. We investigate quantisation and overflow characteristics in various combinations. In the graphical representations, which show the results of the constructive algorithm, black areas in the coefficient plane indicate that at least one extreme matrix lies outside the stability region of the respective filter structure (refer to Fig. 8-38). In this case, we need not apply the algorithm, as stability is not guaranteed. There is in any case one sequence of matrices in A' that is unstable, that is, the sequence consisting of only this unstable extreme matrix. In the grey areas, potential instability is proven by the algorithm. It is interesting to note that the algorithm may come to the result "unstable" even if all extreme matrices are formed by coefficient pairs that lie within the respective stability regions. In the white areas, globally asymptotical stability is guaranteed.

Direct-form filters

With only one quantiser, placed behind the adder as shown in Fig. 8-42, the two possible extreme matrices take on the form

$$\begin{pmatrix} -\alpha_i a_1 & -\alpha_i a_2 \\ 1 & 0 \end{pmatrix} \quad i = 1,2$$

$$\text{with } \alpha_1 = k_q, \quad \alpha_2 = \min(0, k_o) \ .$$

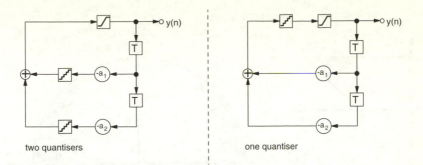

two quantisers   one quantiser

Fig. 8-42 Placement of nonlinearities in the direct form structure

If the sector of the overflow characteristic completely overlaps with the sector of the quantiser, the overflow correction does not contribute anything else to the variation range of the coefficients. Only for negative values of k_o, overflow may reduce the stability range compared to pure quantisation. The latter applies to two's complement overflow and, in certain coefficient ranges, to the triangle characteristic. If we intend to investigate overflow alone, the parameters α_1 and α_2 assume the following values:

$$\alpha_1 = 1, \quad \alpha_2 = k_o .$$

It is assumed in this case that adders, multipliers and memories operate as if they were analog.

If we quantise directly after each coefficient multiplication (Fig. 8-42), we obtain four extreme matrices:

$$\begin{pmatrix} -\alpha_i a_1 & -\beta_j a_2 \\ 1 & 0 \end{pmatrix} \quad i, j = 1,2$$

$$\text{with } \alpha_1 = \beta_1 = k_q, \quad \alpha_2 = \beta_2 = \min\left(0, k_o k_q\right) .$$

Figure 8-43 shows the stability regions for magnitude truncation and rounding with one or two quantisers respectively and without considering overflow. Fig. 8-44 depicts the results for overflow without quantisation. In addition to the stable regions of the coefficients a_1 and a_2, we also show the regions within the unit circle in the z-plane, where the corresponding poles (realised with the coefficients a_1 and a_2) are stable. For reasons of symmetry it is sufficient to show the upper half of the unit circle only. In some cases, the stability regions are considerably restricted compared to the linear filter, which is stable within the whole unit circle. What recommendation can be derived from the figures? Sign-magnitude arithmetic with truncation after summation, which corresponds to the case of one MT quantiser, yields the largest stability region. The two's complement overflow characteristic should be avoided in any case. Saturation is the best choice.

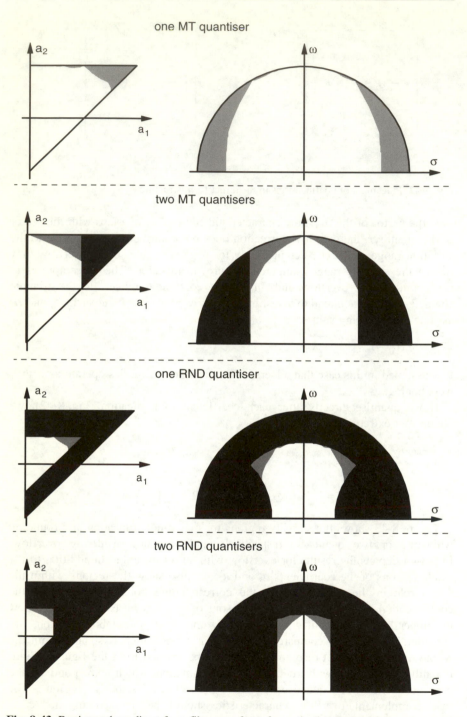

Fig. 8-43 Regions where direct form filters are free of quantisation limit cycles

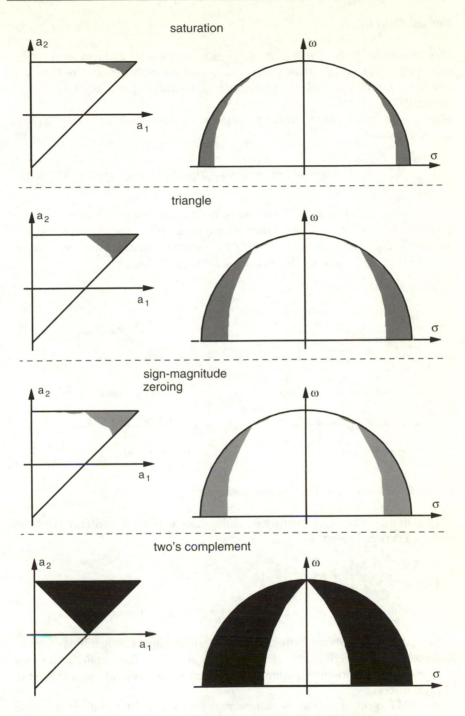

Fig. 8-44 Regions where direct form filters are free of overflow limit cycles

Normal-form filters

Quantisation may be performed after each coefficient multiplication which requires four quantisers. If an adder or accumulator with double wordlength is available, quantisation can take place after summation, which only requires two quantisers (Fig. 8-45).
With two quantisers placed behind the adders, the four possible extreme matrices can be expressed as

$$\begin{pmatrix} -\alpha_i\sigma & -\alpha_i\omega \\ \beta_j\omega & \beta_j\sigma \end{pmatrix} \quad i,j = 1,2 \qquad \text{with } \alpha_1 = \beta_1 = k_\mathrm{q}, \quad \alpha_2 = \beta_2 = \min\left(0,k_\mathrm{o}\right).$$

A negative k_o only occurs in the case of two's complement overflow. All other overflow characteristics do not contribute anything else to the variation range of the coefficients. If we intend to investigate overflow isolated from quantisation, the parameters α_1, α_2, β_1 and β_2 assume the following values:

$$\alpha_1 = \beta_1 = 1, \qquad \alpha_2 = \beta_2 = k_\mathrm{o} \ .$$

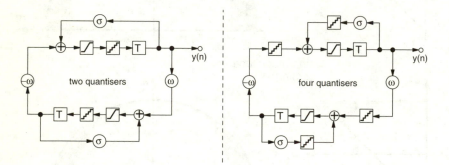

Fig. 8-45 Nonlinearities in the normal-form structure

Quantisation after each coefficient multiplication (four quantisers) yields 16 extreme matrices in total:

$$\begin{pmatrix} -\alpha_i\sigma & -\beta_j\omega \\ \mu_k\omega & \nu_l\sigma \end{pmatrix} \quad i,j,k,l = 1,2$$

$$\text{with } \alpha_1 = \beta_1 = \mu_1 = \nu_1 = k_\mathrm{q}, \quad \alpha_2 = \beta_2 = \mu_2 = \nu_2 = \min\left(0,k_\mathrm{o}k_\mathrm{q}\right) \ .$$

In contrary to the direct form, the normal form exhibits configurations that are absolutely free of overflow and quantisation limit cycles. This applies to the case of sign-magnitude arithmetic with the quantisers placed behind the adders (two MT quantisers).

Four MT quantisers lead to narrow regions close to the unit circle where instabilities my occur for saturation, zeroing and triangle overflow characteristics

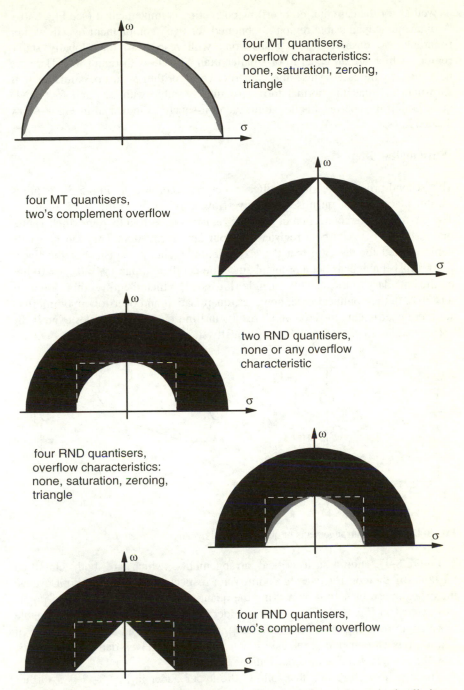

Fig. 8-46 Regions where normal form filters are free of overflow and quantisation limit cycles

as well as for the case that no overflow correction is implemented (see Fig. 8-46). A diamond-shaped stable region is obtained for two's complement overflow. For rounding, we observe severe restrictions with respect to the globally stable regions, which are all equal to or smaller than a circle with radius 1/2. Here we have one of the cases where the constructive algorithm is too pessimistic. It is shown in [2], that the normal form structure is stable with two and four RND quantisers if the coefficients lie within the unit square as indicated in Fig. 8-46 by dashed lines.

Wave digital filter

The second-order wave digital filter structure according to Fig. 5-52 requires careful placement of quantisers and overflow corrections in the block diagram. The crosswise interconnection of the state registers leads to a loop in which values may oscillate between both registers without any attenuation. This can be easily demonstrated for the case that the coefficients γ_1 and γ_2 vanish. For the linear filter, this behaviour is not a problem since the coefficient pair ($\gamma_1 = 0$, $\gamma_2 = 0$) lies on the boundary of the stability triangle (Fig. 8-31) which identifies this system as unstable. In the nonlinear case, however, quantisation and overflow may in effect set these coefficients to zero, which results in limit or overflow cycles even if the corresponding linear system is asymptotically stable.

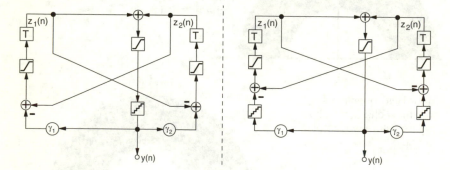

Fig. 8-47 Unstable arrangements of quantisers and overflow corrections

Figure 8-47 shows such critical arrangements, which are both classified unstable by the constructive algorithm. With respect to quantisation limit cycles, the problem can be solved by moving the quantisers behind the summation points as depicted in Fig. 8-48. This requires adders that are able to handle the double wordlength resulting from the coefficient multiplication. If an appropriate type of quantiser is chosen, the previously mentioned loop receives damping elements, which force quantisation limit oscillations to die down.

The overflow correction that follows the upper adder in Fig. 8-47 is a further critical element that has to be considered in the context of overflow limit cycles. It turns out that none of the five overflow characteristics (8.63) leads to filters that ensure freedom from overflow limit cycles. Investigation of various alternatives using the constructive algorithm leads to the conclusion that stable conditions can only be achieved if this overflow correction is completely omitted. So in order to preserve the sum $z_1(n) + z_2(n)$, an extra front bit is required which the coefficient multipliers and the subsequent adders have to cope with.

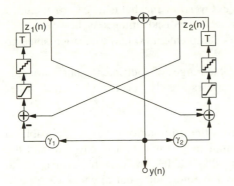

Fig. 8-48
Optimum positioning of quantisers and overflow corrections

With the mentioned special properties of the adders and multipliers, the structure according to Fig. 8-48 possesses four extreme matrices:

$$\begin{pmatrix} -\gamma_1 \alpha_i & (1-\gamma_1)\,\alpha_i \\ (-1+\gamma_2)\,\beta_j & \gamma_2 \beta_j \end{pmatrix} \quad i,j = 1,2$$

$$\text{with } \alpha_1 = \beta_1 = k_q, \quad \alpha_2 = \beta_2 = \min(0,k_o) \ .$$

Application of the constructive algorithm using these extreme matrices yields very contrary results. In the case of rounding, no combination of the coefficients γ_1 and γ_2 exists that is globally asymptotically stable. The situation with MT quantisation is totally different. Stability is guaranteed in combination with any overflow characteristic in the whole coefficient triangle. The only exception is two's complement overflow, which only allows stable filters if both coefficients are positive (Fig. 8-49).

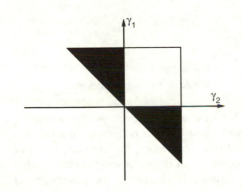

Fig. 8-49
Stable coefficient range with MT quantisation and two's complement overflow

8.7.5 A Search Algorithm in the State Plane

The constructive algorithm, which we introduced in the previous section, proved to be a powerful and universal tool to identify coefficient ranges in real filter implementations that are globally asymptotically stable and thus free of any limit cycles. The computation times to produce graphical representations of the stable regions such as in Fig. 8-43 are moderate. Overflow mechanisms can be included in the stability analysis. The following aspects are disadvantageous, however:

- The method cannot be applied to two's complement (value) truncation since no sector condition is valid in this case.
- The method identifies potentially unstable ranges of coefficients, but is not able to give any indication if limit cycles really occur in actual filter implementations or not.

The method described below overcomes these limitations, but is applicable to quantisation limit cycles only. Two's complement truncation is covered, and the algorithm clearly terminates with the result "stable" or "unstable". In order to come to these concrete results, we finally have no choice but to check each point in the state plane to see if a trajectory starting in this point

- converges into the origin,
- lies on a limit cycle, or
- leads into a limit cycle (Fig. 8-50).

If we find a point that ends up in a cycle, the respective pair of filter coefficients is unstable. Because of the finite precision of the signal representation, the state plane consists of a grid of discrete points (z_1/z_2). If we multiply a state vector $z(n)$ by the system matrix A and apply the respective rounding or truncation rules, we obtain a state vector $z(n+1)$ that again fits into the grid of discrete state values. Before starting the "exhaustive search" [3, 49] for unstable points in the state plane, we must have some idea of the size of the region around the origin that we have to scan and of the maximum period of the limit cycle.

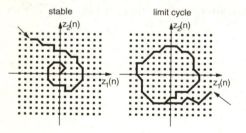

Fig. 8-50
Stable and unstable trajectories

If the bounds on the amplitude of the limit cycles are known, we also know the size of the region to be scanned. Any point outside these bounds cannot lie on a limit cycle and hence must not be considered. The reason for the occurrence of limit cycles is the existence of periodic sequences of error values $e(n)$ that develop from the quantisation process. The mechanism is similar to the generation of quantisation noise as demonstrated in [35], where the behaviour of a filter is described that exhibits limit cycles in the absence of an input signal. The amplitude of the input signal $x(n)$ is raised step by step from zero up by a few quantisation steps. The initial periodic limit cycle gradually turns into a random sequence which is superimposed onto the signal. The amplitudes of the initial limit cycle and of the finally superimposed noise are of the same order of magnitude.

The quantisation error satisfies the following condition for the various types of quantisers:

$$|e(n)| \le \varepsilon \, \Delta x_q$$

with $\varepsilon = 1/2$ for RND quantisation and $\varepsilon = 1$ for MT and VT quantisation.

The maximum amplitude that a limit cycle may assume depends on the transfer function $H(z)$ between the place of the respective quantiser and the state registers (delay elements) of the filter whose output are the state variables $z_1(n)$ and $z_2(n)$. In Sect. 8.4.3, we have derived the following relationship between the maximum values of input and output signal of a LSI system, which is characterised by its unit-sample response $h(n)$:

$$|y|_{\max} = |x|_{\max} \sum_{n=0}^{\infty} |h(n)| \ .$$

A bound M on the amplitude of limit cycles can therefore expressed as

$$M = \varepsilon \, \Delta x_q \sum_{n=0}^{\infty} |h(n)|$$

with $\varepsilon = 1/2$ for RND quantisation and $\varepsilon = 1$ for MT and VT quantisation.

If a filter possesses NQ quantisers, the partial contributions of these quantisers to the overall limit cycle have to be summed up at the output.

$$M = \varepsilon \, \Delta x_q \sum_{i=1}^{NQ} \sum_{n=0}^{\infty} |h_i(n)| \qquad (8.70)$$

According to the results of Sect. 8.5.1, the internal transfer functions between a quantiser and a state register can be represented in the following general form:

$$H(z) = \frac{z^{-1}\left(1 - b_1 z^{-1}\right)}{1 + a_1 z^{-1} + a_2 z^{-2}} = \frac{z - b_1}{z^2 + a_1 z + a_2} = \frac{z - b_1}{(z - z_{\infty 1})(z - z_{\infty 2})} \ . \qquad (8.71)$$

The transfer function (8.71) may be rewritten in three different ways:

$$H_a(z) = \frac{z - b_1}{z - z_{\infty 1}} \frac{1}{z - z_{\infty 2}}$$

$$H_b(z) = \frac{1}{z - z_{\infty 1}} \frac{z - b_1}{z - z_{\infty 2}}$$

$$H_c(z) = \frac{z_{\infty 1} - b_1}{z_{\infty 1} - z_{\infty 2}} \frac{1}{z - z_{\infty 1}} + \frac{z_{\infty 2} - b_1}{z_{\infty 2} - z_{\infty 1}} \frac{1}{z - z_{\infty 2}} \, .$$

$H_a(z)$ and $H_b(z)$ are based on the assumption of a cascade arrangement of the pole terms with two different assignments of the zero to one of the poles. $H_c(z)$ is based on the assumption of a parallel arrangement of the pole terms. For each of the three forms, we will determine an upper bound M on the absolute sum over the respective unit-sample response. The smallest of these three values is then used to calculate the bound on the amplitude of the limit cycles using (8.70).

$$H_a(z) = \frac{z - b_1}{z - z_{\infty 1}} \frac{1}{z - z_{\infty 2}}$$

$$H_a(z) = \left[1 + \frac{z_{\infty 1} - b_1}{z - z_{\infty 1}}\right] \frac{1}{z - z_{\infty 2}}$$

The unit-sample responses of the two partial systems are easily obtained as

$$h_I(n) = \delta(n) + (z_{\infty 1} - b_1) \, z_{\infty 1}^{n-1} \, u(n-1) \ ,$$

$$h_{II}(n) = z_{\infty 2}^{n-1} \, u(n-1) \ .$$

The summation is performed in both cases over a geometric progression which will converge in any case as the magnitude of the poles of stable filters is less then unity.

$$\sum_{n=0}^{\infty} |h_I(n)| = 1 + |z_{\infty 1} - b_1| \sum_{n=0}^{\infty} |z_{\infty 1}|^{n-1} = 1 + \frac{|z_{\infty 1} - b_1|}{1 - |z_{\infty 1}|}$$

$$\sum_{n=0}^{\infty} |h_I(n)| = \frac{1 - |z_{\infty 1}| + |z_{\infty 1} - b_1|}{1 - |z_{\infty 1}|}$$

$$\sum_{n=0}^{\infty} |h_{II}(n)| = \sum_{n=0}^{\infty} |z_{\infty 2}|^{n-1} = \frac{1}{1 - |z_{\infty 2}|}$$

$$M_a = \sum_{n=0}^{\infty} |h_I(n)| \sum_{n=0}^{\infty} |h_{II}(n)| = \frac{1 - |z_{\infty 1}| + |z_{\infty 1} - b_1|}{(1 - |z_{\infty 1}|)(1 - |z_{\infty 2}|)}$$

The corresponding result for the transfer function $H_b(z)$ can be directly down written since $z_{\infty 1}$ and $z_{\infty 2}$ simply reverse roles.

$$M_b = \frac{1 - |z_{\infty 2}| + |z_{\infty 2} - b_1|}{(1 - |z_{\infty 1}|)(1 - |z_{\infty 2}|)}$$

$H_c(z)$ is a partial fraction representation of (8.71) with the following unit-sample response:

$$h(n) = \frac{z_{\infty 1} - b_1}{z_{\infty 1} - z_{\infty 2}} \, z_{\infty 1}^{n-1} \, u(n-1) + \frac{z_{\infty 2} - b_1}{z_{\infty 2} - z_{\infty 1}} \, z_{\infty 2}^{n-1} \, u(n-1) \ .$$

Using the triangle inequality we have

$$h(n) \leq \frac{\left|z_{\infty 1} - b_1\right|}{\left|z_{\infty 1} - z_{\infty 2}\right|} \, \left|z_{\infty 1}\right|^{n-1} \, u(n-1) + \frac{\left|z_{\infty 2} - b_1\right|}{\left|z_{\infty 2} - z_{\infty 1}\right|} \, \left|z_{\infty 2}\right|^{n-1} \, u(n-1) \ .$$

Summation over the geometric progressions finally results in

$$M_c = \sum_{n=0}^{\infty} \left|h(n)\right| = \frac{1}{\left|z_{\infty 1} - z_{\infty 2}\right|} \left(\frac{\left|z_{\infty 1} - b_1\right|}{1 - \left|z_{\infty 1}\right|} + \frac{\left|z_{\infty 2} - b_1\right|}{1 - \left|z_{\infty 2}\right|} \right) \ .$$

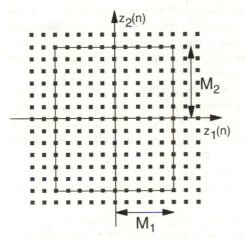

Fig. 8-51
Region to be scanned with respect to the occurrence of limit cycles

For direct-form and normal-form filters, the bounds on the amplitude of the limit cycles M_1 and M_2 are equal for both state variables $z_1(n)$ and $z_2(n)$, which yields a square in the state plane that has to be scanned (Fig. 8-51). For the wave digital filter structure, M_1 and M_2 differ depending on the values of the coefficients γ_1 and γ_2, resulting in a rectangle. The number of points that we have to check for stability thus amounts to $4 \times M_1 \times M_2$.

One point after the other in the bounded region is chosen as a starting point of a trajectory, which is obtained by continued multiplication by the system matrix A and application of the respective rounding or truncation operations. After at least $4 \times M_1 \times M_2$ operations, one of the following termination criteria applies:

- The trajectory leaves the bounded region. This means that the starting point does not lie on a limit cycle. The further calculation of the trajectory can be stopped, and the next point is checked.

- The trajectory returns back to the starting point. This means that a limit cycle has been found within the bounded region. The corresponding pair of filter coefficients is unstable. Further points need not be checked.
- The trajectory converges into the origin. The next point is checked.
- After $4 \times M_1 \times M_2$ operations, none of the above events has taken place. This means that the trajectory has entered a limit cycle since at least one point must have been reached for a second time. The corresponding pair of filter coefficients is unstable. Further points need not be checked.

If all $4 \times M_1 \times M_2$ point are scanned and no limit cycle has been found, then the corresponding pair of filter coefficients is stable.

There are ways to make the described algorithm more efficient. One possibility is to mark all starting points that finally end up in the origin. If such a point is hit later by another trajectory, the further pursuit of the trajectory can be stopped. This requires the establishment of a map of the scanned region and hence the provision of memory. A further improvement is to record each trajectory temporarily in an interim buffer. If the trajectory converges into the origin, all points recorded in the interim buffer are copied into the map. The algorithm becomes more complex, and still more memory is required. A problem is the dimensioning of the arrays in a software implementation of the algorithm as the occurring bounds M_1 and M_2 are not known in advance when the program is started.

A strategy that does not require any additional memory, but still saves a lot of computation time, is based on a special sequence in which the points in the state plane are checked for stability. We start with the 8 points that directly surround the origin. If all these points are stable, we proceed to the next quadratic band which consists of 16 points. If these points are stable, the next band with 24 points is checked, and so on. If a trajectory reaches one of these inner bands that have already been checked and identified as stable, the further calculation of the respective trajectory can be stopped.

The latter method has been applied in calculating the following figures which show the definitely stable coefficient ranges of the various filter structures. The coefficients are chosen in steps according to the VGA screen resolution. In the black regions, quantisation limit cycles have been found, whereas the white areas are asymptotically stable.

Direct-form filter

For RND and MT quantisation, the coefficient ranges that are free of quantisation limit cycles are in all cases larger than predicted by the constructive approach. For one MT and for two RND quantisers, the unstable regions coincide with regions where at least one extreme matrix is unstable (Fig. 8-52). For two MT quantisers, there are even small regions free of quantisation limit cycles in which unstable extreme matrices occur. With one exception, all grey regions in Fig. 8-43 turn out to be stable. We remember that these are regions where finite products of stable extreme matrices exist that are unstable. The mentioned exception is the case of one MT quantiser, where small irregular regions are observed along the upper edge of the stability triangle that exhibit limit cycles. These fall into the grey regions in Fig. 8-43 which predict potential instability. A magnified detail of this area is depicted in Fig. 8-53. The stable regions for value truncation are asymmetrical and have extremely irregular boundaries.

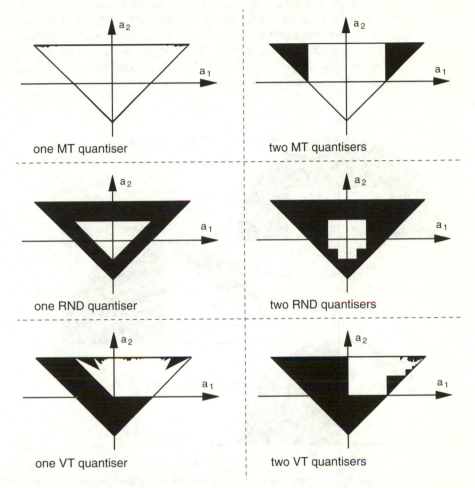

one MT quantiser

two MT quantisers

one RND quantiser

two RND quantisers

one VT quantiser

two VT quantisers

Fig. 8-52 Coefficient regions of direct-form filters that are free of limit cycles

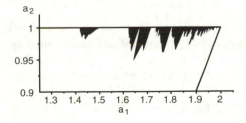

Fig. 8-53
Magnified detail of the case with one MT quantiser (direct form)

Normal-form filter

For the normal-form filter too, the results of the constructive approach are more conservative compared to those of the search algorithm (Fig. 8-54). The normal-form structure, if realised with two or four MT quantisers, is in any case free of quantisation limit cycles. For RND quantisers, the unit square is stable, which is in contrast to the constructive algorithm which predicts the circle with radius ½ to be free of limit cycles. For value truncation, we again obtain asymmetrical regions with irregular boundaries. In the case of two VT quantisers, we observe a narrow linear region that is stable. Under the angle of 45° where both filter coefficients have equal magnitude, quantisation effects obviously cancel out.

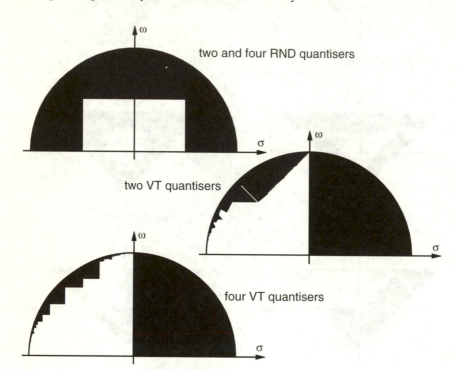

Fig. 8-54 Coefficient regions of normal-form filters that are free of limit cycles

Wave digital filter

The wave digital filter structure according to Fig. 8-48 proves to be free of quantisation limit cycles within the whole stability triangle if MT quantisation is applied. For RND and VT quantisation, however, wide areas of the coefficient plane are unstable. Merely some rectangular regions show asymptotic stability (Fig. 8-55). For VT quantisation, we again observe as a special feature a narrow linear region along the γ_1-axis that is free of limit cycles.

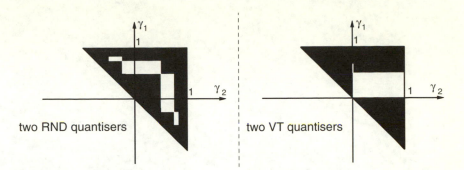

Fig. 8-55 Coefficient regions of the wave digital filter structure that are free of limit cycles

The computational effort of the search algorithm is higher compared to the constructive approach. For the application in filter design programs, this is not important, however, since only a few pairs of coefficients have to be checked for stability. In contrast to that, about 100,000 coefficient pairs have been calculated to obtain the graphs in Fig. 8-52, Fig. 8-54 and Fig. 8-55.

8.7.6 Summary

Numerous combinations of filter structures and quantisation or overflow characteristics have been checked with respect to their stability properties. It turns out that MT quantisation is superior. Normal form and wave digital filter structure are definitely free of quantisation limit cycles if MT quantisation is applied. As for the direct form, the variant with one quantiser should be chosen because only very small coefficient regions are unstable.

As for overflow limit cycles, the two's complement overflow characteristic should be avoided. For all other overflow characteristics, normal form and wave digital filter structures are free of overflow oscillations. In the case of the direct form, saturation yields the largest coefficient region without overflow limit cycles.

The direct form is the only structure that cannot be made free of quantisation or overflow limit cycles by any combination of arithmetic, quantiser and overflow correction. The direct form is nevertheless attractive, as it features the lowest implementation complexity of all structures.

9 Oversampling and Noise Shaping

Analog-to-digital and digital-to-analog conversion are important functions that greatly determine the quality of the signal processing chain. In Chap. 4 we showed that a certain effort is required for filtering in the continuous-time domain in order to avoid errors in the transition process from the analog to the discrete-time world and vice versa. The complexity of the involved filters increases more, the better the frequency range $0 \leq |\omega| < \omega_s/2$, the baseband, is to be utilised and the higher the requirements are with respect to the avoidance of signal distortion. The resolution of the converters determines the achievable signal-to-noise ratio (SNR). Each additional bit improves the SNR by about 6 dB. The costs of the converters, however, increase overproportionally with the resolution.

Since modern semiconductor technologies admit ever-increasing processing speeds and circuit complexities, a trend can be observed to reduce the complexity of analog circuitry as far as possible and shift the effort into the digital domain. In particular, the truncation of the spectra before A/D and after D/A conversion is carried out by digital filters. Additionally, the resolution of the converters can be reduced without degrading the SNR. This can be achieved by choosing sampling frequencies much higher than required to meet Shannon's sampling theorem.

9.1 D/A Conversion

From the viewpoint of the frequency domain, the periodically continued portions of the spectrum of the discrete-time signal have to be filtered away in order to reconstruct the corresponding analog signal. Figure 9-1 shows an example of such a periodic spectrum. In case of a converter that operates without oversampling, the spectrum has to be limited to half the sampling frequency by means of sharp cutoff analog low-pass filters. Converters functioning according to this principle are sometimes referred to as Nyquist converters. The name originates from the fact that half the sampling frequency is also called the Nyquist frequency in the literature.

The disadvantages of sharp cutoff analog filters in this context have already been discussed in Chap. 4. In order to avoid the drawbacks of this filtering, oversampling is applied. All today's CD players, for instance, use this technique. On a compact disc, signals are stored at a sampling rate of 44.1 kHz and a resolution of 16 bits. By appropriate signal processing, the sampling rate is increased by a factor of 4, 8 or even more before the signal is applied to the D/A converter. We denote this factor in the following as the oversampling ratio, OSR.

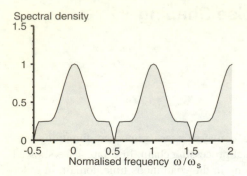

Fig. 9-1
Periodic spectrum of a discrete-time signal

If we aim at reducing the complexity of the analog filter only, *OSR*s of 4, 8 or 16 are common. But oversampling can also be used to simplify or even eliminate completely the D/A converter. In this case, *OSR*s of 128 or higher are needed.

9.1.1 Interpolation

In order to arrive at the higher sampling rate, we start with inserting *OSR*−1 zeros between two adjacent samples of the present sequence (e.g. 3 in the case of four-times and 7 in the case of eight-times oversampling). We will show, in the following, what the consequences of this procedure are in the frequency domain. According to (3.9a), the discrete-time sequence *f(n)* to be converted to analog has the spectrum

$$F\left(e^{j\omega T}\right) = \sum_{n=-\infty}^{+\infty} f(n)\, e^{-j\omega T n}\;.$$

The sequence *f'(n)* that results from supplementing with zeros as described above has the spectrum

$$F'\left(e^{j\omega T}\right) = \sum_{n=-\infty}^{+\infty} f'(n)\, e^{-j\omega T n}\;,\tag{9.1}$$

accordingly. Since only every *OSR*th sample of *f'(n)* is nonzero, we can replace *n* by *OSR*×*n* in (9.1) without affecting the result.

$$F'\left(e^{j\omega T}\right) = \sum_{n=-\infty}^{+\infty} f'(OSR\, n)\, e^{-j\omega T\, OSR\, n}$$

Since *f'(OSR×n)* corresponds to the original sequence *f(n)*, we have

$$F'\left(e^{j\omega T}\right) = \sum_{n=-\infty}^{+\infty} f(n)\, e^{-j\omega T\, OSR\, n}$$

$$F'\left(e^{j\omega T}\right) = F\left(e^{jOSR\omega T}\right) .$$ (9.2)

The shape of the spectrum is not, in principle, influenced by the insertion of the zeros. Only the scale of the frequency axis has been changed. Figure 9-2 illustrates this result under the assumption of a four-times oversampling. A digital filter can be used to filter out the desired baseband. Since the frequency response of this digital filter is periodic, as indicated in Fig. 9-2, the result of this filtering is a periodic spectrum (Fig. 9-3).

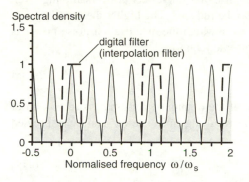

Fig. 9-2
Spectrum of the digital signal after insertion of zeros to achieve four-times oversampling

As a result, we observe a large spacing between the baseband and the periodically continued portions of the spectrum. A relatively simple analog filter is now sufficient to recover the baseband and thus the desired continuous-time signal. A possible frequency response of this analog interpolation filter, which cuts off at 7/8 ω_s, is also shown in Fig. 9-3.

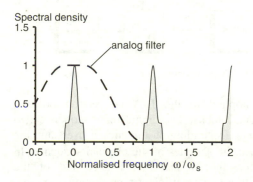
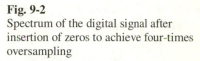

Fig. 9-3
Spectrum of the oversampled signal after low-pass filtering using a digital interpolation filter (dashed line shows frequency response of a subsequent analog filter)

As a result of the digital low-pass filtering, we obtain a new discrete-time sequence featuring a higher temporal resolution than the original. The previously inserted zeros are replaced by values that appropriately interpolate the original samples. One can easily imagine that, with increasing OSR, we achieve an increasingly better approximation of the desired analog signal in the discrete-time domain, which finally reduces the effort for filtering in the continuous-time domain.

Also the equalisation of the $\sin x/x$-distortion becomes easier, since the frequency range of interest only covers one quarter of the range depicted in

Fig. 4-8, leaving only a loss of at most 0.25 dB to be compensated. For eight-times or higher oversampling, as is custom today, this aspect of signal reconstruction can be neglected in practice.

9.1.2 Noise Filtering

The concept of oversampling is not only helpful in reducing the effort for filtering in the signal conversion process. By making use of a side-effect of oversampling, even the complexity of the D/A converter may be reduced. The higher the sampling rate chosen, the larger the frequency range over which the quantisation noise power is spread. According to (8.7), this noise power amounts to

$$N_0 = \Delta x_q^2 / 12 \ ,$$ (9.3a)

where Δx_q is the step size at the output of the quantiser. Assuming a resolution of b bits and a dynamic range of ± 1, the resulting noise power can be expressed according to (8.8) as

$$N_0 = 2^{-2b} / 3 \ .$$ (9.3b)

If the baseband up to the Nyquist frequency is isolated from the oversampled quantised signal by means of an appropriate low-pass filter, a large portion of the broadband quantisation noise is removed.

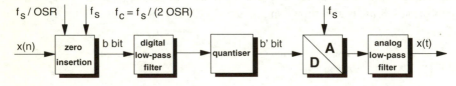

Fig. 9-4 Functional blocks of oversampling D/A conversion

The principle described can be applied to the D/A conversion process. Figure 9-4 shows the involved functional blocks. The digital input sequence $x(n)$ is represented with a resolution of b bits. Depending on the chosen oversampling ratio, $OSR-1$ zeros are inserted between the samples of the original sequence. In the subsequent digital low-pass filter, the baseband is filtered out of the oversampled signal. If an FIR filter with a k-bit coefficient representation is used, we obtain a sequence with a resolution of $k+b$ bits at the output. The subsequent quantiser reduces the wordlength to the b'-bit resolution of the D/A converter, resulting in a broadband quantisation noise power of

$$\sigma_e^2 = N_0 = 2^{-2b'} / 3 \ ,$$

which is also superimposed onto the stepped signal at the output of the D/A converter. The final analog low-pass filter can be optimised in two respects:

If low effort for filtering is an objective, a low-pass filter is chosen with a passband edge at $f_s(0.5/OSR)$ and a stopband edge at $f_s(1-0.5/OSR)$. The resolution b' of the D/A converter has to be chosen to be at least equal to the original resolution b of the input signal $x(n)$ if loss of signal quality is to be avoided.

If the complexity of the converter is to be minimised, on the other hand, an analog low-pass filter is required that cuts off sharply at the Nyquist frequency. Besides the unwanted out-of-band portions of the periodic spectrum (Fig. 9-3), a large amount of the noise power outside the baseband is removed. In case of four-times oversampling, only a quarter of the quantisation noise still exists in the signal, corresponding to a loss of 6 dB. The signal-to-noise ratio is thus improved as if the signal was present at a one-bit higher resolution. With $OSR = 16$, the effective resolution even rises by two bits. Generally speaking, each doubling of the sampling rate leads to a gain in resolution of half a bit, equivalent to a lowering of the quantisation noise by 3 dB. Since the signal is available at a better signal-to-noise ratio after low-pass filtering compared to ordinary Nyquist conversion, the resolution b' of the quantiser and thus of the D/A converter can be reduced below b without a deterioration in the quality of the original signal sequence $x(n)$.

A good idea at first glance to further reduce the complexity of the converter would be to increase the OSR such that a resolution of only one bit would be sufficient. The A/D converter would then consist of a simple sign detector, the D/A converter of a two-level digital output. Noise filtering as described above is not appropriate, however, to reach this desired goal. On the one hand, an OSR of about 10^9 would be required mathematically to achieve a signal-to-noise ratio equivalent to a 16-bit resolution. On the other hand, the complete information about the course of a signal between the zero crossings gets lost and cannot be recovered no matter what OSR is used. It can be stated that the superimposed noise signal $e(n)$ becomes increasingly correlated with the processed signal as the resolution decreases. As a consequence, larger portions of the noise spectrum fall into the baseband. The concept of noise filtering thus becomes ineffective. The strong correlation between the signal $x(n)$ and the error $e(n)$ becomes obvious if we consider the simple case of a one-bit (two-level) quantisation.

$$e(n) = Q[x(n)] - x(n)$$

$$e(n) = \text{sign}(x(n)) - x(n)$$

$$e(n) = \begin{cases} 1 - x(n) & \text{for } x(n) > 0 \\ -1 - x(n) & \text{for } x(n) < 0 \end{cases}$$

9.1.3 Noise Shaping

The portion of the noise spectrum that falls into the baseband can be effectively reduced if the concept of noise shaping that we are already familiar with in the context of noise reduction in IIR filters is applied. It is the strength of this concept that it decorrelates signal and noise so that it is also applicable with very low level resolutions, down to even one bit. In order to implement noise shaping, the bits that have been truncated in the quantiser are delayed and subtracted from the input signal of the quantiser, as depicted in Fig. 9-5a.

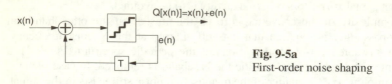

Fig. 9-5a
First-order noise shaping

The dropped bits correspond to the error signal $e(n)$ which is superimposed onto the signal $x(n)$. Since this error signal is added to $x(n)$ once directly and once after delay with inverse sign, it is finally weighted by the transfer function

$$H(z) = 1 - z^{-1} \; .$$

The square magnitude of this transfer function determines the noise power density at the output of the quantiser, which is now frequency-dependent (refer to 8.21).

$$\left| H\left(e^{j\omega T}\right) \right|^2 = 2 - 2\cos\omega T = 2(1 - \cos\omega T)$$

At low frequencies, the noise spectrum is attenuated, while it is boosted close to the Nyquist frequency. It becomes obvious that noise shaping makes no sense without oversampling, since the overall noise power even increases. In case of oversampling, however, the amplified portion of the noise spectrum is later removed by the low-pass filter that decimates the spectrum to the baseband again.

The described concept can be extended to higher-order noise shaping. The error signal passes a chain of delays and is added to the input signal of the quantiser after weighting by integer coefficients. Fig. 9-5b shows a second-order noise shaper.

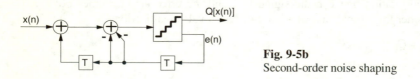

Fig. 9-5b
Second-order noise shaping

In the second-order case, the quantisation noise is weighted by the transfer function

$$H(z) = \left(1 - z^{-1}\right)^2 = 1 - 2z^{-1} + z^{-2} \; .$$

The corresponding square magnitude can be expressed as

$$\left| H\left(e^{j\omega T}\right) \right|^2 = 4(1 - \cos\omega T)^2 = 4\left(1 - 2\cos\omega T + \cos^2\omega T\right) \; .$$

We now calculate the portion of the noise power that falls into the baseband if noise shaping is applied. For first-order error feedback, the integral

$$\sigma_e^2 = N_0 \frac{1}{\pi} \int\limits_0^{\pi/OSR} 2(1 - \cos\omega T)\, \mathrm{d}\omega T$$

has to be evaluated.

$$\sigma_e^2 = N_0\, 2\left[1/OSR - \sin(\pi/OSR)/\pi\right] \tag{9.4}$$

Second-order error feedback yields

$$\sigma_e^2 = N_0\left[6/OSR - 8\sin(\pi/OSR)/\pi + \sin(2\pi/OSR)/\pi\right], \tag{9.5}$$

accordingly. Table 9-1 summarises the achievable reduction of the noise power under the assumption that the spectrum is limited to the baseband $f_s/(2OSR)$ after quantisation, using a sharp cutoff low-pass filter. The figures for noise filtering and first- and second-order noise shaping are compared.

Table 9-1 Possible reduction of the quantisation noise power by oversampling

OSR	Noise filtering	First-order noise shaping	Second-order noise shaping
Nyquist	0.0 dB	+3.0 dB	+7.8 dB
2	-3.0 dB	-4.4 dB	-3.4 dB
4	-6.0 dB	-13.0 dB	-17.5 dB
8	-9.0 dB	-22.0 dB	-32.3 dB
16	-12.0 dB	-31.0 dB	-47.3 dB
32	-15.1 dB	-40.0 dB	-62.4 dB
64	-18.1 dB	-49.0 dB	-77.4 dB
128	-21.1 dB	-58.0 dB	-92.5 dB

For large values of the oversampling ratio, the noise in the baseband (in-band noise) decreases with $1/OSR^3$ for first-order and with $1/OSR^5$ for second-order noise shaping. By appropriate series expansion of the sine function in (9.4) and (9.5), we obtain the following approximations:

$$\sigma_e^2 = N_0/OSR \qquad\qquad \text{filtering} \tag{9.6a}$$

$$\sigma_e^2 \approx N_0\, \pi^2/3OSR^3 \qquad \text{first-order noise shaping} \tag{9.6b}$$

$$\sigma_e^2 \approx N_0\, \pi^4/5OSR^5 \qquad \text{second-order noise shaping}. \tag{9.6c}$$

Each doubling of the oversampling ratio thus reduces the noise by 3 dB, 9 dB or 15 dB and provides 0.5, 1.5 or 2.5 bits of extra resolution respectively.

In the case of a one-bit resolution, second-order noise shaping with $OSR = 128$ results in an achievable signal-to-noise ratio, comparable to a 16-bit resolution. Assuming amplitude limits of ± 1, the maximum power of a sinusoidal signal amounts to 0.5. The step size of the quantiser is $\Delta x_q = 2$.

According to (9.3), this results in a noise power of $N_0 = 1/3$ which is reduced by 92.5 dB by applying oversampling and noise shaping. The resulting signal-to-noise ratio SNR can thus be calculated as

$$SNR = 10 \log(1/2) \, \text{dB} - 10 \log(1/3) \, \text{dB} + 92.5 \, \text{dB} = 94.3 \, \text{dB} \quad .$$

9.2 A/D Conversion

9.2.1 Oversampling and Decimation

Oversampling is also able to simplify the A/D conversion process in terms of filter and converter complexity. The solid line in Fig. 9-6 represents the spectrum of the analog signal to be converted to digital. The shaded portion of the spectrum is the actually desired baseband. The periodically continued portions of the spectrum, developing from four-times oversampling, are represented as dotted lines. These extend into the baseband and distort the original signal. Because of the four-times oversampling, the baseband extends up to $|\omega| < 1/8 \, \omega_s$ rather than to half the sampling frequency, which would be necessary to meet the sampling theorem. It is easy to see that the analog antialiasing filter has to cut off at $7/8 \, \omega_s$ in order to avoid distortions of the baseband signal. A possible magnitude response of this filter is shown in Fig. 9-6 as well.

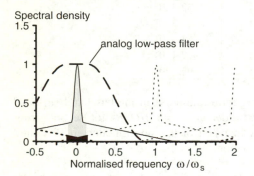

Fig. 9-6
Analog low-pass filtering in the case of four-times oversampling.
The desired baseband is shaded. The black areas represent the distorting spectral portions that are mirrored into the baseband

In the following, we compare the filter complexity required for a Nyquist converter with that required for a converter using four-times oversampling. In the baseband, the assumed Chebyshev filter shall exhibit a passband ripple of 0.1 dB. Alias spectral components in the baseband shall be attenuated by at least 72 dB. With this attenuation, these components are of the order of magnitude of the quantisation noise of a 12-bit A/D converter. In the case of the Nyquist converter, the transition band of the filter shall cover the range $0.45 \, \omega_s$ to $0.5 \, \omega_s$. According to (2.26b), a filter order of $N = 24$ is required to fulfil this specification. In the case of four-times oversampling, the transition band extends between $1/8 \, \omega_s$ and $7/8 \, \omega_s$. This merely requires a filter order of $N = 5$.

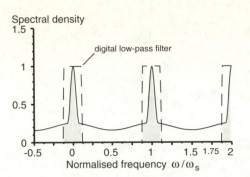
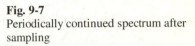

Fig. 9-7
Periodically continued spectrum after
sampling

Figure 9-7 shows the periodically continued spectrum after analog low-pass
filtering (cutoff frequency at 7/8 ω_s) and sampling at a rate of ω_s. The baseband
remains unaffected. The actual limitation of the spectrum to the range up to 1/8 ω_s
is performed in the discrete-time domain by a digital filter whose periodic
magnitude response is depicted in Fig. 9-7. For this purpose, it is preferable to use
FIR filters which allow the implementation of exactly linear-phase filters and
exclude the emergence of group delay distortions on this occasion. Figure 9-8
shows the corresponding spectrum after digital low-pass filtering.

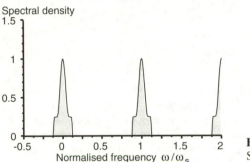

Fig. 9-8
Spectrum after digital low-pass filtering

In order to arrive at the actually desired sampling rate, which is reduced by a
factor of four, each fourth sample is gathered from the digitally filtered signal. In
the following, we investigate what the consequences of this decimation process
are if we consider arbitrary oversampling ratios OSR. According to (3.9b), the
samples $f(n)$ and the associate spectrum $F(e^{j\omega T})$ are related as

$$f(n) = \frac{1}{2\pi} \int_{-\pi}^{+\pi} F\left(e^{j\omega T}\right) e^{j\omega T n} \, d\omega T \,.$$

By taking each OSRth sample out of $f(n)$, we obtain the sequence $f'(n)$.

$$f'(n) = f(OSR\, n) = \frac{1}{2\pi} \int_{-\pi}^{+\pi} F\left(e^{j\omega T}\right) e^{j\omega T\, OSR\, n} \, d\omega T$$

$$f'(n) = \sum_{i=0}^{OSR-1} \frac{1}{2\pi} \int_{-\pi/OSR}^{+\pi/OSR} F\left(e^{j(\omega T + 2i\pi/OSR)}\right) e^{j\omega T \, OSR \, n} \, d\omega T$$

$$f'(n) = \frac{1}{2\pi} \int_{-\pi}^{+\pi} \frac{1}{OSR} \sum_{i=0}^{OSR-1} F\left(e^{j(\omega T/OSR + 2i\pi/OSR)}\right) e^{j\omega T n} \, d\omega T \qquad (9.7)$$

The spectrum $F'(e^{j\omega T})$ of the new sequence can thus be expressed as

$$F'\left(e^{j\omega T}\right) = \frac{1}{OSR} \sum_{i=0}^{OSR-1} F\left(e^{j(\omega T/OSR + 2i\pi/OSR)}\right) . \qquad (9.8)$$

Inspection of (9.8) shows that the gaps in the spectrum according to Fig. 9-8 are filled up and the scale of the frequency axis is changed. Because of the preceding digital low-pass filtering, the periodically continued spectral portions do not overlap (Fig. 9-9). The described procedure finally leads to the same result as expensive analog low-pass filtering and subsequent sampling at double the Nyquist frequency.

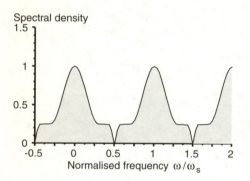

Spectral density

Normalised frequency ω/ω_s

Fig. 9-9
Spectrum of the digital signal after reduction of the sampling rate by a factor of four. The spectrum is normalised to unity again.

9.2.2 Noise Filtering

Oversampling, with subsequent limitation of the spectrum to the baseband, is also advantageous in the context of A/D conversion. Fig. 9-10 shows the functional block diagram of an oversampling A/D converter. The initial analog low-pass filter avoids the occurence of alias spectral components in the baseband after sampling. After A/D conversion with a resolution of b bits, a quantisation noise power N_0 according to (9.3) is superimposed onto the digitised signal. We assume for the time being that the subsequent digital low-pass filter operates with such a high signal resolution that practically no additional noise is induced. Together with the undesired spectral components (Fig. 9-7), the portion of the quantisation noise power outside the basedband is also filtered off. The noise power that is superimposed onto the remaining signal thus amounts to merely N_0/OSR. Assuming $OSR = 4$, the signal-to-noise ratio will be improved as if the analog signal was sampled with a one-bit higher resolution. With $OSR = 16$, the effective resolution is even increased by two bits. We can say in general that each doubling of the oversampling ratio provides 0.5 bits of extra resolution, equivalent to an improvement of the signal-to-noise ratio by 3 dB.

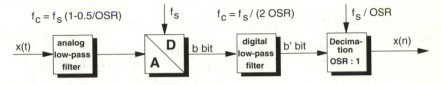

Fig. 9-10 Oversampling A/D conversion

Example
An analog signal is to be sampled with a 14-bit converter and an oversampling ratio of 16. The digital FIR low-pass filter operates with a 12-bit coefficient wordlength. After multiplication by the coefficients, the filtered signal is present with a wordlength of 26 bits. Since the noise power after filtering corresponds to a signal resolution of 16 bits, we can round-off to 16 bits without noticeably degrading the signal quality. We have increased the effective resolution of the A/D converter by two bits at the expense of a four-times-higher sampling rate. The low-pass filter is thus able to correctly interpolate four additional steps between the discrete output levels of the A/D converter.

A cost reduction with respect to the A/D converter can hardly be achieved using the procedure described above. It is true that a given signal resolution can be achieved by means of a less accurate converter, but the much higher processing speed required for oversampling more-or-less blights this advantage. It would be far more convenient if the OSR could be chosen high enough that a one-bit resolution and thus a simple sign detector would be sufficient for the A/D conversion. This is made possible by the introduction of noise shaping as we previously described in the context of the oversampling D/A converter. The concept has to be modified, however, since the signal to be quantised down to one bit is analog in this case, and analog circuitry has to be kept as simple as possible.

9.2.3 Oversampling $\Delta\Sigma$ Conversion

The A/D converter in Fig. 9-10 can be considered a special quantiser that turns analog signals with infinite amplitude resolution into signals with 2^b discrete amplitude steps. This discretisation generates a broadband quantisation noise N_0 according to (9.3). The principle of noise shaping may also be applied to this special quantiser. The procedure is to delay the quantisation error by one clock cycle and to subtract it from the input signal. For the feedback of the error signal, we cannot fall back on the truncated bits as described in the case of D/A conversion. Instead, we provide this error signal by explicitly calculating the difference between the input and output signals of the A/D converter. The circuit in Fig. 9-11 thus reduces the quantisation noise in the baseband as derived in Sect. 9.1.3. If we replace the A/D converter in Fig. 9-10 by the circuit in Fig. 9-11, we can maintain the same signal-to-noise ratio as in the case of simple oversampling conversion with an even smaller resolution b of the A/D converter.

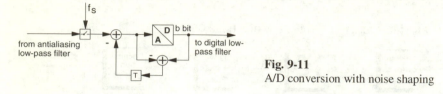

Fig. 9-11
A/D conversion with noise shaping

By choosing a sufficiently high sampling frequency, the A/D converter may be replaced by a one-bit quantiser, equivalent to a simple sign detector. In addition to that, we perform some equivalent rearrangements in the block diagram. Fig. 9-12 shows the result. The recursive structure in front of the sign detector, consisting of a delay and an adder, represents a discrete-time integrator (the value at the input equals the previous output value plus the present input value). The bold paths are carrying analog signals.

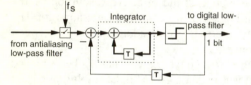
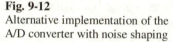

Fig. 9-12
Alternative implementation of the A/D converter with noise shaping

If, in the next step, we shift the sampler behind the quantiser, the discrete-time integrator can be replaced by a common analog integrator. Choosing the time constant of this integrator equal to the sampling period T yields a circuit with approximately identical timing conditions to those in the previous discrete-time implementation according to Fig. 9-12. Figure 9-13 depicts the corresponding block diagram, which includes this modification. The delay in the feedback path may be omitted. Its role in the loop is taken over by the sampler.

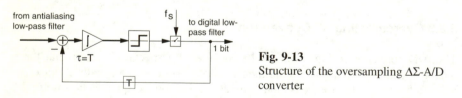

Fig. 9-13
Structure of the oversampling $\Delta\Sigma$-A/D converter

The structure shown in Fig. 9-13 is commonly referred to as $\Delta\Sigma$ modulator [11]. The ouput of this circuit is a binary signal whose pulse width is proportional to the applied analog voltage. Figure 9-14 shows the reaction of the $\Delta\Sigma$ modulator to a sinusoid with amplitude 0.9. If positive values are applied, the comparator provides a "high" level most of the time. In case of negative input values, the output tends more to the "low" level. A zero input yields rapidly alternating "high" and "low" levels with an average value in the middle between the two extreme levels.

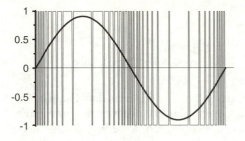

Fig. 9-14
Output of the $\Delta\Sigma$ modulator in case of a sinusoidal input

Since the $\Delta\Sigma$ modulator is still fully equivalent to block diagram 9-11, the spectral contents of the pulse train are composed of the spectrum of the analog input signal and of the spectrum of the quantisation noise, weighted by the transfer function $1-z^{-1}$. The A/D converter can thus be replaced in a practical implementation by an inverting integrator, a comparator and a D flip-flop, which takes over the role of the sampler. The circuit is highly insensitive to nonideal behaviour of the operational amplifier, which is commonly used to realise the integrator. The circuit is inherently stable as long as no extraordinary delay or gain occurs in the loop. For certain constant or slowly varying input values, the noise level may assume marked extreme values. The spectrum may also contain single audible tones. In order to avoid these unfavourable situations, a high frequency noise (dither) is added to the analog input signal, which is later removed again by the digital low-pass filter. This additional noise ensures that the $\Delta\Sigma$ modulator does not stay too long in these unfavourable operational conditions.

For a signal conversion of CD quality, a first-order oversampling $\Delta\Sigma$ converter is not sufficient. We showed in Sect. 9.1.3 that a second-order noise shaper with an *OSR* of at least 128 is required for this purpose. The necessary transfer function $(1-z^{-1})^2$ to provide sufficient attenuation of the quantisation noise in the baseband can be realised by applying the algorithm according to Fig. 9-11 twice, as depicted in Fig. 9-15.

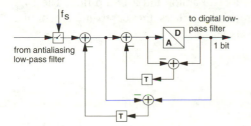

Fig. 9-15
Double application of the noise-shaping algorithm

Equivalent rearrangement of Fig. 9-15 and shift of the sampler behind the comparator leads to the block diagram of a second-order $\Delta\Sigma$ modulator. This circuit features two integrators arranged in series. The binary output signal of the modulator is fed back to the inputs of both integrators with inverse sign (Fig. 9-16).

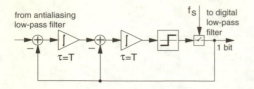

Fig. 9-16
Second-order ΔΣ modulator

The second-order ΔΣ modulator provides a pulse-width modulated signal, too. In the fine structure of the pulse train, however, there are differences compared to the first-order modulator, which result in the desired lowering of the noise spectrum in the baseband. As an example, a constant input value of 0.4 leads to the following periodic sequences at the output:

first-order: 1 1 1 -1 1 1 -1 1 1 -1 ...   average = (7–3)/10
second-order: 1 1 1 -1 1 -1 1 1 1 1 -1 1 -1 1 1 1 -1 1 1 -1 ...   average = (14–6)/20 .

Both sequences have mean values of 0.4.

The second-order ΔΣ modulator is somewhat more complex than the first-order variant. The sensitivity to tolerances of the components is more critical. A 30% increase of loop gain and/or additional delay in the loop will cause instability. An additional dither signal may not be needed, since the correlation between the quantisation error and the input signal is reduced by the additional feedback loop [11].

The explanations have shown that the hardware complexity of the analog front ends of A/D and D/A converters may be considerably reduced by techniques such as oversampling and noise shaping. The interfaces between digital and analog functions are simple binary interfaces operating at high clock rates. The information is coded in these interfaces by means of pulse-width modulation. In the case of high quality audio applications, the required clock frequencies are in the order of 5 to 10 MHz (44.1 kHz × 128 or 256). Considering the A/D converter, the complexity of the analog circuity is reduced to an anti-aliasing filter with moderate performance, two integrators, a comparator and a D flip-flop. The D/A converter is reduced to a low-pass filter that filters the baseband out of the spectrum of the pulse-width modulated signal.

10 Appendix

10.1 Symbols and Abbreviations

a	incident wave
A	system matrix
$a(t)$, $a(n)$	unit-step response of the system
$a(\omega)$	logarithmic magnitude response
a_v	coefficients of the denominator polynomial
A_v	normalised denominator coefficients
B	bandwidth of a bandpass or bandstop filter
b	reflected wave
b	wordlength in bits
$b(\omega)$	phase response
$\boldsymbol{b}, \boldsymbol{c}$	coefficient vectors
b_v	coefficients of the numerator polynomial
B_v	normalised numerator coefficients
C	capacitance
Δ	normalised width of the filter transition band
$\delta(n)$	unit-sample sequence
$\delta(t)$	delta function
DFT	discrete Fourier transform
δ_p, δ_s	ripple in the passband and stopband
Δx_q	step size of the quantiser
ε	design parameter of Chebyshev and Cauer filters
$e(n)$, $e(t)$	error signal (discrete-time and continuous-time)
f	frequency
FIR	finite impulse response
f_s, f_c	sampling frequency, cutoff frequency
γ	coefficient of the wave digital filter
G	conductance
$H(j\omega)$, $H(e^{j\omega T})$	frequency response (continuous-time, discrete-time)
$H(p)$, $H(z)$	transfer function (continuous-time, discrete-time)
$h(t)$, $h(n)$	impulse response, unit sample response
$I_0(x)$	zeroth-order modified Bessel function of first kind
IDFT	inverse discrete Fourier transform
IIR	infinite impulse response
Im	imaginary part of a complex number

$K(k)$	complete elliptic integral of the first kind
k, k_1	design parameters of elliptic filters
L	inductance
M	degree of the numerator polynomial
MT	magnitude truncation
N	degree of the denominator polynomial, order of the filter
N_0	noise power of the quantiser
$O(x)$	overflow characteristic
OSR	oversampling ratio
p	complex frequency
P	signal power
P	normalised complex frequency
p_0, P_0	zero frequency
p_∞, P_∞	pole frequency
Q	quality factor
q	quantisation step of the filter coefficients
$Q(x)$	quantisation operator
$q(x)$	unit rectangle
r	reflection coefficient
R	resistance, port resistance
$r(t), r(n)$	ramp function (continuous-time, discrete-time)
Re	real part of a complex number
RND	rounding
σ	real part of p
σ	variance
S_{11}, S_{22}	reflectance
S_{12}, S_{21}	transmittance
$sn(u,k)$	Jacobian elliptic function
SNR	signal-to-noise ratio
T	sampling period
T	transformation matrix
t, τ	time variables
$\tau_g(\omega)$	group delay
$T_n(x)$	Chebyshev polynomial of the first kind of order n
$u(t), u(n)$	unit-step function (continuous-time, discrete-time)
V	gain parameter
v_p	minimum gain in the passband
v_{norm}	normalised gain
v_s	maximum gain in the stopband
VT	value truncation
ω	angular frequency, imaginary part of p
Ω	normalised angular frequency
$W(\omega)$	weighting function
ω_c	cutoff frequency

$W_n(1/x)$	Bessel polynomial
ω_p	passband edge frequency
ω_s	sampling angular frequency
ω_s	stopband edge frequency
$x(t)$, $x(n)$	input signal (continuous-time, discrete-time)
ψ	frequency variable of the bilinear transform
$y(t)$, $y(n)$	output signal (continuous-time, discrete-time)
z	frequency variable of discrete-time systems
z	state vector

10.2 References

[1]   Avenhaus E (1972) A Proposal to Find Suitable Canonical Structures for the Implementation of Digital Filters with Small Coefficient Wordlength. NTZ 25:377-382

[2]   Barnes CW, Shinnaka S (1980) Stability Domains for Second-Order Recursive Digital Filters in Normal Form with "Matrix Power" Feedback. IEEE Trans. CAS-27:841-843

[3]   Bauer PH, Leclerc LJ (1991) A Computer-Aided Test for the Absence of Limit Cycles in Fixed-Point Digital Filters. IEEE Trans. ASSP-39:2400-2409

[4]   Bellanger M (1984) Digital Processing of Signals. John Wiley & Sons, New York

[5]   Bose T, Brown DP (1990) Limit Cycles Due to Roundoff in State-Space Digital Filters. IEEE Trans. ASSP-38:1460-1462

[6]   Bose T, Chen MQ (1991) Overflow Oscillations in State-Space Digital Filters. IEEE Trans. CAS-38:807-810

[7]   Brayton RK, Tong CH (1979) Stability of Dynamical Systems: A Constructive Approach. IEEE Trans. CAS-26:224-234

[8]   Brayton RK, Tong CH (1980) Constructive Stability and Asymptotic Stability of Dynamical Systems. IEEE Trans. CAS-27:1121-1130

[9]   Bronstein I, Semendjajew K (1970) Taschenbuch der Mathematik. Verlag Harri Deutsch, Frankfurt

[10] Butterweck HJ, Ritzerfeld J, Werter M (1989) Finite Wordlength Effects in Digital Filters. AEÜ 43:76-89

[11] Candy JC, Temes GC (1992) Oversampling Methods for A/D and D/A Conversion. In: Oversampling Delta-Sigma Data Converters. IEEE Press, Piscataway

[12] Chang TL (1981) Suppression of Limit Cycles in Digital Filters Designed with One Magnitude-Truncation Quantizer. IEEE Trans. CAS-28:107-111

[13] Claasen TACM, Mecklenbräuker WFG, Peek JBH (1973) Second-Order Digital Filter With Only One Magnitude-Truncation Quantiser And Having Practically No Limit Cycles. Electron Lett. 9:531-532

[14] Claasen TACM, Mecklenbräuker WFG, Peek JBH (1973) Some Remarks on the Classification of Limit Cycles in Digital Filters. Philips Res. Rep. 28:297-305

[15] Claasen TACM, Mecklenbräuker WFG, Peek JBH (1975) Frequency Domain Criteria for the Absence of Zero-Input Limit Cycles in Nonlinear Discrete-Time Systems, with Applications to Digital Filters. IEEE Trans. CAS-22:232-239

[16] Dorf RC (Editor) (1993) The Electrical Engineering Handbook. CRC Press, London

[17] Ebert PM, Mazo JE, Taylor MG (1969) Overflow Oscillations in Digital Filters. Bell Syst. Tech. J. 48:2999-3020

[18] Erickson KT, Michel AN (1985) Stability Analysis of Fixed-Point Digital Filters Using Computer Generated Lyapunov Functions-Part I: Direct Form and Coupled Form Filters. IEEE Trans. CAS-32:113-132

[19] Erickson KT, Michel AN (1985) Stability Analysis of Fixed-Point Digital Filters Using Computer Generated Lyapunov Functions-Part II: Wave Digital Filters and Lattice Digital Filters. IEEE Trans. CAS-32:132-142

[20] Fettweis A (1971) Digital Filter Structures Related to Classical Filter Networks. AEÜ 25:79-89

[21] Fettweis A (1972) A Simple Design of Maximally Flat Delay Digital Filter. IEEE Trans. AU-20:112-114

[22] Fettweis A (1986) Wave Digital Filters: Theory and Practice. Proc. IEEE 74: 270-327

[23] Fettweis A (1988) Passivity and Losslessness in Digital Filtering. AEÜ 42:1-8

[24] Gold B, Rader CM (1969) Digital Processing of Signals. McGraw-Hill

[25] Hamming RW (1962) Numerical Methods for Scientists and Engineers. McGraw-Hill, New York

[26] Herrmann O (1971) On the Approximation Problem in Nonrecursive Digital Filter Design. IEEE Trans. CT-18:411-413

[27] Herrmann O, Rabiner LR, and Chan DSK (1973) Practical Design Rules for Optimum Finite Impulse Response Low-Pass Digital Filters. Bell Syst. Tech. J.52:769-799

[28] Herrmann O, Schuessler W (1970) Design of Nonrecursive Digital Filters with Minimum Phase. Electron. Lett. 6:185-186

[29] Jackson LB (1969) An Analysis of Limit Cycles Due to Multiplication Rounding in Recursive Digital (Sub)Filters. Proc. 7th Annu. Allerton Conf. Circuit System Theory, pp. 69-78

[30] Jackson LB (1979) Limit Cycles in State-Space Structures for Digital Filters. IEEE Trans. CAS-26:67-68

[31] Jahnke E, Emde F, Lösch F (1966) Tafeln Höherer Funktione. Teubner, Stuttgart

[32] Jinaga BC, Dutta Roy SC (1984) Explicit Formulas for the Weighting Coefficients of Maximally Flat Nonrecursive Digital Filters. Proc. IEEE 72:1092

[33] Kaiser JF (1974) Nonrecursive Digital Filter Design Using The I_0-Sinh Window Function. Proc. IEEE Int. Symp. on Circuits and Syst., pp. 20-23

[34] Kaiser JF (1979) Design Subroutine (MXFLAT) for Symmetric FIR Low Pass Digital Filters with Maximally-Flat Pass and Stop Band. In: Programs for Digital Signal Processing, IEEE Press, pp. 5.3-1 to 5.3-6

[35] Kieburtz RB (1973) An Experimental Study of Roundoff Effects in a Tenth-Order Recursive Digital Filter. IEEE Trans. COM-21:757-763

[36] Kingsbury NG (1972) Second-Order Recursive Digital Filter Element for Poles Near the Unit Circle and the Real z-Axis. Electronic Letters 8:155-156

[37] La Salle J, Lefschetz S (1967) Die Stabilitätstheorie von Ljapunov. Bibliographisches Institut, Mannheim

[38] Laugwitz D (1965) Ingenieurmathematik V. Bibliographisches Institut, Mannheim

[39] Lepschy A, Mian GA, Viaro U (1986) Stability Analysis of Second-Order Direct-Form Digital Filters with Two Roundoff Quantizers. IEEE Trans. CAS-33:824-826

[40] Lightstone M, Mitra SK, Lin I, Bagchand S, Jarske P, Neuvo Y (1994) Efficient Frequency-Sampling Design of One- and Two-Dimensional FIR Filters Using Structural Subband Decomposition. IEEE Trans. CAS-41:189-201

[41] Lueder E (1981) Digital Signal Processing with Improved Accuracy. Proc. 5th European Conf. on Circuit Theory and Design, The Hague, pp. 25-33

[42] Lueder E, Hug H, Wolf W (1975) Minimizing the Round-Off Noise in Digital Filters by Dynamic Programming. Frequenz 29:211-214

[43] McClellan JH, Parks TW, Rabiner LR (1973) A Computer Program For Designing Optimum FIR Linear Phase Digital Filters. IEEE Trans. AU-21:506-526

[44] McClellan JH, Parks TW, Rabiner LR (1979) FIR Linear Phase Filter Design Program. In: Programs for Digital Signal Processing, IEEE Press, pp. 5.1-1 to 5.1-13

[45] Meerkötter K, Wegener W (1975) A New Second-Order Digital Filter Without Parasitic Oscillations. AEÜ 29:312-314

[46] Oppenheim AV, Schafer RW (1975) Digital Signal Processing. Prentice-Hall, Englewood Cliffs

[47] Pfitzenmaier G (1971) Tabellenbuch Tiefpässe, Unversteilerte Tschebyscheff- und Potenz-Tiefpässe. Siemens Aktiengesellschaft, Berlin · München

[48] Pregla R, Schlosser W (1972) Passive Netzwerke, Analyse und Synthese. Teubner, Stuttgart

[49] Premaratne K, Kulasekere EC, Bauer PH, Leclerc LJ (1996) An Exhaustive Search Algorithm for Checking Limit Cycle Behavior of Digital Filters. IEEE Trans. ASSP-44:2405-2412

[50] Rabiner LR, McClellan JH, Parks TW (1975) FIR Digital Filter Design Techniques Using Weighted Chebyshev Approximation. Proc. IEEE 63:595-610

[51] Rabiner LR, McGonegal CA, Paul D (1979) FIR Windowed Filter Design Program - WINDOW. In: Programs for Digital Signal Processing, IEEE Press, pp. 5.2-1 to 5.2-19

[52] Rajagopal LR, Dutta Roy SC (1989) Optimal Design of Maximally Flat FIR Filters with Arbitrary Magnitude Specifications. IEEE Trans. ASSP-37:512-518

[53] Saal R (1979) Handbook of Filter Design. AEG-Telefunken, Backnang

[54] Shannon CE (1949) Communication in the Presence of Noise. Proc. IRE 37:10-21

[55] Shen J, Strang G (1999) The Asymptotics of Optimal (Equiripple) Filters. IEEE Trans. ASSP-47:1087-1098

[56] Vaidyanathan PP (1984) On Maximally-Flat Linear-Phase FIR Filters. IEEE Trans. CAS-31:830-832

[57] Vaidyanathan PP, Mitra SK (1984) Low Pass-Band Sensitivity Digital Filters: A Generalized Viewpoint and Synthesis Procedures. Proc. IEEE 72:404-423

[58] Weinberg L (1962) Network Analysis and Synthesis. McGraw-Hill

[59] Wupper H (1975) Neuartige Realisierung RC-aktiver Filter mit geringer Empfindlichkeit. Habilitationsschrift, Ruhr-University Bochum

[60] Zölzer U (1989) Entwurf digitaler Filter für die Anwendung im Tonstudiobereich. Doctorate dissertation, Technical University Hamburg-Harburg

10.3 Filter Design Tables

This section presents a selection of design tables for the filter types that we introduced in Chap. 2:

- Butterworth
- Chebyshev
- Inverse Chebyshev

- Cauer
- Bessel

All tables show the coefficients for filter orders up to $N = 12$. For Butterworth and Bessel filters, which have the filter order as the only independent design parameter, these tables are complete. Chebyshev and Cauer filters have far more degrees of freedom, so we can only include a representative choice in this textbook. In the case of design problems for which these tables are not sufficient, extensive filter catalogues are available in the literature [47, 53].

All first- and second-order filter blocks that occur in the context of the filter types named above can be expressed in the following general form:

$$H(P) = \frac{1 + B_2 P^2}{1 + A_1 P + A_2 P^2} \quad .$$

Three different forms of transfer functions appear in practice.

Simple real pole

$$B_2 = 0 \quad A_2 = 0 \quad A_1 \neq 0$$

This simple first-order pole occurs in the case of odd-order low-pass filters.

Complex-conjugate pole pair

$$B_2 = 0 \quad A_2 \neq 0 \quad A_1 \neq 0$$

Butterworth, Chebyshev and Bessel filters are composed of these second-order filter blocks.

Complex-conjugate pole pair with imaginary zero pair

$$B_2 \neq 0 \quad A_2 \neq 0 \quad A_1 \neq 0$$

Inverse Chebyshev and Cauer filters have zero pairs on the imaginary $j\omega$-axis.

Filter catalogues often show poles and zeros instead of normalised filter coefficients. The filter coefficients A_1, A_2 and B_2 can be easily calculated from these poles and zeros. For second-order filter blocks we have the following relations:

$$A_1 = -2 \operatorname{Re} P_\infty \big/ |P_\infty|^2$$
$$A_2 = 1 \big/ |P_\infty|^2$$
$$B_2 = 1 \big/ |P_0|^2 \quad .$$

For the first-order pole, A_1 is calculated as

$$A_1 = -1/P_\infty \quad .$$

The following filter tables contain a factor V, which normalises the gain of the cascade arrangement of second-order filters blocks to unity. V equals one in most cases. In the case of even-order Chebyshev and Cauer filters, V depends on the given passband ripple.

$$V = v_p = 1\Big/\sqrt{1+\varepsilon^2}$$

Butterworth filters

$V = 1$
$B_2 = 0$

N	A_1	A_2	N	A_1	A_2
1	1.000000		9	0.347296	1.000000
2	1.414214	1.000000		1.000000	1.000000
3	1.000000	1.000000		1.532089	1.000000
	1.000000			1.879385	1.000000
4	0.765367	1.000000		1.000000	
	1.847759	1.000000	10	0.312869	1.000000
5	0.618034	1.000000		0.907981	1.000000
	1.618034	1.000000		1.414214	1.000000
	1.000000			1.782013	1.000000
6	0.517638	1.000000		1.975377	1.000000
	1.414214	1.000000	11	0.284630	1.000000
	1.931852	1.000000		0.830830	1.000000
7	0.445042	1.000000		1.309721	1.000000
	1.246980	1.000000		1.682507	1.000000
	1.801938	1.000000		1.918986	1.000000
	1.000000			1.000000	
8	0.390181	1.000000	12	0.261052	1.000000
	1.111140	1.000000		0.765367	1.000000
	1.662939	1.000000		1.217523	1.000000
	1.961571	1.000000		1.586707	1.000000
				1.847759	1.000000
				1.982890	1.000000

Chebyshev filters

$B_2 = 0$

V equals one for odd filter orders. In the case of even filter orders, V depends on the passband ripple:

    0.25 dB: $V = 0.971628$
    0.50 dB: $V = 0.944061$
    1.00 dB: $V = 0.891251$
    2.00 dB: $V = 0.794328$

Chebyshev 0.25 dB

N	A_1	A_2	N	A_1	A_2
1	0.243421		9	0.080466	0.974280
2	0.849883	0.473029		0.294846	1.239844
3	0.573140	0.747032		0.775649	2.128887
	1.303403			2.575534	5.762661
4	0.365795	0.860621		4.205054	
	2.255994	2.198549	10	0.065459	0.979263
5	0.245524	0.912880		0.231071	1.191128
	1.318006	1.864219		0.553748	1.832684
	2.288586			1.512316	3.972113
6	0.175678	0.940542		6.018950	14.261358
	0.809742	1.586787	11	0.054269	0.982923
	3.535069	5.071206		0.186303	1.155992
7	0.130987	0.956845		0.417320	1.642616
	0.543394	1.416677		0.989240	3.031036
	1.959450	3.535167		3.181532	8.546939
	3.250989			5.155207	
8	0.101215	0.967243	12	0.045709	0.985689
	0.390423	1.310156		0.153607	1.129805
	1.171842	2.627537		0.327217	1.512940
	4.783087	9.092020		0.700890	2.486660
				1.843984	5.617915
				7.248804	20.579325

Chebyshev 0.5 dB

N	A_1	A_2
1	0.349311	
2	0.940260	0.659542
3	0.548346	0.875314
	1.596280	
4	0.329760	0.940275
	2.375565	2.805743
5	0.216190	0.965452
	1.229627	2.097461
	2.759994	
6	0.151805	0.977495
	0.719120	1.694886
	3.691705	6.369532
7	0.112199	0.984148
	0.471926	1.477359
	1.818204	3.938897
	3.903658	
8	0.086211	0.988209
	0.335124	1.348920
	1.036706	2.788231
	4.980968	11.356882
9	0.068277	0.990873
	0.251348	1.266842
	0.671707	2.209747
	2.385021	6.396216
	5.040188	
10	0.055392	0.992718
	0.196118	1.211094
	0.474268	1.880381
	1.335834	4.203190
	6.259890	17.768641
11	0.045831	0.994050
	0.157651	1.171413
	0.355133	1.673930
	0.855617	3.139405
	2.943283	9.468583
	6.173405	
12	0.038544	0.995044
	0.129707	1.142112
	0.277330	1.535091
	0.599686	2.547068
	1.627510	5.935974
	7.533733	25.605025

Chebyshev 1 dB

N	A_1	A_2
1	0.508847	
2	0.995668	0.907021
3	0.497051	1.005829
	2.023593	
4	0.282890	1.013680
	2.411396	3.579122
5	0.181032	1.011823
	1.091107	2.329385
	3.454311	
6	0.125525	1.009354
	0.609201	1.793016
	3.721731	8.018803
7	0.0920921	1.007375
	0.391989	1.530326
	1.606177	4.339334
	4.868210	
8	0.070429	1.005894
	0.275575	1.382088
	0.875459	2.933762
	5.009828	14.232606
9	0.055600	1.004790
	0.205485	1.289680
	0.556611	2.280179
	2.103365	7.024249
	6.276263	
10	0.045006	1.003957
	0.159743	1.227864
	0.389282	1.921118
	1.126613	4.412333
	6.289486	22.221306
11	0.037176	1.003317
	0.128092	1.184305
	0.289916	1.700378
	0.708287	3.233739
	2.593589	10.382028
	7.681621	
12	0.031226	1.002818
	0.105202	1.152365
	0.225632	1.553671
	0.491819	2.598631
	1.371714	6.223779
	7.565021	31.985093

Chebyshev 2 dB

N	A_1	A_2	N	A_1	A_2
1	0.764783		9	0.042558	1.015849
2	0.976619	1.214978		0.157778	1.307956
3	0.416333	1.128547		0.432087	2.337937
	2.710682			1.723645	7.602867
4	0.225886	1.076803		8.289826	
	2.285707	4.513278	10	0.034388	1.012859
5	0.141700	1.050236		0.122298	1.241205
	0.898462	2.543558		0.299870	1.953980
	4.580677			0.887534	4.589612
6	0.097258	1.035249		5.913813	27.587958
	0.481606	1.876388	11	0.028368	1.010640
	3.508721	10.007393		0.097873	1.194521
7	0.070933	1.026046		0.222356	1.721517
	0.304860	1.573834		0.549392	3.311060
	1.317946	4.708418		2.124017	11.223492
	6.437500			10.140090	
8	0.054045	1.020012	12	0.023804	1.008948
	0.212583	1.408883		0.080271	1.160468
	0.690371	3.057181		0.172584	1.568437
	4.714502	17.698915		0.378610	2.640205
				1.080071	6.467695
				7.109976	39.674546

Inverse Chebyshev filters

$V = 1$

The coefficients B_2 are independent of the stopband attenuation of the filter.

N	B_2	N	B_2
1	---	9	0.969846
2	0.500000		0.750000
3	0.750000		0.413176
4	0.853553		0.116978
	0.146447	10	0.975528
5	0.904508		0.793893
	0.345492		0.500000
6	0.933013		0.206107
	0.500000		0.024472
	0.066987	11	0.979746
7	0.950484		0.827430
	0.611260		0.571157
	0.188255		0.292292
8	0.961940		0.079373
	0.691342	12	0.982963
	0.308658		0.853553
	0.038060		0.629410
			0.370590
			0.146447
			0.017037

30 dB stopband attenuation

N	A_1	A_2
1	31.606961	
2	5.533785	15.811388
3	1.866438	4.233593
	1.866438	
4	0.943411	2.372921
	2.277597	1.665815
5	0.573384	1.765237
	1.501139	1.206220
	0.927755	
6	0.386924	1.491737
	1.057095	1.058725
	1.444018	0.625712
7	0.279334	1.344439
	0.782677	1.005215
	1.131001	0.582210
	0.627658	
8	0.211428	1.255566
	0.602097	0.984968
	0.901101	0.602284
	1.062921	0.331686
9	0.165740	1.197594
	0.477229	0.977748
	0.731157	0.640924
	0.896897	0.344725
	0.477229	
10	0.133490	1.157571
	0.387403	0.975935
	0.603395	0.682043
	0.760322	0.388150
	0.842823	0.206514
11	0.109859	1.128721
	0.320677	0.976405
	0.505516	0.720132
	0.649401	0.441267
	0.740675	0.228348
	0.385972	
12	0.092017	1.107207
	0.269779	0.977798
	0.429157	0.753654
	0.559288	0.494835
	0.651304	0.270691
	0.698936	0.141282

40 dB stopband attenuation

N	A_1	A_2
1	99.995000	
2	9.959874	50.000000
3	2.838494	8.807047
	2.838494	
4	1.337350	3.906721
	3.228648	3.199614
5	0.784536	2.515899
	2.053941	1.956882
	1.269406	
6	0.518865	1.937759
	1.417566	1.504746
	1.936431	1.071734
7	0.369949	1.641491
	1.036575	1.302267
	1.497894	0.879262
	0.831268	
8	0.277721	1.468564
	0.790882	1.197966
	1.183639	0.815282
	1.396197	0.544684
9	0.216469	1.358346
	0.623297	1.138499
	0.954947	0.801675
	1.171416	0.505477
	0.623297	
10	0.173632	1.283518
	0.503901	1.101883
	0.784844	0.807990
	0.988961	0.514097
	1.096272	0.332462
11	0.142458	1.230251
	0.415834	1.077935
	0.655521	0.821662
	0.842102	0.542797
	0.960461	0.329878
	0.500504	
12	0.119043	1.190908
	0.349015	1.061499
	0.555203	0.837355
	0.723555	0.578536
	0.842597	0.354392
	0.904218	0.224982

50 dB stopband attenuation

N	A_1	A_2
1	316.226185	
2	17.754655	158.113883
3	4.233618	18.673520
	4.233618	
4	1.842787	6.650654
	4.448881	5.943547
5	1.037435	3.722221
	2.716039	3.163204
	1.678604	
6	0.669948	2.608071
	1.830333	2.175059
	2.500281	1.742046
7	0.470579	2.068541
	1.318534	1.729317
	1.905337	1.306312
	1.057382	
8	0.349764	1.765502
	0.996045	1.494904
	1.490686	1.112221
	1.758384	0.841623
9	0.270736	1.577551
	0.779554	1.357705
	1.194346	1.020881
	1.465083	0.724683
	0.779554	
10	0.216072	1.452477
	0.627065	1.270841
	0.976676	0.976948
	1.230684	0.683056
	1.364224	0.501420
11	0.176614	1.364773
	0.515534	1.212457
	0.812689	0.956184
	1.044004	0.677319
	1.190740	0.464399
	0.620505	
12	0.147160	1.300743
	0.431453	1.171334
	0.686342	0.947190
	0.894458	0.688371
	1.041619	0.464227
	1.117795	0.334817

60 dB stopband attenuation

N	A_1	A_2
1	999.999500	
2	31.606961	500.000000
3	6.259920	39.936595
	6.259920	
4	2.501934	11.539482
	6.040202	10.832375
5	1.345577	5.644660
	3.522766	5.085643
	2.177189	
6	0.845771	3.602658
	2.310690	3.169645
	3.156462	2.736632
7	0.583966	2.672247
	1.636236	2.333023
	2.364430	1.910018
	1.312160	
8	0.429063	2.171174
	1.221868	1.900576
	1.828654	1.517893
	2.157045	1.247295
9	0.329439	1.869655
	0.948583	1.649809
	1.453313	1.312985
	1.782752	1.016787
	0.948583	
10	0.261378	1.673459
	0.758548	1.491823
	1.181466	1.197931
	1.488734	0.904038
	1.650274	0.722402
11	0.212706	1.538214
	0.620885	1.385898
	0.978764	1.129625
	1.257349	0.850760
	1.434071	0.637840
	0.747307	
12	0.176634	1.440779
	0.517863	1.311369
	0.823801	1.087225
	1.073599	0.828406
	1.250232	0.604262
	1.341665	0.474853

Cauer filters

The tables of the Cauer filters contain an additional column indicating the width of the transition band ω_s/ω_p of the respective filter.

Passband ripple: 1 dB

$V = 1$ for odd N

$V = 0.891251$ for even N

Cauer filter, 1-dB passband ripple, 30-dB stopband attenuation

N	A_1	A_2	B_2	ω_s/ω_p
1	0.508847			62.114845
2	0.962392	0.893097	0.031688	4.004090
3	0.404019	0.984053	0.262019	1.732505
	1.787125			
4	0.165088	0.995394	0.581107	1.250383
	1.754288	2.297846	0.139656	
5	0.067497	0.998247	0.801280	1.095539
	0.579010	1.440900	0.463510	
	2.219443			
6	0.027596	0.999292	0.913451	1.037991
	0.212903	1.162894	0.731929	
	1.925770	2.751622	0.169694	
7	0.011281	0.999711	0.963679	1.015356
	0.083238	1.063737	0.880383	
	0.603380	1.551367	0.501672	
	2.304612			
8	0.004611	0.999882	0.984993	1.006248
	0.033405	1.025582	0.949274	
	0.217117	1.198555	0.755958	
	1.955920	2.837955	0.175090	
9	0.001885	0.999952	0.993839	1.002549
	0.013551	1.010378	0.978949	
	0.084134	1.076951	0.892081	
	0.607276	1.571091	0.508125	
	2.319263			
10	0.000770	0.999980	0.997477	1.001041
	0.005522	1.005229	0.991342	
	0.033642	1.030770	0.954408	
	0.217706	1.204760	0.759916	
	1.961000	2.852703	0.176001	
11	0.000315	0.999992	0.998968	1.000425
	0.002254	1.001726	0.996452	
	0.013627	1.012464	0.981110	
	0.084233	1.079226	0.893987	
	0.607922	1.574423	0.509205	
	2.321723			
12	0.000129	0.999997	0.999578	1.000174
	0.000921	1.000705	0.998549	
	0.005549	1.005075	0.992236	
	0.033661	1.031660	0.995241	
	0.217801	1.205804	0.760576	
	1.961850	2.855176	0.176154	

Cauer filter, 1-dB passband ripple, 40-dB stopband attenuation

N	A_1	A_2	B_2	ω_s/ω_p
1	0.508847			196.512846
2	0.985058	0.902474	0.010126	7.044820
3	0.452106	0.994690	0.131433	2.416184
	1.909413			
4	0.210869	1.001457	0.386003	1.515484
	2.013945	2.764201	0.080466	
5	0.099952	1.001110	0.636119	1.218682
	0.733855	1.674652	0.321263	
	2.595082			
6	0.047599	1.000574	0.806052	1.098870
	0.303672	1.282199	0.585449	
	2.356753	3.734747	0.113924	
7	0.022695	1.000279	0.902280	1.045995
	0.135465	1.126194	0.775128	
	0.796829	1.934861	0.378329	
	2.796049			
8	0.010825	1.000133	0.952133	1.021682
	0.062612	1.058360	0.885644	
	0.317801	1.373784	0.632778	
	2.443707	4.012650	0.122697	
9	0.005163	1.000064	0.976873	1.010285
	0.029425	1.027430	0.943723	
	0.139262	1.163888	0.804391	
	0.811036	2.002385	0.391919	
	2.844769			
10	0.002463	1.000030	0.988900	1.004893
	0.013937	1.012992	0.972748	
	0.063819	1.075113	0.901436	
	0.320594	1.396459	0.643513	
	2.464013	4.079467	0.124755	
11	0.001175	1.000014	0.994690	1.002331
	0.006626	1.006177	0.986906	
	0.029869	1.035156	0.951713	
	0.139888	1.173013	0.810872	
	0.814260	2.018229	0.395040	
	2.856020			
12	0.000560	1.000007	0.997463	1.001111
	0.003156	1.002942	0.993732	
	0.014120	1.016619	0.976668	
	0.063975	1.079126	0.904893	
	0.321206	1.401720	0.645953	
	2.468661	4.094866	0.125227	

Cauer filter, 1-dB passband ripple, 50-dB stopband attenuation

N	A_1	A_2	B_2	ω_s/ω_p
1	0.508847			621.456171
2	0.992304	0.905568	0.003213	12.484568
3	0.475795	1.000427	0.063303	3.460613
	1.969598			
4	0.240441	1.006120	0.239692	1.908184
	2.177965	3.086060	0.045962	
5	0.125733	1.004054	0.470461	1.407231
	0.849859	1.870794	0.215871	
	2.847722			
6	0.066518	1.002272	0.670150	1.198914
	0.381611	1.398334	0.448369	
	2.711769	4.680949	0.077468	
7	0.035320	1.001225	0.808326	1.101265
	0.186351	1.195534	0.653917	
	0.967594	2.339835	0.282982	
	3.231412			
8	0.018777	1.000654	0.892993	1.052645
	0.094878	1.099700	0.797976	
	0.413323	1.575551	0.517273	
	2.890996	5.311008	0.088636	
9	0.009986	1.000348	0.941574	1.027667
	0.049332	1.051861	0.886913	
	0.196415	1.273942	0.705474	
	1.002200	2.503109	0.303867	
	3.343619			
10	0.005312	1.000185	0.968482	1.014624
	0.025930	1.027262	0.938161	
	0.098546	1.137504	0.830830	
	0.421486	1.633180	0.537209	
	2.944416	5.508338	0.091991	
11	0.002825	1.000099	0.983108	1.007753
	0.013707	1.014411	0.966614	
	0.050838	1.070945	0.906151	
	0.198629	1.298530	0.719802	
	1.012099	2.552223	0.309923	
	3.376373			
12	0.001503	1.000052	0.990978	1.004117
	0.007267	1.007641	0.982099	
	0.026610	1.037135	0.948932	
	0.099213	1.149131	0.839763	
	0.423725	1.650156	0.542875	
	2.959761	5.565840	0.092956	

Cauer filter, 1-dB passband ripple, 60-dB stopband attenuation

N	A_1	A_2	B_2	ω_s/ω_p
1	0.508847			1965.225746
2	0.994603	0.906560	0.001017	22.176712
3	0.487098	1.003271	0.029897	5.021121
	1.998311			
4	0.258381	1.009197	0.142774	2.460783
	2.276772	3.289813	0.026091	
5	0.144391	1.006476	0.330087	1.671161
	0.931913	2.019421	0.141961	
	3.072028			
6	0.082338	1.003943	0.529063	1.343537
	0.443972	1.498899	0.332159	
	2.990838	5.508885	0.052929	
7	0.047295	1.002313	0.692997	1.185474
	0.231599	1.262563	0.531840	
	1.112351	2.728504	0.210158	
	3.601562			
8	0.027238	1.001341	0.809410	1.103082
	0.126530	1.143868	0.695302	
	0.499674	1.784259	0.415274	
	3.289942	6.644266	0.065191	
9	0.015702	1.000775	0.885187	1.058218
	0.070824	1.080607	0.811008	
	0.251636	1.396267	0.604619	
	1.177950	3.042259	0.236698	
	3.806594			
10	0.009055	1.000447	0.932070	1.033179
	0.040168	1.045732	0.886205	
	0.134654	1.212283	0.748612	
	0.517389	1.900155	0.444653	
	3.398181	7.086114	0.069688	
11	0.005223	1.000258	0.960233	1.019007
	0.022949	1.026130	0.932690	
	0.074465	1.117437	0.846299	
	0.257159	1.447926	0.628906	
	1.200468	3.157678	0.245952	
	3.878535			
12	0.003013	1.000149	0.976864	1.010920
	0.013165	1.014991	0.960602	
	0.041937	1.066146	0.908245	
	0.136565	1.237964	0.765807	
	0.523176	1.941448	0.454589	
	3.435250	7.241268	0.071232	

Bessel filters

<table>
<tr><td colspan="3" align="center">$V = 1$</td><td colspan="3" align="center">$V = 1$</td></tr>
<tr><td colspan="3">Normalised to t_0</td><td colspan="3">Normalised to ω_{3dB}</td></tr>
</table>

N	A_1	A_2	N	A_1	A_2
1	1.000000		1	1.000000	
2	1.000000	0.333333	2	1.361654	0.618034
3	0.569371	0.154812	3	0.999629	0.477191
	0.430629			0.756043	
4	0.366265	0.087049	4	0.774254	0.388991
	0.633735	0.109408		1.339664	0.488904
5	0.256073	0.055077	5	0.621595	0.324533
	0.469709	0.070065		1.140177	0.412845
	0.274218			0.665639	
6	0.189781	0.037716	6	0.513054	0.275641
	0.358293	0.047955		0.968607	0.350473
	0.451926	0.053188		1.221734	0.388718
7	0.146771	0.027325	7	0.433228	0.238072
	0.281315	0.034558		0.830363	0.301095
	0.370779	0.038961		1.094437	0.339457
	0.201135			0.593694	
8	0.117236	0.020647	8	0.372765	0.208745
	0.226517	0.025927		0.720236	0.262125
	0.306756	0.029468		0.975366	0.297924
	0.349492	0.031272		1.111250	0.316161
9	0.096041	0.016118	9	0.325742	0.185418
	0.186326	0.020085		0.631960	0.231049
	0.256809	0.022911		0.871017	0.263562
	0.302019	0.024637		1.024356	0.283414
	0.158805			0.538619	
10	0.080289	0.012913	10	0.288318	0.166512
	0.156045	0.015968		0.560356	0.205909
	0.217637	0.018235		0.781532	0.235149
	0.261565	0.019770		0.939275	0.254934
	0.284462	0.020548		1.021499	0.264964
11	0.068245	0.010565	11	0.257940	0.150928
	0.132690	0.012969		0.501515	0.185268
	0.186582	0.014805		0.705206	0.211495
	0.227721	0.016132		0.860698	0.230458
	0.253569	0.016940		0.958389	0.241998
	0.131193			0.495859	
12	0.058816	0.008797	12	0.232862	0.137889
	0.114304	0.010723		0.452546	0.168086
	0.161652	0.012226		0.640003	0.191640
	0.199526	0.013363		0.789953	0.209464
	0.226028	0.014132		0.894879	0.221511
	0.239675	0.014520		0.948908	0.227595

10.4 Filter Design Routines

This part of the Appendix contains some useful subroutines for the design of IIR and FIR filters. They have been developed in Turbo Pascal 6.0 and can be easily included in the reader's own filter design programs. A special feature is the method of passing arrays to the subroutines. These are passed as pointers, which makes the subroutines independent of the array size. Each subroutine contains tips how these arrays have to be declared in the calling program and an example of a subroutine call.

Cauer

The subroutine Cauer allows the calculation of the poles and zeros of Cauer (elliptic) filters. Input parameters are the parameters k, k_1, ε and N, as introduced in Sect. 2.5. This parameter set must satisfy Eqn. (2.42).

```
{ Subroutine to calculate poles and zeros of elliptic filters. The subroutine
delivers: Nz = N DIV 2 conjugate complex poles and zeros plus one real pole if
N is odd, in total Np = (N+1) DIV 2 pole values.
INPUT:     k       transition width wp/ws,        Double
           k1      vsnorm/vpnorm,                 Double
           e       1/vpnorm,                      Double
           N       filter degree,                 Integer
RETURN: ZerR   real part of the zeros,           pointer to an Array of Double
           ZerI   +-imaginary part of the zeros,  pointer to an Array of Double
           PolR   real part of the poles,          pointer to an Array of Double
           PolI   +-imaginary part of the poles,   pointer to an Array of Double
REQUIRED EXTERNAL ROUTINES: -
The arrays ZerR,ZerI,PolR,PolI have to be declared in the main program as
Array[1..Max] of Double with Max corresponding to half the highest filter degree
used in the program. Max has to fulfil the condition Max >= (N+1) DIV 2. In the
actual subroutine call the arrays have to be referenced by their address.
Example:  Cauer(k,k1,e,N,@ZerR,@ZerI,@PolR,@PolI);
The parameters N, k and k1 cannot be chosen independently. They are related
by a formula making use of complete elliptic integrals (refer to Chap. 2.5).
Subroutine CauerParam calculates these parameters from the tolerance
scheme of the desired low-pass filter.}


Procedure Cauer(k,k1,e:Double;n:Integer;ZerR,ZerI,PolR,PolI:Pointer);

Type ADouble  =  Array[1..2] of Double;
        PADouble = ^ADouble;

Var   u0,s1,c1,d1,dx,sn,cn,dn,de : Double;
         kc,k1c                          : Double;
         Lm,L,Nz,Np                      : Integer;
```

```
Function ArcSin(x:Double):Double;
Begin
ArcSin:=ArcTan(x/sqrt(1.0-x*x));
End;

Function Cei1(x:Double):Double;
Var    a,a1,b,b1,Test : Double;
       I              : Integer;
Begin
a:=1.0;
b:=Sqrt(1.0-x*x);
Test:=1e-12;
For I:=1 To 25 Do
   Begin
   a1:=0.5*(a+b);
   b1:=Sqrt(a*b);
   a:=a1;
   b:=b1;
   If (a-b) < Test Then I:=25;
   End;
Cei1:=0.5*Pi/a;
End;

Function InverseSn(x,k:Double):Double;
Var    p,q,tt,tw         : Double;
       a,b,a1,b1,f,Test  : Double;
       kq,kh,J           : Byte;
Begin
a:=1.0;b:=k;
Test:=1e-12;
p:=ArcTan(x);
tw:=1.0;
kq:=1;
For J:=1 to 25 do
   Begin
   f:=p/a/tw;
   tw:=tw*2.0;
   kq:=2*kq-1;
   q:=b*Sin(p)/a/Cos(p);
   tt:=ArcTan(q);
   if tt < 0 then kq:=kq+1;
   kh:=Trunc(kq/2.0);
   p:=p+tt+pi*kh;
   a1:=a;b1:=b;
   a:=0.5*(a1+b1);b:=Sqrt(a1*b1);
   If (a-b) < Test Then J:=25;
   End;
InverseSn:=f;
End;
```

```
Procedure EllipFunctions(x,k:Double;Var sn,cn,dn:Double);
Var    p,q,kc,Test      : Double;
       l,limit,j1,jk,nn  : Integer;
       a,b,c            : Array[1..25] of Double;
Begin
kc:=Sqrt(1.0-k*k);
a[1]:=1.0;
b[1]:=kc;
c[1]:=k;
Test:=1e-12;
For l:=1 to 24 do
   Begin
   a[l+1]:=0.5*(a[l]+b[l]);
   b[l+1]:=sqrt(a[l]*b[l]);
   c[l+1]:=0.5*(a[l]-b[l]);
   If c[l+1] < Test then
      Begin
      limit:=l+1;
      l:=24;
      End;
   End;
j1:=limit-1;
nn:=j1-1;
p:=Exp(nn*Ln(2.0))*a[j1]*x;
For l:=1 to nn do
   Begin
   jk:=j1+1-l;
   q:=c[jk]*Sin(p)/a[jk];
   p:=0.5*(p+ArcSin(q));
   End;
sn:=Sin(p);
cn:=Cos(p);
dn:=Sqrt(1.0-k*k*sn*sn);
End;

Begin
Nz:=N DIV 2;
Np:=(N+1) DIV 2;
kc:=Sqrt(1.0-k*k);
k1c:=Sqrt(1.0-k1*k1);
u0:=Cei1(kc)/Cei1(k1c)*InverseSn(1.0/e,k1);
EllipFunctions(u0,kc,s1,c1,d1);
dx:=Cei1(k)/N;
For Lm:=1 to Nz do
   Begin
   L:=N+1-Lm-Lm;
   EllipFunctions(L*dx,k,sn,cn,dn);
   PADouble(ZerR)^[Lm]:=0;
   PADouble(ZerI)^[Lm]:=1.0/k/sn;
   de:=c1*c1+k*k*sn*sn*s1*s1;
```

```
    PADouble(PolR)^[Lm]:=-cn*dn*s1*c1/de;
    PADouble(PolI)^[Lm]:=sn*d1/de;
    End;
if Odd(N) then
    Begin
    PADouble(PolR)^[Np]:=-s1/c1;
    PADouble(PolI)^[Np]:=0;
    End;
End;
```

CauerParam

The subroutine **CauerParam** allows the calculation of a harmonised set of the parameters k, k_1, ε and N from the characteristic data of a given tolerance scheme. Passband ripple and stopband attenuation are specified in dB. The edge frequencies may be specified in arbitrary units since only the quotient of both frequencies is of interest.

```
{ Subroutine to calculate the filter parameters of elliptic filters from the data of
the tolerance scheme
INPUT: wd      passband cutoff frequency, Double
       ws      stopband cutoff frequency, Double
       ad      passband ripple in dB, Double
       as      stopband ripple in dB, Double
RETURN:  k      transition width wp/ws, Double
         k1     vsnorm/vpnorm, Double
         e      1/vpnorm, Double
         N      filter order, Integer
REQUIRED EXTERNAL ROUTINES: -
EXAMPLE: CauerParam(wd,ws,ad,as,k,k1,e,N) }

Procedure CauerParam(wd,ws,ad,as:Double;var k,k1,e:Double;var N:Integer);
Var    vpnorm,vsnorm,q,kc,k1c:Double;

Function Modul(q:Double):Double;
Var    Theta2,Theta3,qp,qf,Test    : Double;
Begin
Test:=1e-20;
Theta2:=1.0;
Theta3:=1.0;
qp:=1.0;
qf:=1.0;
    Repeat
    qp:=qp*q;
    qf:=qf*qp;
    Theta3:=Theta3+2.0*qf;
    qf:=qf*qp;
    Theta2:=Theta2+qf;
    Until qf < Test;
```

```
Theta2:=2.0*Sqrt(Sqrt(q))*Theta2;
Modul:=(Theta2*Theta2)/(Theta3*Theta3);
End;

Function Cei1(x:Double):Double;
Var   a,a1,b,b1,Test  : Double;
      I               : Integer;
Begin
a:=1.0;
b:=Sqrt(1.0-x*x);
Test:=1e-12;
For I:=1 to 25 Do
   Begin
   a1:=0.5*(a+b);
   b1:=Sqrt(a*b);
   a:=a1;
   b:=b1;
   If (a-b) < Test Then I:=25;
   End;
Cei1:=0.5*Pi/a;
End;

Begin
vpnorm:=1/Sqrt(Exp(Ln(10)*ad/10)-1);
vsnorm:=1/Sqrt(Exp(Ln(10)*as/10)-1);
k1:=vsnorm/vpnorm;
k1c:=Sqrt(1-k1*k1);
k:=wd/ws;
kc:=Sqrt(1-k*k);
N:=Trunc(Cei1(k1c)*Cei1(k)/Cei1(k1)/Cei1(kc))+1;
q:=Exp(-Pi*Cei1(k1c)/Cei1(k1)/N);
k:=Modul(q);
e:=1/vpnorm;
End;
```

Bessel

Subroutine **Bessel** calculates the coefficients of analog Bessel filters (Chap. 2.6). The input parameter **Param** must be set to zero in this case. Moreover, the coefficients of analog reference filters for the design of digital Bessel filters by means of the bilinear transform can be determined by equating **Param** with the reciprocal of the design parameter μ (refer to Sect. 6.4.4).

```
{ Routine to calculate the filter coefficients of Bessel filters
INPUT:   N          filter order, Integer
         Param      design parameter, Double
For Param = 0, the subroutine calculates the Bessel coefficients of analog
filters. For Param = 1/mu the subroutine calculates the Bessel coefficients for
the design of digital filters with mu = 2*t0/T and mu > N-1.
```

RETURN: Coeff   pointer to an Array of Double containing the filter
                  coefficients
REQUIRED EXTERNAL ROUTINES: -
The array Coeff has to be declared in the main program as Array[0..Max] of
Double with Max being the maximum filter degree used in the program. In the
subroutine call, the array has to be referenced by its address.
EXAMPLE: Bessel(@Coeff,N,1/mu); }

```
Procedure Bessel(Coeff:Pointer;N:Integer;Param:Double);

Type    ADouble = Array[0..1] of Double;
        PADouble = ^ADouble;

Var     I,J   : Integer;
        g,h   : Double;

Begin
PADouble(Coeff)^[0]:=1;
For J:=1 to N Do
   Begin
   PADouble(Coeff)^[J]:=0;
   End;
For I:=N DownTo 1 Do
   Begin
   g:=2*I-1;
   h:=1-(I-1)*(I-1)*Param*Param;
   For J:=N DownTo 1 Do
   If Odd(J)   Then PADouble(Coeff)^[J]:=PADouble(Coeff)^[J-1]*h
               Else PADouble(Coeff)^[J]:=PADouble(Coeff)^[J-1]
                  +PADouble(Coeff)^[J]*g;
   PADouble(Coeff)^[0]:=PADouble(Coeff)^[0]*g;
   End;
End;
```

FirFreqResp

Subroutine FirFreqResp calculates the frequency response of FIR filters in the
frequency range 0 to Range. Range is normalised to the sampling period
(Range $= f_{max} * T$). NP is the number of equidistant points to be calculated
within the specified frequency range. The array Coeff contains the filter
coefficients. N specifies the filter order. The calculated zero phase frequency
response is returned back to the calling program in the array H0. Typ specifies
the kind of symmetry of the impulse response. Typ $= 0$ applies to Case 1 and
Case 2 filters, Typ $= 1$ to Case 3 and Case 4 filters accordingly.

{ Procedure to calculate the zero phase frequency response of FIR filters from
the coefficients of the filter with an arbitrary resolution.
INPUT: NP   number of computed frequency points, Integer

Coeff   pointer to an Array of Double containing the filter coefficients
N   degree of the filter, Integer
Typ   symmetry of the impulse response, Integer
         Typ   0 : impulse response with even symmetry
                1 : impulse response with odd symmetry
Range desired normalised frequency range fmaxT, Double
RETURN: H0   pointer to an Array of Double containing the computed
                frequency response
REQUIRED EXTERNAL ROUTINES: -
H0 has to be declared in the main program as Array[0..NP] of Double. Coeff
has to be declared as Array[0..Nmax] with Nmax being the largest filter degree
used in the main program.
EXAMPLE: FirFreqResp(@H0,NP,@Coeff,N,0,0.5) }


```
Procedure FirFreqResp(H0:Pointer;NP:Integer;Coeff:Pointer;N,Type:Integer;
                      Range:Double);

Type    ADouble = Array[0..2] of Double;
        PADouble = ^ADouble;


Var     IMax,I,J          : Integer;
        NHalf,Dw,W,Sum    : Double;

Begin
IMax:=(N-1) DIV 2;
NHalf:=N/2.0;
DW:=2*Pi*Range/(NP-1);
If Type = 0 Then
   Begin
   For J:=0 To NP-1 Do
      Begin
      Sum:=0;
      W:=J*Dw;
      For I:=0 To IMax Do
         Begin
         Sum:=Sum+PADouble(Coeff)^[I]*Cos(W*(I-NHalf));
         End;
      Sum:=Sum*2.0;
      If Not Odd(N) Then Sum:=Sum+PADouble(Coeff)^[N DIV 2];
      PADouble(H0)^[J]:=Sum;
      End;
   End
Else
   Begin
   For J:=0 To NP-1 Do
      Begin
      Sum:=0;
      W:=J*Dw;
      For I:=0 To IMax Do
         Begin
```

```
        Sum:=Sum+PADouble(Coeff)^[I]*Sin(W*(I-NHalf));
        End;
      Sum:=Sum*2.0;
      PADouble(H0)^[J]:=Sum;
        End;
    End;
  End;
```

CoeffIDFT

The procedure **CoeffIDFT** calculates the N+1 coefficients of an Nth-order FIR filter from values of the zero phase frequency response taken at N+1 equidistant frequency points. These values are sampled at the frequencies $\omega_i = 2*\pi*i/T/(N+1)$ with $i = 0...N$. The algorithm is based on an inverse discrete Fourier transform (IDFT) and makes use of the special symmetry conditions of linear-phase filters. The array **H** passes the samples of the frequency response to the subroutine. The filter coefficients are returned back in the array **Coeff**. **Typ** again specifies the kind of symmetry of the impulse response. **Typ** = 0 applies to Case 1 and Case 2 filters, **Typ** = 1 to Case 3 and Case 4 filters accordingly.

```
{ Procedure to calculate the N+1 coefficients of an Nth-order FIR filter from
N+1 equally spaced samples of the zero-phase frequency response. The
samples are taken at the frequencies wi = 2*pi*i/T/(N+1) with i=0 ... N . The
algorithm is based on an inverse discrete Fourier transform (IDFT) and makes
use of the special symmetry conditions of linear phase filters.
INPUT:    H         pointer to an Array of Double containing the samples
                    of the frequency response
          N         order of the filter, Integer
          Typ       symmetry of the impulse response, Integer
                    Typ  0 : impulse response with even symmetry
                         1 : impulse response with odd symmetry
RETURN:  Coeff     pointer to an Array of Double containing the calculated
                    filter coefficients
REQUIRED EXTERNAL ROUTINES: -
H and Coeff have to be declared in the main program as Array[0..Nmax] with
Nmax being the maximum filter order used in main program.
EXAMPLE:  CoeffIDFT(@H,@Coeff,N,Typ) }

Procedure CoeffIDFT(H,Coeff:Pointer;N,Typ:Integer);

Type  ADouble = Array[0..2] of Double;
      PADouble = ^ADouble;

Var   I,J,IMax          : Integer;
      Sum,NHalf,DW      : Double;

Begin
IMax:=N DIV 2;
```

```
NHalf:=N/2.0;
For J:=0 To N Do
   Begin
   DW:=2*pi*(J-NHalf)/(N+1);
   If Typ = 0 Then Sum:=PADouble(H)^[0]
   Else
      Begin
      Sum:=0;
      If Odd(N) Then Sum:=PADouble(H)^[IMax+1];
      If Odd(J-IMax-1) Then Sum:=-Sum;
      End;
   For I:=1 To IMax Do
      Begin
      If Typ = 0   Then Sum:=Sum+2*PADouble(H)^[I]*cos(DW*I)
                   Else Sum:=Sum+2*PADouble(H)^[I]*sin(DW*I);
      End;
   Sum:=Sum/(N+1);
   PADouble(Coeff)^[J]:=Sum;
   End;
End;
```

MaxFlatPoly

The function MaxFlatPoly returns values of the MaxFlat polynomial according to Sect. 7.6 (Eq. 7.52). Input values are the argument x, the degree of the polynomial Nt and the parameter L.

```
{ Function to calculate values of the maximum flat polynomial PNt,L(x). This
algorithm is an optimised version of the formula given by Kaiser (1979). The
polynomial is defined in the range x=0 ... 1. L is in the range L=1 ... Nt.
INPUT:     x              argument of the polynomial, Double
           Nt             Nt=N/2 (half order of the filter), Integer
           L              degree of tangency at x=0, Integer
RETURN:  MaxFlatPoly   result of the polynomial calculation, Double
REQUIRED EXTERNAL ROUTINES: -
EXAMPLE:  y = MaxFlatPoly(x,L,Nt)  }

Function MaxFlatPoly(x:Double;L,Nt:Integer):Double;

Var   j,jj        : Integer;
      Sum,y,z    : Double;

Begin
If (L = 0) Or (L = Nt+1) Then
   Begin
   If L = 0 Then Sum:=0;
   If L = Nt+1 Then Sum:=1;
   End
Else
```

```
Begin
Sum:=1.;
If L<>1 Then
   Begin
   y:=x;
   For j:=1 To L-1 Do
      Begin
      z:=y;
      If Nt+1-L<>1 Then
         Begin
         For jj:=1 To Nt-L Do
            Begin
            z:=z*(1.0+j/jj)
            End;
         End;
      y:=y*x;
      Sum:=Sum+z;
      End;
   End;
y:=1.0-x;
For j:=1 To Nt+1-L Do Sum:=Sum*y;
End;
MaxFlatPoly:=Sum;
End;
```

MaxFlatDesign

The procedure MaxFlatDesign determines the parameters of a MAXFLAT filter from the characteristic data of a low-pass tolerance scheme. Input data to the procedure are the passband edge frequency wp, the stopband edge frequency ws, the minimum gain in the passband vp and the maximum gain in the stopband vs. As a result, we obtain half the filter order Nt (Nt = $N/2$) and the parameter L.

```
{ Procedure to calculate the filter parameters Nt and L for a MAXFLAT FIR
filter. This routine is based on the algorithm introduced by Rajagopal and Dutta
Roy (1989). Input is the tolerance contour of the low pass. The cutoff
frequencies are normalized to the sampling frequency of the filter.
INPUT:     wp           passband cutoff frequency, Double
           ws           stopband cutoff frequency, Double
           vp           minimum passband gain (linear), Double
           vs           maximum stopband gain (linear), Double
RETURN: Nt              half the order of the filter (Nt=N/2), Integer
        L               degree of tangency at w=0, Integer
REQUIRED EXTERNAL ROUTINES: MaxFlatPoly
EXAMPLE:  MaxFlatDesign(wp,ws,vp,vs,Nt,L);  }

Procedure MaxFlatDesign(wp,ws,vp,vs:Double;Var Nt,L:Integer);

Const Lim = 200;
```

```
Var   xp,xs  : Double;
      Exit   : Boolean;

Function Found(vp,vs,xp,xs:Double;L,Nt:Integer):Boolean;
Var Flg : Boolean;
Begin
Flg:=True;
If MaxFlatPoly(xp,L,Nt) < vp Then Flg:=False;
if MaxFlatPoly(xs,L,Nt) > vs Then Flg:=False;
Found:=Flg;
End;

Function InRange(vp,xp:Double;L,Nt:Integer):Boolean;
Begin
If (vp > MaxFlatPoly(xp,L,Nt)) And (vp <= MaxFlatPoly(xp,L+1,Nt))
   Then Inrange:=True
   Else InRange:=False;
End;

Begin
xp:=(1-cos(2*pi*wp))/2.0;
xs:=(1-cos(2*pi*ws))/2.0;
L:=0;Nt:=1;Exit:=False;
While (Not Exit) And (Nt < Lim) Do
   Begin
   Nt:=Nt+1;
   While Not InRange(vp,xp,L,Nt) Do L:=L+1;
   If Found(vp,vs,xp,xs,L,Nt) Then Exit:=True;
   If Found(vp,vs,xp,xs,L+1,Nt) Then
       Begin Exit:=True;L:=L+1;End;
   End;
End;
```

Using the function MaxFlatPoly and the procedure MaxFlat Design, it is extremely easy to calculate the coefficients of an FIR low-pass filter with maximally flat frequency response. The following listing shows a short example how these routines may be applied to calculate the coefficients of a MAXFLAT filter.

```
Program MaxFlat;

Var   wp,ws,vp,vs,x   : Double;
      I,L,Nt,N        : Integer;
      H,Coeff         : Array[0..100] Of Double;

{$I MaxFlatPoly, MaxFlatDesign, CoeffIDFT}

{ Start of the main program }
Begin
```

```
ReadLn(wp);
ReadLn(ws);
ReadLn(vp);
ReadLn(vs);
MaxFlatDesign(wp,ws,vp,vs,Nt,L);
N:=2*Nt;

For I:=0 to N do
   Begin
   x:=(1.0-Cos(2*pi*i/(N+1)))/2.;
   H[I]:=MaxFlatPoly(x,L,NT)
   End;

CoeffIDFT(@H,@Coeff,N,0);

{ The filter coefficients can be found in the array Coeff. }

End.
```

Index

Printing: Mercedes-Druck, Berlin
Binding: Buchbinderei Lüderitz & Bauer, Berlin